T0229381

Mobile Social Networking and Computing

A Multidisciplinary Integrated Perspective

Mobile Social Networking and Computing

A Multidisciplinary Integrated Perspective

Yufeng Wang and Jianhua Ma

CRC Press
Taylor & Francis Group
Boca Raton   London   New York

CRC Press is an imprint of the
Taylor & Francis Group, an **Informa** business

CRC Press
Taylor & Francis Group
6000 Broken Sound Parkway NW, Suite 300
Boca Raton, FL 33487-2742

© 2015 by Taylor & Francis Group, LLC
CRC Press is an imprint of Taylor & Francis Group, an Informa business

Printed on acid-free paper
Version Date: 20140313

International Standard Book Number-13: 978-1-4665-5275-3 (Hardback)

Library of Congress Cataloging-in-Publication Data

Wang, Yufeng (Computer scientist)
  Mobile social networking and computing : a multidisciplinary integrated perspective / Yufeng Wang and Jianhua Ma.
      pages cm
  Summary: "This book introduces mobile social networking (MSN) and computing from a multi-disciplinary perspective. It covers fundamental theory and key problems in MSN, including characteristics, inner structural relationship, incentive mechanisms, resource allocating, information diffusion, search, ranking, privacy, trust and reputation in MSN. It reviews various applications and includes analysis on related platforms"-- Provided by publisher.
      Includes bibliographical references and index.
      ISBN 978-1-4665-5275-3 (hardback)
      1. Mobile computing. 2. Online social networks. I. Ma, Jianhua, 1962- II. Title.

QA76.59.W36 2014
006.7'54--dc23   2014009746

Visit the Taylor & Francis Web site at
http://www.taylorandfrancis.com

and the CRC Press Web site at
http://www.crcpress.com

Contents

SECTION I MSN BASIC CONCEPTS, APPLICATIONS, AND CHALLENGES

SECTION II FUNDAMENTAL THEORY AND KEY PROBLEMS IN MSNs

Preface

Over the past few years, online social networks (OSNs) have been steadily growing and have become ubiquitous on the Internet, greatly improving social connectivity and collaboration. Despite the wide spread of OSNs, the flexibility and sociability of these networks are questionable. First, access to OSN services is still not widely available upon user demand. Furthermore, human communication is still highly embedded in the physical contact and closeness provided by the physical environment. Unfortunately, there is no automated means to facilitate communication in the physical environment, which leads to the issue of sociability. Thus, people with shared interests and backgrounds fail to leverage interpersonal affinities for personal benefits.

On the contrary, we are living in a mobile device–focused society (most people today would find it very difficult to live without a mobile phone). Perhaps, this is because people love to socialize. With the great popularity of smartphones and the recent availability of open mobile platforms, we believe that there is a significant latent impact in the convergence of distributed content sharing, OSNs, sensor networks, and pervasive computing on the mobile phone platform.

Now, with numerous advancements in mobile phone technologies (especially smartphones and new location technologies), this is the right time to connect the virtual community with physical space and to tightly bind the rich social context with the local environmental context by providing personalized services to people who interact. This is the key idea behind mobile social networks (MSNs).

In a sense, MSNs could be regarded as part of an ecosystem of social devices. This book explicitly distinguishes MSNs from OSNs as follows:

- First, MSN applications are not just extensions of computer-based social computing because the usage patterns of mobile devices significantly differ from those in a wired environment. It is expected that user involvement in mobile social computing will foster undiscovered usages and interactions. In practical terms, because mobile devices have rich sensing capabilities, they allow users to augment the real-world commons with the Internet. In short, the mobile device will become the natural tool to bridge the gap between the physical world surrounding us and the wealth of information on the Internet.

■ Second, we argue that it is the user, through the mobile device, who will be placed at center stage, and the potential success of MSNs lies in active collaboration among participants. The importance of people's participation, their feedback, and finding ways to satisfy their needs brings with it many interdisciplinary challenges. For example, a possible collaboration could be supported through peer-to-peer (P2P) networking that occurs when users are close to each other, and their mobile devices store and share location-based information. This would allow data to be shared easily among users who are working together.

In summary, user empowerment will have a determinant role in the mobile social networking ecosystem. Users are no longer passive consumers; they have the opportunity to create content or to contribute to social networks. Users' real-life situations will be at the core of their mobile usage, and they will use their mobile devices as tools in both the real and the information/content/application domains.

This work differs from most existing books related to MSNs in the following aspects.

■ First, most other books look at MSNs only from a technical standpoint, but we believe that focusing only on the technical viewpoint is far from sufficient to thoroughly characterize MSNs. Our discussions on multidisciplinary perspectives and approaches will shed new light on various related issues in mobile networks and systems, will foster new applications, and will inspire novel economic and business models, forming the crux of our book *Mobile Social Networking and Computing: A Multidisciplinary Integrated Perspective*. This volume outlines incentive mechanisms inspired by classical economics, behavioral economics, and social psychology, and, perhaps for first time, summarizes economic and business models of MSNs.

■ Second, this book not only deals with the theoretical aspects of mobile social networking and computing but also throws light on the applications viewpoint. This book is an attempt to thoroughly introduce and categorize various existing applications related to mobile social networking and computing, aimed at inspiring potentially interesting social networking applications and suggesting important research opportunities. Both location-based service (LBS), one of the most popular MSN applications, and mobile social networks in proximity (MSNPs), an emerging application, have been presented in a comprehensive manner.

■ Finally, we also address popular mobile social networking development platforms, such as Android and iOS, and some cross-platform development frameworks that would be of immense help for engineers and developers of MSN applications. The latest generation of mobile devices supports more application capabilities with fewer vendor-imposed restrictions. Openness and computational power combined with diverse capabilities of modern smartphones and cross-platform development tools could significantly facilitate the emergence of new MSN applications that may ease communication and enhance intelligent social life.

In addition, this book includes the following features:

- An emphasis on the interaction between the macrolevel structure and the local rational behaviors (microlevel) in MSN: These two perspectives are interrelated—local interactions lead to the emergence of high-level properties, which, in turn, affect the local behaviors of rational participants.
- Two distinct (but related) viewpoints about MSN applications: Broadly speaking, we think that mobile social networking and computing include two different areas. The first is socially inspired networking technology, which adopts the concepts and methods in social fields to address the problems in networking areas, such as mobile search and ranking. The second is networking technology, which uses various recent advancements to facilitate our social life, thereby serving the society.
- Economic viewpoint of social networking and computing: Making the leap from fixed-line social networking to mobile networking is a lot trickier than it might appear. The distinguishing features of mobile terminals and their users, such as tiny screens, low bandwidth, different requirements, preferred experience, and so on, will render mobile terminal–oriented applications of social networks and computing very different from traditional computing. This book will discuss various issues from multidisciplinary viewpoints, including technology, economics, social sciences, psychology, and so on.
- Cross-platform development frameworks of MSN application development: Present day developments in MSN applications are much easier to operate than earlier mobile application development platforms but still have some of the same complexities and issues. Each platform is unique and exhibits different features, capabilities, and behaviors. Developing an app for separate mobile platforms requires in-depth knowledge of those platforms and their Software Development Kits (SDKs). This increases the cost of development and the time to market an app and decreases the ease of updating. This is where cross-platform development tools come into picture. Our book summarizes several existing multiplatform developer tools in the market today.

Briefly, we attempt to organize the book contents from two viewpoints: the structural viewpoint and the application viewpoint. From the structural viewpoint, multidisciplinary inspired research on open mobile networking and computing has two distinctive levels. First, from a high-level (macrolevel) perspective, we conduct some social and economic analysis of network and information infrastructure; then, from a relatively low-level (microlevel) perspective, we adopt some concepts and theories from other research fields to properly address or alleviate the detailed problems in the mobile networking and computing fields. Naturally, these two perspectives are interrelated; that is, local interactions lead to the emergence of high-level properties, which, in turn, affect the local behaviors of participants. From the application viewpoint, we argue that mobile social networking includes two different areas. The first area includes the socially inspired networking technologies, which adopt the concepts and methods in social fields to address the problems in networking areas. The second area includes the social networks serving society, which use various advanced technologies to facilitate our social life.

Certain fundamental issues run through all MSN research and applications, including incentive mechanisms, trust and reputation, energy efficiency, and so on. The new generation of mobile devices has more applications that run with few vendor-imposed restrictions. The combination of openness, computational power, and the diverse capabilities of modern smartphones and MSN cross-platform development frameworks could be used to construct applications that may ease communication and enhance intelligent social life.

This book exhaustively presents various aspects of MSN research and applications: fundamental theory, key problems, typical applications, development frameworks, and so on. It is organized into four parts comprising 13 chapters.

Section I introduces basic concepts, applications, and challenges related to MSNs (Chapter 1).

Section II presents the fundamental theory and key problems encountered in MSNs. Three chapters address the "fundamental theory": Chapter 2 details the multidimensional (temporal–spatial–social) structural characteristics of MSNs; Chapter 3 focuses on the interaction between network structure and local autonomous interaction; and Chapter 4 discusses incentive mechanisms in mobile networking and computing. Five chapters address "key problems": Chapter 5 highlights information diffusion; mobile search and ranking are covered in Chapter 6; Chapter 7 focuses on energy efficiency schemes in MSNs; Chapter 8 highlights issues of privacy, trust, and reputation in mobile networking and computing; the section ends with a discussion of the economic and business model in MSNs (Chapter 9).

Section III features a discussion on MSN applications, which is subdivided into two parts: socially inspired mobile networking (Chapter 10) and enhanced social life with mobile technologies (Chapters 11 and 12).

The book concludes with a final section (IV) on MSN development platforms and pertinent examples (Chapter 13).

Acknowledgments

First of all, this work would not have been possible without the great help and patience of Ruijun He and Laurie Schlags of CRC Press, who have been very supportive and encouraging throughout the process of preparing this book for publication. And moreover, I would appreciate many anonymous editors for their great efforts to improve the presentation quality of this book.

I wish to thank my students, Jing Tang, Jing Xu, Shaobing Fu, Qicai Zhou, Jiabing Cheng, Li Wei, Minmin Cai, and Xiaohong Cheng, for spending a great deal of effort in seeking and collecting potential material for this book.

I especially would like to express my gratitude again to Jing Tang for carefully checking and correcting typos and the format of some of the book's chapters and to Qicai Zhou and Li Wei for perfectly completing the tedious task of acquiring permissions from corresponding organizations and authors.

We are also indebted to Professor Hongbo Zhu, vice president of Nanjing University of Posts and Telecommunications (NJUPT), and Professor Longxiang Yang, vice dean of the College of Telecommunications and Information Engineering at NJUPT, for their considerable and continuing supports in book writing.

Finally, this book is dedicated to my loving family— Bo Zhang, my wife; Yichun Wang, my daughter; and Jinmin Wang and Xiurong Liu, my parents. Without their endless love, patience, and motivation, I would never have been able to complete this long journey of book writing.

Yufeng Wang

Author Biographies

Yufeng Wang received his doctoral degree from the State Key Laboratory of Networking and Switching Technology, Beijing University of Posts and Telecommunications (BUPT), China, in July 2004. From July 2006 to April 2007, he worked as a postdoctoral researcher at Kyushu University, Japan. In May 2007, he became an associate professor at Nanjing University of Posts and Telecommunications (NUPT), China. From February 2008 to March 2011, he was an expert researcher at the National Institute of Information and Communications Technology (NICT), Japan. Since 2013, he has been a full professor at NUPT and is also a guest researcher at the State Key Laboratory of Networking and Switching Technology and at the Media Lab at Waseda University, Japan.

Professor Wang's research interests include multidisciplinary inspired research on networking and systems, specifically new generation networks, peer-to-peer (P2P), wireless ad hoc network and sensor networks, trust and reputation systems, and mobile social networking and computing.

Dr. Wang has published more than 40 academic papers in journals and conference proceedings and has organized two special issues on multidisciplinary networks and systems for the *Telecommunication Systems Journal* and the *Journal of Computer and System Sciences*.

Jianhua Ma received his bachelor's and master's degrees in communication systems from the National University of Defense Technology (NUDT), China, in 1982 and 1985, respectively. In 1990, he received his doctoral degree in information engineering from Xidian University, China. He has been on the faculty of Hosei University, Japan, since 2000. Presently, he is a professor of computer and information sciences in the Digital Media Department. Prior to joining Hosei University, Dr. Ma had 15 years of teaching and/or research experience at NUDT, Xidian University, and the University of Aizu, Japan.

Dr. Ma's main research interest is ubiquitous computing, especially devoted to what he calls "smart worlds" filled with smart/intelligent ubiquitous things, or u-things, including three kinds of essential elements: smart objects, smart spaces/ hyperspaces, and smart systems. These are based on his vision for the future: ubiquitous intelligence (UI, u-intelligence) or pervasive intelligence (PI), solving the crucial problems caused by intelligence pervasion due to the fast progress of semiconductors, microelectromechanical systems (MEMS), nanoelectromechanical systems (NEMS), sensors, radio frequency identifications (RFIDs), embedded devices, ubiquitous computers, pervasive networks, universal services, and so forth.

Dr. Ma is a member of Institute of Electrical and Electronics Engineers (IEEE) and Association for Computing Machinery (ACM). He has edited 10 books/proceedings and has published more than 180 academic papers in journals, books, and conference proceedings. He has delivered more than 10 keynote speeches at international conferences, and he has given invited talks at more than 30 universities/institutes. (More detailed information about Prof. Jianhua Ma can be found at http://cis.k.hosei.ac.jp/~jianhua.)

MSN BASIC CONCEPTS, APPLICATIONS, AND CHALLENGES

I

Chapter 1

Introduction to Mobile Social Networking and Computing

1.1 Introduction

Over the past years, online social networks (OSNs) have steadily been growing and have become ubiquitous on the Internet, which greatly improves social connectivity and collaboration. Despite the predominance of OSNs, the flexibility and sociability of these networks can be questioned. First, access to OSN services was not available upon a user's demand, as had occurred exclusively while using a desktop computer. Further, human communication is still highly embedded in physical contact and closeness provided by the physical environment. Unfortunately, there is no automated way to facilitate communication in the physical environment, which leads to a sociability issue. Thus, people with shared interests and backgrounds fail to leverage interpersonal affinities for personal benefits.

On the contrary, presently we are living in a mobile device–dependent society (most people would find it difficult to live without a mobile phone). With the great adoption of smartphones and the recent availability of open mobile platforms, we believe that the convergence of distributed content sharing, OSNs, sensor networks, and pervasive computing on the mobile phone platform has significantly greater impact. New services for mobile phones have been developed in many OSN sites. For example, at the time of writing this manuscript, Facebook claimed to have more than 100 million mobile users and more than

200 mobile operators in 60 countries working to deploy and promote Facebook mobile products.* However, most approaches simply extend the Web interfaces of social networks to mobile devices (i.e., you can view the social networks through your mobile phone). Currently, people living or working in the same places have a lack of awareness about nearby places, interpersonal affinities, and social events, which they would normally like to visit or attend.

Historically, the relationship between people and places has been addressed by research on "community," which indicated that people's interactions and social relations were highly local, grounded in and organized around shared physical space. In the early 1960s, community theorists took these place-mediated interactions as essential to social relations/a community [1]. The emergence of "online communities" in the 1960s (with the early Internet) showed that social relations could be separated from shared physical space. And it is possible to develop relationships and form a virtual community based on computer-mediated technologies. Now, with the great development of mobile technologies (especially smartphones and new location technologies), it is the right time to reconnect the virtual community with physical space and to bind the rich social context tightly with the local environmental context of people interacting to provide personalized services. For example, location-based applications are one type of applications that help set mobile social network (MSN) applications apart from OSN applications. In their simplest form, location-based service (LBS) can include providing users with relevant maps, driving directions, and proximity searches for points of interest. Briefly, location-based MSNs can straightforwardly extend existing OSNs by allowing a user to know when his or her friends are around and by providing that user with data about new people to meet who share his or her interests. It is estimated that 20% of users will make use of LBS by 2012, and the size of the mobile LBS market will exceed $12 billion by 2014.

Broadly speaking, social networking on mobile networks (including the current narrow meaning of MSN) was launched as chat services in Japan, Scandinavia, Italy, France, and the United States from 1999 and evolved into chat rooms and texting community services. By 2004, camera phones and 3G networks introduced a second generation of platforms, primarily for dating services. In 2006–2007, a third generation emerged offering richer services predominantly based on WAP 2.0 and MMS. In 2008, a fourth generation of MSNs provided users with a high level of control over their information broadcast via their profiles or active handset services (location awareness, for example). Technologies such as Web 2.0 widgets, Flash Lite, Open Social, and the Open Handset Alliance (OHA) operating system, coupled with advanced social media capture and transfer systems, have delivered a higher level of functionality to MSNs [2]. Table 1.1 summarizes the evolution of MSNs, including features, technologies and business model, and so on.

---

* "Facebook statistics," https://www.facebook.com/press/info.php?statistics

Table 1.1 History of Social Networking on Mobile Networks

The first generation	• Began in 1999/early 2000, continues to be offered • Features: Text-only chat via chat room; most people anonymous • Technology: Application based, preinstalled on mobile terminals • Business model: Pay as you go (prepay); subscription based
The second generation	• Began in 2004–2006, based on region; usually coinciding with launches of 3G and camera phones; continues to be offered • Features: Uploading of photos, mobile search for person based on simple profile (gender, type of relationship sought, zip code), anonymous contact with people, rating/voting • Technology: SMS for purchase confirmation, preinstalled handset, and user-downloaded applications Applications: mostly dating • Business model: Pay as you go (prepay); subscription based
The third generation	• Experiment trails in 2006, reaches widespread adoption in 2009 • Features: Richer user experience, automatic publishing to Web profile and status update, some Web 2.0 features, searches by group/join interest groups, alerts of updates to favorite profiles, Location Based Service (LBS) emerging, free/ad-supported content (games, ringtones, etc.), User Generated Content (UGC) content ratings, and so on • Technologies: General interest, music, mobile-specific content distribution • Business models: Advertising and ad-supported content becoming increasingly important; pay as you go (prepay) and subscription based still popular; networks gain scale to become content distribution platform
The fourth generation	• Begun in 2008, and significantly popular now • Features: In addition to the above, presence, privacy, multiplayer mobile gaming, promoting offline and online social interactions; mobile/online network consolidation: borderline between those two communities is blurring • Technologies: Web widgets, open social, opportunistic network, open handset alliance • Business model: All of the above plus virtual currency purchase and trade of virtual goods

The increasing popularity of MSN applications is obvious and is due to new and interesting applications, systems, and services so that more and more people are engaging in social interaction and collaboration through mobile devices. Indeed, in addition to locating and being alerted about friends and communities, users can also use LBSs (such as those that recommend nearby points of interests) and data-sharing services (such as videos). Most MSNs have been extended from PC-based social networking sites to be available on mobile devices almost everywhere, anytime, by following their location- and proximity-aware facilities. A broad insight into state-of-the-art MSN applications is provided in Ref. [3]. Table 1.2 offers typical MSN services and their providers and links.

In a sense, MSNs could be regarded as part of an ecosystem of social devices. For the following considerations, this book explicitly distinguishes MSNs from OSNs:

- First, MSN applications are not just an extension of computer-based social computing, as the usage patterns of mobile devices significantly differ from those in a wired environment. It is expected that user involvement in mobile social computing will open previously undiscovered usages and interactions. In practical terms, as mobile devices have rich sensing capabilities, they allow augmenting real-world experiences with the Internet. Briefly, the mobile device will then be the natural tool to bridge the physical world around us with the wealth of information on the Internet.
- Second, we argue that it would be the user, through the mobile device, that will be placed at the center stage. The potential success of MSNs lies in the active collaboration of participants. This stems from people's participation, their feedback, and satisfying their needs. From this will naturally arise many interdisciplinary challenges. For example, one possible collaboration could be supported through peer-to-peer (P2P) networking when users are close to each other, in which mobile devices themselves may store and share location-based information. This would allow data to be shared easily between users who are working together.

In summary, user empowerment will have a determinant role in the mobile social networking ecosystem, not only because users are no longer passive consumers and have the possibility of becoming creators of content or of contributing to social networks but mainly because they will put the many situations of their real daily lives at the core of mobile usage, using the mobile device as a tool between the real and the information/content/application domains.

The chapter is organized as follows: In Section 1.2, the Mickey Mouse–like figure is designed to schematically characterize and correlate the components of research and applications in MSNs. Section 1.3 characterizes the structural properties of MSNs from economic viewpoints and introduces the dynamic evolution of

Table 1.2 Typical MSN Services and Their Providers

Main Service Categories	Providers and Links
Services Provided by Centralized Servers	
The MSN provider has a remote server with which mobile users interact to get services, including recommendations (such as Yelp), finding friends (such as Friendfinder), seeking and exchanging information and goods (such as PeopleNet), exchanging messages, viewing profiles, reading and sending bulletins, and viewing photos (such as Facebook, MySpace, and Twitter), downloading and playing games (such as Zynga), streaming video (such as YouTube), making professional contacts (such as LinkedIn), and learning (such as Pinterest).	• Facebook (http://www.facebook.com/mobile/) • Pinterest (https://www.pinterest.com/teachingideas/) • LinkedIn (http://touch.www.linkedin.com/mobile.html) • Friendfinder (www.friendfinder.com) • MySpace (http://www.myspace.com/) • PeopleNet (http://www.peoplenet.com) • Twitter (http://twitter.com/#!/twittermobile) • Yelp (http://www.yelp.com/) • YouTube (http://www.youtube.com/mobile) • Zynga (http://company.zynga.com/)
Service based on location-aware devices	
The MSN provider gets the locations of its users (from their Global Positioning System (GPS)–enabled mobile devices or other localization technologies) and uses the information to locate nearby friends, to provide recommendations, and allow users to discover their surroundings.	• Foursquare (https://foursquare.com/) • Facebook places (http://facebook.com/places) • FourWhere (http://fourwhere.com/)

Continued

Table 1.2 (*Continued*)   Typical MSN Services and Their Providers

Main Service Categories	Providers and Links
Services based on proximity-aware devices	
The MSN provider allows individuals to use their Bluetooth-enabled devices (such as Bluedating, MobiClique, Proxidating, Speck) or Wi-Fi connectivity (such as FaceTime and Open Garden) to find and communicate with nearby friends and others with similar profiles and interests.	• Bluedating (http://www.bluedating.com/) • FaceTime (http://www.apple.com/iphone/built-in-apps/ facetime.html) • Open Garden (https://opengarden.com/) • Proxidating (http://www.proxidating.com) • Speck (http://speck.randomfoo.net/)

local interaction. In Section 1.4, we discuss two related categories of MSNs from an applications perspective: (1) socially inspired mobile networking and computing, and (2) the enhanced social life with mobility and location technologies. The latter includes two types of applications: people centric and place centric. In Section 1.5, we summarize the fundamental issues in all MSN applications: incentive mechanisms, trust and reputation, identity management and privacy, energy efficiency and location technologies, and some issues in human–computer interaction (HCI), such as tiny screen, easy-to-use, nonintrusive, and so on. Most of those issues also will be dealt in other chapters in this book. Finally, we briefly conclude the chapter.

1.2 Research and Application Framework of MSNs

In MSNs, we assume that users carry mobile locatable devices, which can access services across the Internet using either short-range wireless interfaces (Wi-Fi, Bluetooth) or cellular interfaces; moreover, mobile users can share their profile information (including location) with trusted applications under certain privacy constraints.

Figure 1.1 depicts the basic design principles, fundamental issues, and the MSN development platforms that run through all MSN applications and research [4]. MSN application and research are systematically organized according to two different perspectives: structure and interaction as well as application. The components and their relationships are shown in Figure 1.1.

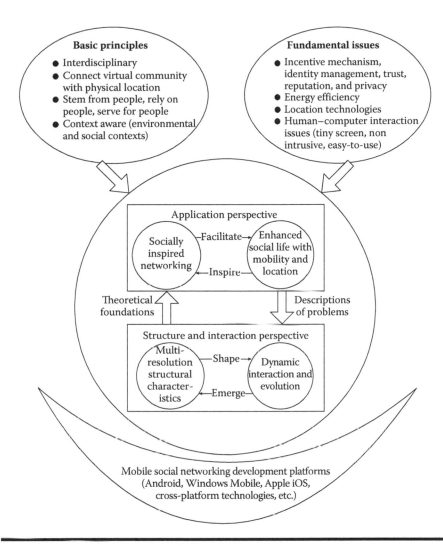

Figure 1.1 Research and application framework of mobile social networks (MSNs). (With kind permission from Springer Science+Business Media: *Telecommunications Systems Journal*, Mobile social networking: reconnect virtual community with physical space, 52, 2003, 91–110, Zhang, B. et al.)

We argue that every aspect of MSN research and applications should follow several principles:

■ The most important features of MSNs are reconnecting the virtual community to the physical region and moving users between them in a way that enhances both.

■ To achieve social networking between mobile devices, it is necessary to cross disciplines in order to deeply understand and exploit the traits, formation, and maintenance of physical and virtual social networks.

■ The locus of control in creation and configuration of content has been shifting to the grassroots: stemming from people, relying on people, and servicing for people. This principle also implies that, for the huge user-generated contents, mobile social search and ranking engineering are also indispensable.

■ MSNs' context-aware principle includes not only environmental contexts, which are naturally adopted in mobile computing and wireless technologies (such as location, time, presence, handset capability, etc.) but also interaction-oriented social contexts, which refer to the users' changing relationship and structure with respect to the particular task or goal.

In general, the new generation of mobile devices expose more application capabilities that run on them with fewer vendor-imposed restrictions. The combination of openness, computational power, and diverse capabilities of modern smartphones could be used to construct applications that may ease and enhance intelligent social life [5–7].

As noted earlier, MSNs are organized from two distinct perspectives: structure and interaction, and application. The former provides a theoretic foundation for the latter and, in turn, the latter will describe and shed light on the technological problems in MSNs. The structure and interaction perspective of MSNs is composed of two different levels: the high level (macrolevel) investigates the multidimensional structural characteristics (spatio–temporal–social); meanwhile, the low-level (microlevel) models the evolutionary dynamics of local interactions among all participants. Naturally, those two-level viewpoints are interrelated: local interactions lead to the emergence of high-level properties, which, in turn, affect the local behaviors of participants. From an application perspective, we argue that MSNs include two different but related areas: first, the socially inspired mobile networking technologies, which analyze, design, and implement communications and information technologies by taking social contexts into account; and the second focuses on the applications to enhance real social life with mobility and location technologies, which use various advanced mobile technologies to facilitate and enhance our social life. Naturally, as we know more about our society, we could design more advanced socially inspired mobile networking technologies to improve our social life.

1.3 Structure and Interaction Perspective about MSNs

1.3.1 Multidimensional Structure—Economic and Social Characteristics of MSNs

The foundations of MSNs are centered on two building blocks: understanding complex structures of social connectivity and exploiting those structures for interaction opportunities. The classic analysis of social and technological network study would follow a static or aggregated graph, that is, networks that do not change over time or are built as the result of aggregation of information over a certain period of time, which, in a sense, could be regarded as single snapshots taken at a particular point of time that presents a coarse-grained view even in networks where temporal dynamics are essential components.

Although most social networking sites allow only a binary state of friendship, it has been unsurprisingly observed that not all links are created equally. The relationships within a social network can be classified into "strong tie" and "weak tie" [8]. "Strong tie" is associated with a high frequency of interactions; "weak tie" implies that the members are loosely bound to their main social network. Members in a "weak tie" social network are more likely to form new connections outside their main social network or to bring new members into the existing social network. Thus, weak ties appear to be crucial for maintaining the network's structural integrity, and strong ties play an important role in maintaining local communities. A weak tie was shown to play an important role in the dissemination of content updates over an MSN [9]. However, most traditional static metrics, such as characteristic path length, clustering coefficient and centrality, and so on, are all based on the binary state (existing or no existing relationship). Thus, we need more general metrics to characterize the structural properties of static snapshot social networks. Two leading concepts were proposed in Ref. [10] for general weighted networks— efficiency (global efficiency and local efficiency) and cost. Efficiency measures how well information propagates over the social network G, and cost measures how expensive it is to build a network. Specifically, the global efficiency is defined as proportional to the sum of the inverse of the shortest path length between any pair of users. On the other hand, the local properties of social network graph G can be characterized by evaluating the efficiency of G_i for each user i, where G_i is defined as the subgraph of the neighbors of user i (the set of i's neighbors, except i itself). And the local efficiency of social network G is an average of the efficiency in each subgraph G_i. Note that, in a sense, global efficiency corresponds to a traditional characteristic path length, and local efficiency plays a role similar to a traditional clustering coefficient. Based on the concept of network efficiency, the information centrality of edge (or node) was proposed in Ref. [11], which is defined as the relative drop in the network efficiency caused by the removal of the edge (node) from G. An algorithm of hierarchical clustering was developed to find community structure in social, biological, and technological networks, which consists of finding and removing iteratively the edge with the highest information centrality [12]. The book's authors also have proposed Vickrey–Clarke–Groves (VCG) overpayment-based centrality and show that there exists a high correlation between global efficiency–based centrality and our proposed VCG-based centrality [13].

Note that these measurements are more or less related to the concept of the shortest path length. A random walk–based betweenness centrality was proposed in Ref. [14], which does consider random walks connecting all couples of nodes instead of the shortest paths. Although the author did not mention it, the random walk–based between centrality is significantly similar to the eigenvector centrality [15,16], which is based on the following social concept: your centrality also depends on your neighbors' centralities. Similar ideas are successfully used in well-known ranking systems, such as PageRank in the Google search engineer [17] and EigenTrust in (P2P) systems [18]. The book's authors also improve the P2P ranking algorithm through the personalized eigenvector centralities calculated along two directions [19,20].

Watts and Strogatz have identified two main attractive properties of a small-world network: highly cluster-like regular lattices and a small characteristic path length (like a random graph) [21]. The combination of these factors leads us to introduce the concept of economic small worlds, which formalizes the idea that networks are "cheap" (cost-effective) to build but nevertheless efficient in propagating information, on both global and local scales [10].

Until now, a majority of prior empirical OSN studies have focused on the characterization of the OSN's friendship structure inferred from a single snapshot taken at a particular point of time. But, social connections inherently vary over time and exhibit more dimensionality than static analysis can capture. Thus, it is imperative to investigate the OSNs' spatio–temporal–social metrics using multiple snapshots of the system taken over different periods of time. Inspired by the aforementioned definition of network efficiency, Tang et al. [22] proposed temporal distance metrics to quantify and compare the speed (delay) of information diffusion processes, in which all temporal distance metrics are based on the temporal hops among users. As described earlier, the distinguishing feature of MSNs is reconnecting the virtual community with the physical community by allowing a user to know when his or her friends are around and by providing the ability to meet new people who share interests. How users are connected in MSNs was only preliminarily investigated in Ref. [23]. The authors propose a three-layer friendship model that simultaneously considers users' mobility characteristics, social graph properties, and tag profiles, and that shows that the model-based friend recommendation is very effective.

Online location-based social networks (LBSNs) have recently attracted millions of users and are experiencing a huge increase in popularity over a short period of time. Thanks to the widespread adoption of location-sensing mobile devices, users can share information about their location with their friends. Among the biggest providers are Foursquare and Gowalla, although other hugely popular social networking services such as Facebook and Twitter have also introduced location-based features. Location is increasingly becoming a crucial facet of many online services; people appear more willing to share information about their geographic position with friends, although companies can customize their services by taking into account where the user is located. As a consequence, service providers have access to a valuable source of data on the geographic location of users, as well as to online friendship connections among them. The combination of these two factors offers a groundbreaking opportunity to understand and exploit the spatial properties of the social networks arising among online users as well as a potential window on real human sociospatial behavior.

Briefly, the distinguished features of MSNs yield many open interesting research issues:

How can the ever-changing multidimensional structural characteristics of MSNs be analyzed, including social link structures and their spatiotemporal properties and moreover, can constructive models be designed that could generate appropriate MSN network structures compatible with the observed and analyzed properties?

Based on people-to-people and people-to-places affinities, how can previously unknown emergent temporal–geo–social patterns be discovered? How can communities integrate participants who do and do not share a geographical location? How do such "mixed-reality" communities coexist with nonmembers within the same geographical locale?

How can we quantitatively measure the financial effects of the MSN community?

1.3.2 Evolutionary Interaction

Although advances in understanding the structure and dynamics of MSNs have been enormous, a conceptual framework for studying this issue from the perspective of interaction design is still lacking. Generally, the global-level phenomenon in social networks is affected by local behaviors and by local properties of substructures. Thus, one important research field includes various approaches to modeling the social network evolution and formation. Basically, there exist two different categories of approaches, that is, the so-called mechanical models and the economic models [24]. Mechanical models have their roots in the random graph literature and model formation by specifying either some stochastic process or an algorithmic process through which the links in a network are formed. The second approach is game theoretic and stems from the economic literature. It has mainly focused on models where the links are formed at the discretion of the nodes that derive benefits and face costs associated with various links and network configurations, the so-called economic model. Specifically, Matthew and Rogers [25] analyzed how the small-world features can be traced to variations in costs and benefits. In brief, random network models lean too heavily on change (to answer the problem of "how"), and at the other extreme, the economic approaches lean too heavily on choice (to answer the problem of "why"). Reality is clearly a mix of those two. This book's authors made a first step in this direction by proposing an evolutionary game theory (EGT)–based overlay topology evolution scheme to drive a given overlay into the small-world structure [26].

Due to the dynamic nature of social interactions, an online social network always keeps participants and information content evolving. Several works experimentally investigate the users' interaction evolution with time in OSNs. For example, in Flickr, only a very small fraction of users are active, and active users appear to form a core in the interaction graph that is responsible for the vast majority of social interactions [27]. Considering that the level of user activity in OSNs is likely to be highly dynamic, researchers have suggested examining the activity network that is based on the actual interaction among users, rather than mere friends [28]. Specifically, Wilson et al. [29] revealed that the interaction graphs derived from Facebook user traces exhibit significantly lower levels of "small-world" properties than those shown in their social graph counterparts. And, the authors qualify the impact of their observations on two social-based applications. Viswanath et al. [30] studied the evolution of activity among users in the Facebook social network to the capture growth of social links (becoming stronger or weaker), finding that links in the activity network tend to come and go rapidly over time and the strength of ties exhibits a general decreasing trend of activity as the social network link ages.

Intuitively, MSNs would facilitate, enhance, and develop the social interaction because MSNs will reconnect the virtual community with the physical region, which is implicitly shown in Ref. [31]. But, few studies thoroughly conduct an analysis of multiple snapshots taken over different periods of time and space in order to examine the dynamic nature of real-world MSNs in great detail. The general lack of attention to user interactions in prior studies of MSNs (and OSNs) is mainly due to the difficulties associated with capturing user interactions through measurement. Reality mining [32] conducted experiments on mobile phones, and their experiments mostly revolved around understanding social networks and understanding information flow. Their sanitized collected data are available for download by researchers.

Briefly, the basic social state of the physical–virtual community is composed of people profiles, place profiles, social ties among people, and associations between people and places. This state evolves continuously over time as new user profiles, social ties, place-related information, and events are created. Naturally, in MSNs, we should deeply investigate how physical places shape behaviors, and, on the other hand, we should examine how to label semantic tags for places according to participants' long-term behaviors in these very places. The high-level relationships among people, place, and time are shown in Figure 1.2. Specifically, people–place–people are circularly related to each other, and the structure and relationships may evolve as time goes by.

The aforementioned empirical studies on interaction in OSNs have examined the growth of these systems, but little is known about the patterns of decline in user population or user activity. Torkjazi et al. [33] examined the evolution of user population and user activity in MySpace. The main findings are that a significant fraction of accounts have been deleted, and a large fraction of valid accounts have not been visited for more than three months. The authors conclude that the departure of users after OSNs become very popular appears to be a common phenomenon and not specific to a particular OSN, and they raise the following question: What are the main forces that enable some systems to compete and strive in the Internet's OSN ecosystem, while others decline and ultimately die out? We think, the results in Ref. [34] partially answer this question from four perspectives: location and purpose, monitor social activity, provide feedback, and organize and maintain the space.

Finally, one important aspect that was neglected in MSNs is that almost all related work only focuses on collecting the interaction data of real MSNs (or OSNs) and

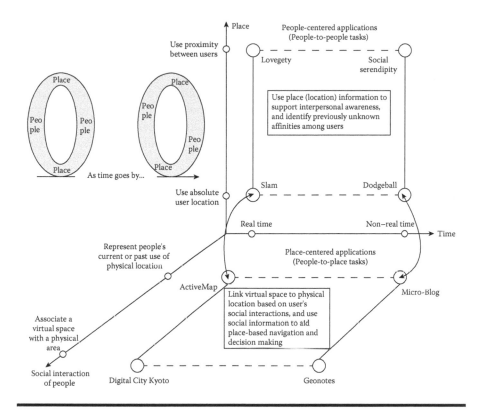

Figure 1.2 The categorization of mobile social networking techniques and typical applications. (With kind permission from Springer Science+Business Media: *Telecommunications Systems Journal*, **Mobile social networking: reconnect virtual community with physical space, 52, 2003, 91–110, Zhang, B. et al.)**

on inferring some phenomenal results, but those neither try to formally model user interaction and evolutionary behaviors nor go further to design some rules to guide users' behaviors to fulfill the designer' specific goals. We argue that an EGT model would especially help to formally model user interaction behaviors in MSNs [35,36], and mechanism design (MD) would focus on the "game rule" to achieve the global goals when facing rational and autonomous participants [37].

1.4 MSN from an Application Perspective

1.4.1 Socially Inspired Mobile Networking

The mobile handset is, by its own nature, a social artifact—an object carried by people to connect with other people. This is the reason why the next big development in mobility and mobile service involves social behaviors (in some form or fashion) to enable better ways to find, communicate, and share with friends and family; to learn about nearby places; and to consume information—all while on the go. Briefly, the goal of socially inspired mobile networking is to harness adaptive human social structures (people and their social practices) as critical resources to design successful information and communication systems. Contrary to conventional Internet design, socially inspired mobile networking technologies do not seek to enforce or maintain end-to-end connectivity; instead, they seek to create an underlying structure between devices where opportunistic interactions could be harnessed to convey relevant content. There is an enormous amount of data about socially inspired networking technologies, including opportunistic communications [38], epidemic information propagation [39], mobile search and ranking [40], and so on. We especially argue that the mobile P2P networking field has many topics that can be investigated by socially inspired methods, including topology construction, resource search (content-based routing) and dissemination, and so on, for, in mobile P2P, mobile users could take advantage of encounters with other users to share content in a P2P fashion. Due to space limitations, we have omitted this subsection.

1.4.2 *Enhanced Social Life with Mobile Technologies*

In real social life, the most common questions asked by people could be listed as: Who are you? Where are you at? Naturally, MSNs techniques could be categorized into people centric and place centric [41]. Instead of applications or systems, the reason for using "techniques" is that a particular application may naturally implement multiple techniques.

People-centered tasks (people-to-people tasks) put people at the center; this is facilitated by providing information about location of or proximity between users. In other words, people-centered techniques use place (location) information to support interpersonal awareness, to enable informal communication, and to identify previously unknown affinities between users, such as social matching, and so on.

Place-centered tasks shift the focus to making decisions about and acting in specific locations, in which social information is used as a resource for these tasks; in other words, these are people-to-place tasks, with (information from and about other) people used as a resource.

Both people-centered and place-centered techniques can be divided further. For example, some people-centered techniques use absolute user locations, while others use relative location or proximity between users. The difference is between applications that tell users where their buddies are and those that only tell users which buddies are close by. Some place-centered techniques represent people's current or past use of a physical location (e.g., showing who is on a university campus now) and others associate a virtual space with a physical location, such as classical MSNs.

The final dimension is real-time and non-real-time interaction. Besides the traditional implication of real time and non-real time, we also imply that real-time techniques are more related to synchronously providing information about current user location or about activity within a certain place, than to asynchronously providing digested information based on huge historical data about users' behaviors.

It is important to note that these classifications are not stringent at all. A particular application may simultaneously and naturally implement multiple techniques, and almost no existing MSN applications solely rely on asynchronous techniques. In other words, MSN applications always combine synchronous information exchange with asynchronous processing. But these classifications are still meaningful for organizing existing applications into proper categories and for structuring the design space for potentially interesting social networking applications, which will help outline additional sociotechnical challenges and opportunities and suggest new ways of understanding and addressing the fundamental issues associated with a location-aware community system. For example, this could help conceive and implement specific MSN applications that satisfy various users' requirements; this implies that, in a place-centered task, an interactive map is the very basic component. For absolute location-focused applications, maintaining privacy is more challenging than with proximity-based applications; the latter could use some indirect methods to infer the proximity relationships among users.

Figure 1.2 illustrates the categorization of MSN techniques and lists typical applications. We will briefly introduce them here. The pioneering system in location-based social networking application, ActiveBadge,* provides real-time information about user location to facilitate communication. (For example, a phone call could be routed to the phone nearest to an individual based on the last reported location.) The Social Location Annotation Mobile (SLAM) project [42], conducted by the Microsoft social computing group,† is a mobile device–based application that enables lightweight, group-centric real-time communication, location awareness, and photo sharing. By "slam" it means a group of people with whom you can exchange messages and photos. Generally, absolute user location techniques can enhance various applications. For instance, the "status" information for a buddy of an instant messaging (IM) client can be augmented with the buddy's location.

Dodgeball is a mobile service that distributes location-based information about users so that people can meet up at venues within cities. Users can also be alerted when friends of friends who have checked in to Dodgeball are within a 10-block radius. Note that Dodgeball does not use tracking signals such as GPS to determine the location of its members. Instead of an automatic location method, users must actively tell Dodgeball where they are by sending a text message to Dodgeball [43]. On the other hand, Loopt‡ leverages GPS and signal triangulation technologies to automatically

---

* ActiveBadge, http://www.cl.cam.ac.uk/research/dtg/attarchive/ab.html
† Microsoft Social Computing Group, http://research.microsoft.com/en-us/groups/scg/
‡ On Mar. 2013, Loopt, one of the early location-based services, is announcing that it has been bought by Green Dot Corp., a provider of retailer pre-paid cards which will use Loopt to develop mobile wallet and payment services. Green Dot is paying $43.4 million for Loopt, http://ir.greendot.com/phoenix.zhtml?c=235286&p=irol-newsArticle&ID=1671195

sense device location without requiring manual location updates. Ulocate* is designed to let family members track each other's locations on a map. Nokia has launched Lifeblog†, which allows users to create a time line of their activities (photos, short message service, video, etc.), thereby creating a kind of digital recording of one's own life.

Synchronous, proximity-based, people-centric social networking applications have been significantly explored by quite a few commercial and research groups, almost all of which support social awareness among colocated people. Colocation can also be used specifically for social matching to bring people together for interaction and potentially new relationships. Most social matching systems support dating and romance rather than general interaction. In Lovegety,‡ when a blue (male) Lovegety and a pink (female) Lovegety are within 15 feet of one another, they beep and flash, telling the user that another Lovegety owner is close by.

Social serendipity [44] provides the infrastructure to leverage the social networking context within a local physical proximity using mobile smartphones that rely on Bluetooth device discovery for locating nearby users and on a central server for matching the user profiles. Whozthat [45] implements similar infrastructure that shares social networking IDs locally, using wireless technology, while also leveraging a wireless connection to the Internet's social network to bind identity with location.

Now, we move on to place-centered techniques, which link virtual spaces to physical locations (or label semantic meanings to physical spaces); that is, they use social information to aid place-based navigation and decision making.

Some systems provide visualization of the current use of defined locations and areas. For example, ActiveCampus Explorer Map [46] overlays campus maps with avatars showing the location of a user's buddies.

It is also possible to visualize the history of use of a physical space. In Micro-Blog [47], mobile phone users are encouraged to record multimedia blogs on the fly, enriched with inputs from other physical sensors. The blogs are geotagged and uploaded to a remote server that positions these blogs on a spatial platform (e.g., Internet maps). Internet users can zoom onto any part of the map and browse multimedia blogs at those locations. Moreover, users may query selected regions for desired information. Queries are serviced either through explicit human participation or automatic physical sensing. This is similar to the people-centered asynchronous system technique; however, here the focus is on a defined physical place rather than a defined set of people (say, one's buddies).

Synchronous online interaction spaces used by digital cities aim to support interaction among geographical neighbors. Digital cities integrate urban information

---

* Where, Inc. was founded in 2004 under the name uLocate Communications, Inc. And moreover, PayPal, a subsidiary of eBay, announced its acquisition of Where, Inc. in 2011 for $135 million, https://advertising.paypal.com/#home
† Lifeblog, http://en.wikipedia.org/wiki/Nokia_Lifeblog
‡ "Love: Japanese Style", Wired News, 11 Jun 1998, http://archive.wired.com/culture/lifestyle/news/1998/06/12899

and create public online spaces for people living or visiting those cities. For example, the Digital City project in Japan uses high-fidelity, Internet-based simulacra of cities, updated continuously via cameras and other sensors to provide data and support community life [48].

Geonotes [49] allows digital messages to be linked to physical locations. These digital notes behave like electronic Post-its® that are visible to authorized users on their mobile devices remotely or when they enter the vicinity. Moreover, these messages are indexed by locations and then searched or accessed by navigating online maps. Place-its [50] proposes a similar location-aware reminder system; users can schedule reminders on their phones that are triggered only when they reach a location.

As we emphasized earlier, a particular system may simultaneously implement multiple techniques, and almost all MSN applications always combine synchronous information exchange and asynchronous processing. Location-based games on mobile terminals are a typical example, introducing an element missing in interactive console games—"the physical effort of sport." They have great potential to be commercially successful [51,52].

1.5 Fundamental Issues

The mass adoption of MSN applications not only represents another means of generating revenue that could revitalize the mobile market but also a chance for social life. However, there are many challenges that can arise. Generally, the following fundamental issues exist for MSNs: incentive mechanisms, trust and reputation, identity management and privacy, location technology and energy efficiency, and HCI issues (tiny screen, easy-to-use, nonintrusive, etc.).

1.5.1 Incentive Mechanisms

We argue that the locus of control in the creation and configuration of MSN content has been shifting to grassroots. The existence and prosperity of MSNs will depend on their achieving a critical mass of users who share their profiles, places, and real-time location information. Generally, incentive mechanisms may include methods of direct and indirect reciprocation (micropayment or reputation) and social-based methods (i.e., users responded positively when they felt that they had uniquely contributed to the space). Intuitively, MSN incentive mechanisms are inherently multidisciplinary. Nowak reviews five mechanisms for the evolution of cooperation [53]. Douceur and Moscibroda [54] adopted a lottery psychology to disproportionally motivate people to contribute to a developing system. Vassileva and Sun [55] proposes motivating users to participate by visualizing the community and the levels of participation of all community members, which is inspired by the theory of social comparison in psychology.

1.5.2 Trust and Reputation

Trust and reputation each play a pivotal role in MSNs and can assist other parties in deciding whether to transact with a given party through aggregating ratings about the party that reveal a trust or reputation score. Obviously, the concept of reputation is closely linked to that of trustworthiness, but there is a clear and important difference. The most distinguished difference is that trust systems produce a score that reflects the trusting entity's subjective view about the trusted entity's trustworthiness, whereas reputation is a single value (more technically, a social evaluation) that represents what the community as a whole thinks about a certain user. Briefly, reputation can be considered as a collective measure of trustworthiness based on referrals or ratings from members in a community, and an individual's subjective trust can be derived from a combination of a partner's reputation and that partner's personal experience [56,57].

As far as MSNs are concerned, the reconnection of virtual community with physical region will facilitate trust among users, especially those within proximity. In addition, the general tendency to form social networks with those we trust can act as a strong incentive for cooperation.

1.5.3 Identity Management and Privacy

MSNs share environmental and social context data such as location, time, presence information, activity, and social relationships. The disclosure of such information provides higher social awareness and greater flexibility in managing one's MSN, but the highly sensitive nature of this information raises privacy concerns that must be addressed in the design solution. Currently, most MSNs allow users to assess the implications of revealing personal information and to decide whether, when, where, and how to reveal those information.

In addition to privacy issues for desktop-based social computing applications, MSNs must also cope with location privacy, which is highly sensitive for mobile users. The MSNs need to ensure that users cannot track each other. The simple way is to replace identifying attributes with synthetic identifiers. We refer to this procedure as naïve anonymization. Counts and Fisher [58] argued that the relationship between virtual identity in OSNs and identity in physical regions should be investigated deeply; the authors found that SLAM users who knew each other well offline had difficulty recognizing one another online due to alias confusion and lack of visual cues.

However, an even more difficult privacy problem than tracking is the inference of social ties based on location. For instance, if Bob is allowed to see John's location (or social interaction), and he sees that John is too often located at Alice's office (or that he has many close social interactions), Bob might infer that they may have a romantic relationship. Note that Bob, John, and Alice are all anonymized names. That is, an entity's connection (i.e., the network structure around it) can be distinguishing, and it may be used to reidentify an anonymous individual. Hay et al. [59] investigated the threat of structural reidentification in an anonymized network (which consists of entities connected by links representing relations such as friendship, communication, or a shared activity) and proposed a novel approach to anonymizing network data.

Consolvo et al. [60] and Vihavainen et al. [61] conducted an interesting study whether—and to what extent—users are willing to disclose their location to social relations. Their results show that the most important factors are: who is requesting, why the requester wants the location information, and what details would be the most useful to the requester.

Finally, the reconnection of virtual community with physical space in MSNs would alleviate privacy concerns. For example, people located in proximity would like to expose some privacy to some extent. On the contrary, the reconnection will raise some new issues: Do community members desire the enforcement of different access rights to members based on their semantic and physical relationship to the community? To what extent can such rights be enforced?

1.5.4 Location Technologies and Energy Efficiency

Location information is central to MSNs. Not only is content tagged with locations but phones must periodically report their own locations to the remote server so that they can be suitably queried. Research from PlaceLab [62] has proposed a variety of solutions using Wi-Fi–based or Global System for Mobile Communications (GSM) cell tower–based localization. However, energy-location accuracy tradeoffs exist in all these techniques. For example, although GPS-based localization provides good accuracy (around 8 meters), measurements demonstrate an unacceptable battery life of less than 7 hours when GPS is used continuously. Energy consumption with Wi-Fi–based localization, although better than GPS, is significantly worse than GSM schemes. The positioning error or Wi-Fi–based location ranges from 10 to 50 meters, and GSM's location error is within several hundred meters [63]. Thus, the adaptive selection needs to be performed based on the application needs, the residual battery power, and a terminal's mobility characteristics that meet the application's accuracy/energy requirements.

1.5.5 HCI Issues

Specific features of mobile terminals can raise many HCI issues, such as being nonintrusive and easy-to-use, having a tiny screen, and so on. Interestingly, the MIT media lab's SixthSense could provide valuable inspiration.* SixthSense is a wearable gestural interface that augments the physical world around us with digital information and lets us use natural hand gestures to interact with that information.

---

* http://www.youtube.com/watch?v=blBohrmyo-I

Many systems allow users to pick the preferred method of communication according to a person's status and location (e.g., in person, e-mail, home/work phone) [64,65]. Project Satire [66] has the possibility of discriminating among human actions based on their accelerometer signatures. CenceMe [67] combines the inference of the presence of individuals using off-the-shelf, sensor-enabled mobile phones with sharing this information through social networking applications such as Facebook and MySpace and presents the design, implementation, evaluation, and user experiences. Peopletones [68] has designed different vibration patterns to denote various social events.

Because a mobile terminal's screen is so small, there is low tolerance for displaying results that are not of high relevance and quality. Unlike PC screens that can easily accommodate loads of advertisements (which are critical for their financial survival), the matchbox-size screens of cell phones and other mobile devices are not hospitable for the ads that are the lifeblood of traditional OSNs. Thus, for MSNs, advertisement approaches should be treated with great care; otherwise, customers could be offended or even lost.

Psychologists are uncovering the overwhelming extent to which our behaviors are governed by automatic processes. People automatically evaluate other people's personalities when they first meet them. Ambady and her colleagues have labeled our ability to form consensual impressions from very little behavioral information as "thin slicing." Impressions after a "thin slice" of behavior are said to be "accurate": (1) if they match impressions formed after more detailed behavioral information, and (2) if raters agree in their judgments [69]. The social cognitive aspects of human users' complex interactions were examined in Ref. [70]. The authors argue that personality traits can be inferred spontaneously from online profiles (but they are facilitated only if a trait is implied in the online profile), and they are extracted preferentially to other contents, implying more efficient processing of social information. The aforementioned psychological factors could help MSNs arrange the contents shown in mobile terminals.

1.6 Conclusion

Nowadays, we witness more and more online social services expanding their reach into the mobile domain. We argue that MSNs are not just extensions of computer-based OSNs. They should take advantage of mobile computing algorithms, wireless technologies, and real-time location systems to help people reconnect the virtual community with their physical spaces and surroundings and to enhance both. In fact, the most successful MSNs will leverage mobile elements by providing added value to users (e.g., answering queries and providing recommendations about people, places, and events of interest anytime, anywhere). This chapter has organized the sporadic topics and systems of existing MSN research and applications into correlated components including the basic design principles, fundamental issues, structure and interaction perspective, and application perspective, which should help to structure the design space, identify social-technical challenges, and inspire potentially interesting social networking applications.

References

1. Nelson, L., C. E. Ramsey, and C. Verner. *Community Structure and Change*. New York: The Macmillan Company; 1960.
2. Mobile social networking. White paper. Available from: http://www.telecoms.com/files/2009/05/buongiorno_final-fmt_nl-3110-f.pdf
3. Jabeur, N., S. Zeadally, and B. Sayed. Mobile social networking applications. *Communications of the ACM* 2013; 56(3): 71–79.
4. Zhang, B., Y. Wang, A. V. Vasilakos, and J. Ma. Mobile social networking: Reconnect virtual community with physical space. *Telecommunications Systems* 2013; 54: 91–110.
5. Alazzawe, A., D. Wijesekera, and R. Dantu. A testbed for mobile social computing. In: Proceedings of the 5th TRIDENTCOM, Washington, DC, 2009.
6. Oliver, E. A survey of platforms for mobile networks research. *ACM SIGMOBILE Mobile Computing and Communications Review* 2009; 12(4): 56–63.
7. Raento, M., A. Oulasvirta, R. Petit, and H. Toivonen. ContextPhone: A prototyping platform for context-aware mobile applications. *IEEE Pervasive Computing* 2005; 4(2): 51–59.
8. Granovetter, M. The strength of weak ties. *American Journal of Sociology* 1973; 78(6): 1360–1380.
9. Ioannidis, S. and A. Chaintreau. On the strength of weak ties in mobile social networks. In: Proceedings of the Second ACM Workshop on Social Network Systems, Nuremberg, Germany, March 31–April 3, 2009; pp. 19–25.
10. Latora, V. and M. Marchiori. Economic small-world behavior in weighted networks. *The European Physical Journal B* 2003; 32: 249–263.
11. Latora, V. and M. Marchiori. A measure of centrality based on the network efficiency. *New Journal of Physics* 2007; 9: 188.
12. Fortunato, S., V. Latora, and M. Marchiori. A method to find community structures based on information centrality. *Physical Review E* 2004; 70(2): 056104.
13. Wang, Y. and A. Nakao. On novel economic-inspired centrality measures in weighted networks. In: Proceedings of Asia-Pacific Services Computing Conference, Yilan, December 9–12, 2008; pp. 291–296.
14. Newman, M. E. J. A measure of betweenness centrality based on random walks. *Social Networks* 2005; 27: 39–54.
15. Carreras, I. et al. Eigenvector centrality in highly partitioned mobile networks: Principles and applications. *Advances in Biologically Inspired Information Systems* 2007; 69: 123–145.
16. Canright, G. S. and K. Engø-Monsen. Introducing network analysis. *Telektronikk*, 2008; 1: 4–18.
17. Page, L., S. Brin, R. Motwani, and T. Winograd. The PageRank citation ranking: Bringing order to the web. Available from: http://dbpubs.stanford.edu/pub/1999-66
18. Kamvar, S. D., M. T. Schlosser, and H. GarciaMolina. The EigenTrust algorithm for reputation management in P2P networks. In: Proceedings of the 12th International Conference on World Wide Web, Budapest, Hungary, May 20–24, 2003.

19. Wang, Y. and A. Nakao. PoisonedWater: An improved approach for accurate reputation ranking in P2P networks. *Future Generation Computer Systems* 2010; 26(8): 1317–1326.
20. Wang, Y., A. Nakao, and A. V. Vasilakos. DoubleFace: Robust reputation ranking based on link analysis in P2P networks. *Cybernetics and Systems* 2010; 41(2): 167–189.
21. Watts, D. J. and S. Strogatz. Collective dynamics of 'small-world' networks. *Nature* 1998; 393: 440–442.
22. Tang, J., M. Musolesi, C. Mascolo, and V. Latora. Temporal distance metrics for social network analysis. In: Proceedings of ACM SIGCOMM Workshop on WOSN, Barcelona, Spain, August 17, 2009; pp. 31–36.
23. Li, N. and G. Chen. Multi-layered friendship modeling for location-based mobile social networks. In: Proceedings of MobiQuitous, Toronto, ON, July 13–16, 2009; pp. 1–10.
24. Wang, Y. and A. Nakao. Research issues and overview of incentive compatible topology evolution in autonomous overlay networks. In: Proceedings of the 3rd ChinaGrid Dunhuang, Gansu, August 20–22, 2008; pp. 345–351.
25. Matthew, J. O. and B. W. Rogers. The economics of small worlds. *Journal of the European Economic Association* 2005; 3(2–3): 617–627.
26. Wang, Y. and A. Nakao. On cooperative and efficient overlay network evolution based on group selection pattern. *IEEE Transactions on Systems, Man, and Cybernetics, Part B (Cybenetics)* 2010; 40(2): 493–504.
27. Valafar, M., R. Rejaie, and W. Willinger. Beyond friendship graphs: A study of user interactions in Flickr. In: Proceedings of ACM SIGCOMM Workshop WOSN, Barcelona, Spain, August 17, 2009; pp. 25–30.
28. Chun, H., H. Kwak, Y. Eom, Y. Ahn, S. Moon, and H. Jeong. Comparison of online social relations in terms of volume vs. interaction: A case study of cyworld. In: Proceedings of ACM SIGCOMM Conference on Internet Measurement, Vouliagmeni, Greece, October 20–22, 2008; pp. 57–70.
29. Wilson, C., B. Boe, A. Sala, K. Puttaswamy, and B. Zhao. User interactions in social networks and their implications. In: Proceedings of European Conference on Computer Systems, Nuremberg, Germany, March 31–April 3, 2009; pp. 205–218.
30. Viswanath, B., A. Mislove, M. Cha, and K. P. Gummadi. On the evolution of user interaction in Facebook. In: Proceedings of ACM SIGCOMM Workshop on WOSN, Barcelona, Spain, August 17, 2009.
31. Pietiläinen, A., E. Oliver, J. LeBrun, G. Varghese, and D. Christophe. MobiClique middleware for mobile social networking. In: Proceedings of ACM SIGCOMM Workshop on WOSN, Barcelona, Spain, August 17, 2009; pp. 49–54.
32. Eagle, N. and A. S. Pentland. Reality mining: Sensing complex social systems. *Personal and Ubiquitous Computing* 2006; 10: 255–268.
33. Torkjazi, M., R. Rejaie, and W. Willinger. Hot today, gone tomorrow: On the migration of MySpace users. In: Proceedings of ACM SIGCOMM Workshop on WOSN, Barcelona, Spain, August 17, 2009.
34. Fisher, D., T. C. Turner, and M. Smith. Space planning for online community. In: Proceedings of the 2nd International AAAI Conference on Weblogs and Social Media (ICWSM), Seattle, WA, March 30–April 2, 2008.
35. Lee, C., J. Suzuki, and A. V. Vasilakos. iNet-EGT: An evolutionarily stable adaptation framework for network applications. In: Proceedings of the 4th International Conference on Bio-Inspired Models of Network, Information, and Computing Systems (BIONETICS), Avignon, France, December 9–11, 2009.

36. Wang, Y., A. Nakao, A. V. Vasilakos, and J. Ma. P2P soft security: On evolutionary dynamics of P2P incentive mechanism. *Computer Communications* 2010; 34(3): 241–249.
37. Feigenbaum, J. and S. Shenker. Distributed algorithmic mechanism design: Recent results and future directions. In: Proceedings of the 6th International Workshop on Discrete Algorithms and Methods for Mobile Computing and Communications, Atlanta, GA, September 28, 2002.
38. Chaintreau, A. et al. Impact of human mobility on opportunistic forwarding algorithms. *IEEE Transactions on Mobile Computing* 2007; 6(6): 606–620.
39. Vojnović, M., V. Gupta, T. Karagiannis, and C. Gkantsidis. Sampling strategies for epidemic-style information dissemination. In: Proceedings of the IEEE INFOCOM, Phoenix, AZ, April 13–18, 2008.
40. Wang, Y., A. Nakao, and J. Ma. Socially-inspired search and ranking in mobile social networking: Concepts and challenges. *Journal of Frontiers of Computer Science in China* 2009; 3(4): 435–444.
41. Jones, Q. et al. People-to-people-to-geographical places: The P3 framework for location-based community systems. *Computer Supported Cooperative Work* 2004; 13(4): 249–282.
42. Counts, S. Group-based mobile messaging in support of the social side of leisure. *Computer Supported Cooperative Work* 2007; 16(1–2): 75–97.
43. Humphreys, L. Mobile social networks and social practice: A case study of dodgeball. *Journal of Computer-Mediated Communication* 2007; 13(1): 341–360.
44. Eagle, N. and A. Pentland. Social serendipity: Mobilizing social software. *IEEE Pervasive Computing* 2005; 4(2): 28–34.
45. Beach, A. et al. Whozthat? Evolving an ecosystem for context-aware mobile social networks. *IEEE Network* 2008; 22(4): 50–55.
46. Griswold, W. G. et al. ActiveCampus: Experiments in community-oriented ubiquitous computing. *Computer* 2004; 37(10): 73–81.
47. Gaonkar, S., J. Li, R. Choudhury, L. Cox, and A. Schmidt. Micro-blog: Sharing and querying content through mobile phones and social participation. In: Proceedings of the ACM MobiSys, Breckenridge, CO, June 17–20, 2008; pp. 174–186.
48. Ishida, T. Activities and technologies in digital city Kyoto. In: Proceedings of the Third International Digital Cities Workshop, Amsterdam, The Netherlands, September 18–19, 2003.
49. Perrson, P., F. Espinoza, P. Fagerberg, A. Sandin, and R. Cöster. Designing information spaces: The social navigation approach. *GeoNotes: A Location-Based Information System for Public Spaces.* Springer, London, 2002; pp. 151–173.
50. Sohn, T., K. Li, G. Lee, L. Smith, J. Scott, and W. Griswold. Place-its: A study of location-based reminders on mobile phones. In: Proceedings of the Seventh International Conference on Ubiquitous Computing (UbiComp), Tokyo, Japan, September 11–14, 2005.
51. Marek, B., C. Matthew, B. Louise, H. Malcolm, and S. Scott. Interweaving mobile games with everyday life. In: Proceedings of ACM CHI, Montréal, Canada, April 22–27, 2006; pp. 417–426.
52. Christoph, S. et al. Geogames: Designing location-based games from classic board games. *IEEE Intelligent Systems* 2006; 21(5): 40–46.
53. Nowak, M. A. Five rules for the evolution of cooperation. *Science* 2006; 314(5805): 1560–1563.

54. Douceur, J. R. and T. Moscibroda. Lottery trees: Motivational deployment of networked systems. In: Proceedings of SIGCOMM, Kyoto, Japan, August 27–31, 2007; pp. 121–132.

55. Vassileva, J. and L. Sun. Using community visualization to stimulate participation in online communities. *e-Service Journal* 2007; 6(1): 3–40.

56. Jøsang, A., R. Ismail, and C. Boyd. A survey of trust and reputation systems for online service provision. *Decision Support Systems* 2007; 43(2): 618–644.

57. Hoffman, K., D. Zage, and C. Nita-Rotaru. A survey of attack and defense techniques for reputation systems. *ACM Computing Surveys* 2007; 5(14): 1–34.

58. Counts, S. and K. Fisher. Mobile social networking as information ground. In: Proceedings of the HICSS, Hawaii, USA, January 7–10, 2008; p. 153.

59. Hay, M., G. Miklau, D. Jensen, D. Towsley, and P. Weis. Resisting structural re-identification in anonymized social networks. In: Proceedings of VLDB, Auckland, New Zealand, August 23–28, 2008; pp. 102–114.

60. Consolvo, S., I. Smith, T. Matthews, A. LaMarca, J. Tabert, and P. Powledge. Location disclosure to social relations: Why, when, & what people want to share. In: Proceedings of the ACM CHI, Portland, OR, April 2–7, 2005; pp. 81–90.

61. Vihavainen, S., A. Oulasvirta, and R. Sarvas. "I can't lie anymore"—The implications of location automation for mobile social applications. In: Proceedings of MobiQuitous, Toronto, CA, July 13–16, 2009.

62. Cheng, Y.-C. Accuracy characterization for metropolitan-scale Wi-Fi localization. In: Proceedings of the MobiSys, Seattle, WA, June 6–8, 2005.

63. Constandache, I., S. Gaonkar, M. Sayler, R. R. Choudhury, and L. Cox. EnLoc: Energy efficient localization for mobile phones. In: Proceedings of the IEEE INFOCOM, Rio de Janeiro, Brazil, April 19–25, 2009 (Mini).

64. Nakanishi, Y., K. Takahashi, T. Tsuji, and K. Hakozaki. iCAMS: A mobile communication tool using location and schedule information. *IEEE Pervasive Computing* 2004; 3(1): 82–88.

65. Marmasse, N., C. Schmandt, and D. Spectre. WatchMe: Communication and awareness between members of a closely-knit group. In: Proceedings of the 6th International Conference on Ubiquitous Computing (UbiComp), Nottingham, UK, September 7–10, 2004.

66. Ganti, R. K., P. Jayachandran, T. F. Abdelzaher, and J. A. Stankovic. Satire: A software architecture for smart attire. In: ACM MobiSys, Uppsala, Sweden, June 19–22, 2006, pp. 110–123.

67. Miluzzo, E. et al. Sensing meets mobile social networks: The design, implementation and evaluation of the CenceMe application. In: Proceedings of the SenSys, Raleigh, NC, November 5–7, 2008.

68. Li, K. A., T. Y. Sohn, S. Huang, and W. G. Griswold. Peopletones: A system for the detection and notification of buddy proximity on mobile phones. In: Proceedings of the MobiSys, Breckenridge, CO, June 17–20, 2008.

69. Ambady, N., F. Bernieri, and J. A. Richeson. Towards a histology of social behavior: Judgmental accuracy from thin slices of behaviors. *Advances in Experimental Social Psychology* 2000; 32: 201–271.

70. Stecher, K. and S. Counts. Spontaneous inference of personality traits and effects on memory for online profiles. In: Proceedings of the 2nd International AAAI Conference on Weblogs and Social Media (ICWSM), Seattle, WA, March 30–April 2, 2008.

FUNDAMENTAL THEORY AND KEY PROBLEMS IN MSNs

II

Chapter 2

Multidimensional (Temporal–Spatio–Social) Structural Characteristics of Mobile Social Networks

2.1 Introduction

The analysis of real social, biological, and technological networks has attracted a lot of attention as technological advances have given us a wealth of empirical data. Classic studies looked at analyzing static or aggregated networks (i.e., networks that do not change over time or that are built as the result of aggregation of information over a certain period of time). Given the soaring collections of measurements related to very large, real network traces, researchers are quickly starting to realize that connections inherently vary over time and exhibit more dimensionality than static analysis can capture.

The popularity of social Web sites that allow us to keep in touch with friends has exploded over the past decade. The convenience of maintaining friendship networks online, sending messages to friends, arranging events, uploading photos, and sharing locations and thoughts has produced household brands such as Facebook, Twitter, and Foursquare. On those popular online social networks (OSNs), in addition to social relationships, temporal and spatial information are also intrinsic.

Facebook is a very popular OSN, and there are several different crawled data sets on Facebook [1]. In particular, the most comprehensive data set is provided by Wilson et al. [2], which includes both the social network and the interactions between users. First, an individual user profile provides personal information, and the profiles of the concerned individual's friends provide their information. This means a social network of people (nodes) and their relationships (edges) can be constructed. Second, individuals can post messages on one another's profile pages; this again creates an interaction network (nodes represent people and a directed edge represents a message being sent). Because users add new friends and delete ex-partners, former friends, and others over time, information on the social network topology can be obtained as it changes over time. In the case of interactions, messages are time-stamped and so information is available about interactions at different times.

Twitter is a service where users can share short, 140-character messages, known as "tweets" with friends. Users subscribe to (or "follow") any other user's profile to access their tweets but, unlike other OSNs such as Facebook, friendship does not need to be reciprocated. Recent data sets have crawled the entire corpus of tweets over a one-month period [3] and a subset of tweets that included tweet location information over a 12-day period [4]. The former data set contained 41.7 million user profiles, 1.47 billion directed links, and 106 million tweets; the latter data set, which filtered out users with geographic information, contained 400,000 user profiles, 183 million directed links, and 334.5 million tweets. Both these data sets allow users to construct two different graphs: a graph of followers and a graph of tweets. In the aforementioned data sets, temporal information is reflected as follows. First, both data sets contain time stamps of each tweet; hence, we can trace the dynamic spread of tweets and retweets as it cascades through the user network. Second, because users can constantly follow and unfollow users, the topology of followers changes over time.

In addition to temporal information including user check-ins and time-stamped contents, location-based social networks (LBSNs) also help to maintain friendships online. The latest feature is the ability of a user to update his or her current location either manually or by using a Global Positioning System that is built into many devices; this allows people to know where their friends are currently located. Numerous data sets exist in several popular location-based network (LBN) services such as Foursquare [5] and Gowalla [6]. With this user location information, we can construct a graph of user colocations and copresences (i.e., users who report that they are at the same location at the same time). Therefore, this gives us spatiotemporal information, meaning that the topology of colocated users changed over time.

For mobile social networks (MSNs) in proximity, the study of close-range human contacts has received attention from epidemiologists [7], who have studied the spread of viruses, and from technologists, who are interested in opportunistic routing in pocket switched networks and in mining the daily routine of users. This has resulted in several experiments that aim to record participants' meetings with other people. In Haggle study [8], participants were asked to carry Bluetooth-enabled devices that could scan and record other Bluetooth devices in proximity (within a 30 meter range). Different environments and numbers of participants were used, ranging from an office with 12 users to a conference with 78 users, with the intention of investigating decentralized routing of messages between mobile devices. In the Reality Mining study [9], 100 participants were given a Bluetooth-enabled smartphone to carry on campus over the course of nine months with the goal of mining human social behavior, such as predictability. Again, the devices would record other Bluetooth devices that were in proximity. In the EmotionSense study [10], social psychologists and computer scientists asked 18 participants to carry Bluetooth- and other sensor-enabled devices to record their interactions with other people and their emotions, sensed through the device microphone. The SocioPatterns project* used Radio Frequency IDentification (RFID) tags on necklaces to record face-to-face proximity (1 to 1.5 meters) colocations, to study the spread of airborne viruses. All the studies noted here allow us to infer when a pair or even a group of people are in proximity (for either radio communication or to transmit a biological virus). From this, a graph of people (nodes) and their contacts (edges) can be generated. Because we have time stamps when a device comes into and out of range of another device, we have information on the duration of a meeting; conversely, there is information on the time between successive meetings for the same pair of devices and, potentially, on periodic patterns between user colocations. The topology of the graph also changes over time as people move into and out of the range of one another.

In brief, for various MSNs (or OSNs), the structural characteristics are intrinsically multidimensional, which serves as the motivation for this chapter: to summarize new tools used to analyze the temporal–spatial–social structural properties of real mobile social networks that inherently change over time and space. This chapter presents network structures in two distinct ways: the traditional aggregate network of connections and interactions and a nontraditional, dynamic time-ordered series of interactions. Formal definitions of both will be given here.

The chapter is organized as follows: In Section 2.2, we briefly introduce basic concepts and key metrics in a static social network, which, in a sense, can be viewed as an aggregate view of the time-varying dynamic social networks. Section 2.3 thoroughly characterizes time-varying MSNs, including classifying temporal information, key measures, and so on. The spatial–social characteristics of MSNs are presented in Section 2.4. Finally, we briefly conclude this chapter.

2.2 Background on Static Social Network Characteristics and Measurements

Our purpose here is to briefly overview what is known about (static) social networks in terms of their basic structure and how they can be usefully quantified. First, some definitions and terminology are introduced, which will allow us to talk

---

* http://www.sociopatterns.org/

about network structure. Much of this terminology emerges from standard graph theory, with some variation in terms across disciplines [11].

The (static) aggregate social network is defined as the 2-tuple graph $G = (V, E)$ of individuals V and their interactions (edges) E observed over a period of time. In this representation, an edge exists between pairs of individuals if they have ever interacted during the observed time period. Multiple interactions between a pair of individuals over time are represented as a single weighted (by frequency) edge between them or as multiple edges between them (multigraph). This representation provides an aggregate view of the interactions where the timing and order of interactions is neglected. The dynamic, time-respecting social graph is presented in the following section.

A graph can be represented as an N-by-N adjacency matrix A, where $N = |V|$ represents the total number of nodes in a social system, and the value a_{ij} at row i and column j is non-zero if an edge exists from node i to j. In the case of an unweighted graph, $a_{ij} = 1$ if there is an edge, otherwise it is 0; in the case of a weighted graph, a_{ij} can be any real number. A graph is undirected if A is required to be symmetric so that $a_{ij} = a_{ji}$, and it is directed otherwise. Whether a network is directed or undirected depends on the application. In applications where mutual consent is required to maintain a relationship (friendships, alliances, partnerships, contracts, and so forth), it will often be most appropriate to represent these as an undirected graph, although there are other applications where unilateral relationships are possible (such as one author citing another or a Web page linking to another).

It is generally useful to use the notation $ij \in A$ to indicate $a_{ij} = 1$ (i.e., there exists an edge [interaction] between nodes i and j) and $ij \notin A$ to indicate that $a_{ij} = 0$.

Walk: A walk between a pair of nodes u and v in a network (V, E) refers to a sequence of nodes, $u = i_1, i_2, i_3, \ldots, i_{K-1}, i_K, i_{K+1} = v$, such that $(i_k, i_{k+1}) \in E$, for each k from 1 to K. The length of the walk is the number of links in it, or K.

Path: A simple path between a pair of nodes u and v in a network (N, E), denoted as P_{uv}, is a walk in (V, E), $u = i_1, i_2, i_3, \ldots, i_{K-1}, i_K, i_{K+1} = v$, such that all the nodes are distinct.

Component: A component of a network (N, E) is a subnetwork (N', E') (so $N' \subseteq N$ and $E' \subseteq E$) such that there is a path in E' from every node $i \in N'$ to every other node $j \in N'$ ($j \neq i$), and there exists for every node $l \in N$ such that $l \notin N'$ has no link in E to any node in N'. Thus, a component of a network is a maximal connected subgraph, such that the subgraph is connected and there is no way of expanding the set of nodes in the subgraph and still having it be connected.

2.2.1 Degree Distribution

Although the information contained in a full specification of all relationships (V, E) is sometimes very useful, it is generally too cumbersome when there are many nodes; therefore, descriptive statistics that capture facets of the network are useful.

For instance, knowing the average degree in the network gives some idea of the density of the connections in a network. However, often we need richer information, and the distribution of the degrees of the nodes provides more substantial information about network structure.

The degree distribution of a network (V, E) is the frequency distribution P of the degrees in the network. $P(d)$ indicates the fraction of nodes that have degree d. Degree distributions vary across applications. One extreme distribution corresponds to a regular network such that all nodes have the same degree. A useful benchmark is a network where each link is formed at random with the same probability p and independently of all other links in the network. In that case, the probability that a given node has degree d has a binomial distribution described by:

$$\binom{n-1}{d} p^d \left(1-p\right)^{n-1-d}$$

For large n and relatively small p, a standard approximation of a binomial distribution by a Poisson distribution applies, and the probability that a node has d links is approximately

$$p(d) = \frac{e^{-(n-1)p} \left((n-1)p\right)^d}{d!} \tag{2.1}$$

Such networks where all nodes are formed uniformly at random with the same probability have been studied extensively in random graph theory, including seminal papers by Erdős and Rényi and many others. They are often referred to as "Poisson random graphs," due to the (approximate) degree of distribution. They serve as a useful benchmark and exhibit many properties that are common to many random graph models:

■ When p is very low (well below $1/n$), most nodes are completely isolated and only a few nodes are linked as pairs.
■ As p increases (above $1/n$), a network begins to emerge in the sense that some nodes have more than one link, and a large component (referred to as the giant component) begins to emerge and dominate the network, and cycles begin to occur.
■ As p increases further—beyond $\log(n)$—the isolated nodes disappear and the network begins to coalesce into a single connected component.

Another useful distribution benchmark is a power distribution such that $P(d) = c \cdot d^{-\gamma}$.

For some parameters γ and normalizing constant c, the distribution is generally truncated at some upper boundary. In settings where such degree distributions are prevalent, it is often said that a power law is satisfied, and the distributions are referred to as being scale-free. The scale-free property refers to the fact that $P(d)/P(d') = P(k \cdot d)/P(k \cdot d')$, for any rescaling by a factor k. Such distributions have been found in a variety of settings, with prominent examples being the distributions of wealth noted by Pareto (for whom the related Pareto distribution is named) and also including word usage and city sizes (often referred to as Zipf's law). For power distributions, the frequency distribution can be rewritten as $\log(P(d)) = \log(c) - \gamma \cdot \log(d)$, and so they are linear when viewed on a log–log plot. An important feature of such a distribution is that it has "fat tails" relative to a Poisson distribution. Thus, the frequency of very high and very low degree nodes is greater than if links were formed uniformly at random, and correspondingly, the frequency of nodes with degrees near the center of the distribution is lower than if links were formed uniformly at random. This distinction can lead the network to have very different properties because very high degree nodes can serve as "hubs" and play prominent roles in different contexts.

Indeed, there are many examples of networks with degree distributions that have fat tails, and so it sometimes said that a power law is satisfied by many networks. Nevertheless, social networks exhibit a full spectrum of degree distributions across different applications, ranging from one extreme, where the distribution of links is nearly as if they were formed uniformly at random (e.g., matched well, such as by distributions of romances among high school students in the Add-Health data set), to the other extreme, where the distribution is nearly scale-free. Thus, although many networks have fatter tails than one would see uniformly at random, with statistically fitting degree distributions, they can come out somewhere between the extremes of scale-free and being formed uniformly at random, as discussed by Jackson and Rogers [12].

2.2.2 Characteristic Path Length

Between a pair of nodes, u and v, in same component in a network graph, there may be many different paths of different lengths that we refer to as the set $\mathcal{P}_{uv}$. Also, all paths are acyclic, in that there are no cycles or repeated nodes in a path.

The shortest (or geodesic) path length, d_{uv} from u to v, is defined as the minimum path length over all paths $P_{uv} \in \mathcal{P}_{uv}$. From this, the characteristic (or average) path length, L, is defined as:

$$L = \frac{1}{N(N-1)} \sum_{u \neq v \in V} d_{uv} \qquad (2.2)$$

This captures the global characteristics of a graph because transitive paths can connect every pair of nodes.

2.2.3 Clustering Coefficient

The clustering coefficient measures the number of nodes that are also neighbors to one another. More formally, for a node i, its clustering coefficient C_i is calculated as the fraction of links that exists between the neighbors of a node k_i of node i, over the total possible number of edges $k_i(k_i - 1)/2$. That is, C_i, the local clustering coefficient of node i, is defined as:

$$C_i = \frac{\text{Number of edges in } G_i}{\text{Maximum possible number of edges in } G_i}$$

$$= \frac{\text{Number of edges in } G_i}{k_i(k_i - 1)/2}$$

(2.3)

where G_i is the subgraph of neighbors of i, and k_i is the number of neighbors of peer i. Then most $k_i(k_i - 1)/2$ edges can exist in G_i; this occurs when the subgraph G_i is completely connected. C_i denotes the fraction of these allowable edges that actually exist, and the average clustering coefficient $C(G)$ of graph G is defined as the average of C_i over all the vertices i of G, that is:

$$C(G) = \frac{1}{N} \sum_{i \in G} C_i$$

(2.4)

Note that Newman also proposed another definition of the clustering coefficient [13].

2.2.4 Network Efficiency, E(G)

Obviously, characteristic path length (CPL) is only meaningful for connected networks in which there exists at least one path connecting any couple of nodes. The general metric network efficiency, E, was introduced in Ref. [14] to measure how efficiently network G exchanges information, which is the extension of the measurement of CPL. CPL is defined as the average of the shortest path lengths between two generic nodes. Obviously, CPL is only meaningful for connected networks in which there exists at least one path connecting any couple of nodes. The global network efficiency is a good measure of the parallel systems' performance and is based on the assumption that the information/communication in a network travels along the shortest routes. Specifically, the efficiency $\varepsilon(G, u, v)$ in the communication between two points u and v is defined as the inverse of the shortest path length $d(G, u, v)$, and the global efficiency of G, $E_{\text{Global}}(G)$ is the average of $\varepsilon(G, u, v)$:

$$E_{\text{Global}}(G) = \frac{1}{N(N-1)} \sum_{u \neq v \neq G} \varepsilon(G, u, v) = \frac{1}{N(N-1)} \sum_{u \neq v \neq G} \frac{1}{d(G, u, v)}$$

(2.5)

The quantity $E_{\text{Global}}(G)$ is perfectly defined in the case of nonconnected graphs; in fact, when there is no path between two points j and k, we assume $d(G, u, v) = +\infty$ and consistently $\varepsilon(G, u, v) = 0$.

In the same vein, to parallel the local dynamics that $C(G)$ captures, a local efficiency E_{loc} metric is defined as:

$$E_{\text{loc}} = \frac{1}{N} \sum_{i \in N} E(G_i) \qquad (2.6)$$

where G_i is the neighbor subgraph of a node i. Intuitively, the larger clustering coefficient is, the larger the local efficiency.

2.2.5 Small-World Behavior

The particular class of networks named "small-world" because of its similarity to the small-world phenomenon (popularly known as "six degrees of separation") observed more than 40 years ago in social systems. In his famous experiment, Milgram showed that the human acquaintance network has a diameter in the order of six, leading to the small-world qualification.

The term "small-world network" is used by Watts and Strogatz to denote those systems that have a high average clustering coefficient, such as regular lattices, yet also have a small CPL, such as random graphs. Later, Watts and Strogatz introduced a model of small-world phenomenon in static graphs [15]. From a regular ring lattice, they randomly rewired edges in this graph with a probability varying from 0 (i.e., leading to a regular network) to 1 (i.e., leading to a random graph). During this process, they observed an abrupt decrease in the average shortest path length, leading to a short path of the same order of magnitude as observed in random graphs, although the clustering coefficient is still of the same order of magnitude as that of a regular graph. This feature suggested the emergence of the small-world phenomenon.

Kleinberg [16] extended the model to two-dimension lattices and introduced a new rewiring process. The edges are not uniformly rewired but follow a power law $1/d^\alpha$, where d is the distance on the lattice from the starting node of the edge and α is the parameter of the model.

The intuition is that small-world networks exhibit strong clusters of nodes such as groups of friends that are more likely to be linked from certain nodes to distant clusters, providing a "shortcut." It is the combination of these close-knit clusters of nodes (which can interact locally) and these shortcut links (which aid in global interactions) that help in reducing the number of transitive hops between any two nodes in a large network.

From an economic viewpoint, in a small-world network, the global and local efficiency is much higher compared to that of a random graph. In the context of static networks, this is equivalent to the fact that the clustering coefficient is high and the average shortest path length is low (short and small). This special structure allows information to spread as fast as in a random network. Many real static networks exhibit these properties.

2.2.6 Centrality

In complex network and social network analysis, centrality refers to the identification of the most "important" nodes in a network. Clearly, node importance is an ambiguous term and could be interpreted in many different ways depending on the application. For example, one could interpret importance as being equal to popularity (e.g., a person with the most friends); one might argue that a person who can deliver a message quickly to the most people in a network is important; or perhaps, one might give precedence to a person that bridges the most communication channels and therefore is key to mediating among different parties. In fact, all three interpretations have been well studied in social network analysis and are more commonly known, respectively, as degree, closeness, and betweenness centrality.

2.2.7 Degree Centrality

One of the simplest measures in network analysis is node degree d_i, which measures the number of neighbors of a node, where $d_i = \sum_{j \in V} a_{ij}$. Because the degree is defined for each node, deriving a measure of centrality based on popularity is not difficult.

2.2.8 Closeness Centrality

Closeness centrality measures how quickly a node can communicate with all other nodes in a network. This is calculated for a node i as the average shortest path length d_{ij} to all other nodes j in the network. Formally, this can be defined in terms of shortest path lengths:

$$C_i^{\text{clo}} = \frac{1}{N-1} \sum_{j \neq i \in V} d_{ij} \qquad (2.7)$$

or in terms of efficiency to handle disconnected nodes:

$$C_i^{\text{eff}} = \frac{1}{N-1} \sum_{j \neq i \in V} \frac{1}{d_{ij}} \qquad (2.8)$$

2.2.9 Betweenness Centrality

Betweenness centrality measures the shortest paths that pass through a node and can be thought of as the proportional flow of data through each node. The betweenness of node i is calculated as the proportional number of shortest paths between all node pairs in the network that pass through i. More formally, this is defined as:

$$C_i^{\text{bet}} = \sum_{j \neq i, k \neq i \in V} \frac{p_{jk}(i)}{p_{jk}} \tag{2.9}$$

where p_{jk} is the number of shortest paths starting from source node j and destination node k, and $p_{jk}(i)$ is the number of those paths that pass through node i. A key point is that betweenness also takes into account alternative shortest paths, which is meaningful in measuring the robustness of a node to attack. If a node i is the only bridging node on all those paths, then its removal would be highly detrimental, whereas if there were another path that did not include i, then its role would be less critical.

2.2.10 Eigenvector Centrality

In eigenvector centrality, the importance of a vertex depends on how much its neighbors are important themselves. The idea is that ties with vertices with high importance values contribute to the importance value of a vertex more than ties with vertices that are not so important. There are different eigenvector centrality measures. In general, the computation of such measures involves finding the eigenvector corresponding to the first eigenvalue of the graph adjacency matrix, taking it as a centrality measure. Unlike degree centrality, not only are direct connections taken into account but also undirected connections in the entire graph; the centrality of each vertex is proportional to the sum of the centralities of the vertices to which it is connected.

Formally, given the adjacency matrix A of a social graph, the eigenvector centrality is defined as the eigenvector x such that:

$$Ax = \lambda x$$

where λ is the largest eigenvalue of A.

There are two main approaches when designing eigenvector methods for exploiting the structure of a (directed) graph. The first approach is to introduce the following two distinct notions:

- Authorities, which are nodes with several in-links
- Hubs, which are nodes with several out-links

The first algorithm that introduced these notions is hypertext-induced topic search (HITS) [17], which defines an authority score x_i and a hub score y_i for each node i. The main idea behind HITS is that good authorities are pointed to by good hubs, and good hubs point to good authorities. Let us consider a directed graph $G = (V, E)$ with adjacency matrix A. The authority and hub score distribution can be found by the following iterative computing, until reaching a convergence or the desired level of accuracy:

$$x(k) = A^T y(k - 1) \quad \text{and} \quad y(k) = Ax(k) \tag{2.10}$$

where k represents the number of iterations.

The main strengths of HITS are its dual centrality measures scores; it computes on the graph two separate ranked lists of nodes in terms of authority scores and hub scores, respectively; one could be more interested in one or the other depending on the applications. On the contrary, it is simple for "spamming nodes" to greatly influence their own hub scores and consequently the authority scores of nodes in their local vicinity. This is a weakness in some possible applications, such as in Web search information retrieval (for which it was originally designed). The second approach is to see links between nodes in a graph as recommendations from one node to another and to consider just one notion of "importance" of a node.

This is the approach used by PageRank, which is a link analysis algorithm for measuring the importance of nodes in a graph with respect to the structure of the graph itself. It was first introduced by Google founders Larry Page and Sergey Brin, and it is the basis of Google's success as a Web search engine [18]. PageRank tries to model the behavior of a user navigating the Web; in this so-called random surfer model, the user performs a random walk on the graph (representing the Web) by starting from a random page and, at each step, clicking on one of the links contained in the page, choosing uniformly at random. In addition, at each page, the user has a given probability $1 - d$ of jumping to a random page in the graph; the value $0 \le d \le 1$ is the so-called damping factor. Given a node v and the set $N_{in}(v)$ of nodes linking to v, and defining $N_{out}(v)$ as the set of nodes to which a node u links, the recursive definition of the PageRank is

$$PR(v) = \frac{1-d}{n} + d \sum_{u \in N_{in}(v)} \frac{PR(u)}{|N_{out}(u)|} \qquad (2.11)$$

where n is the total number of nodes in the graph.

A PageRank score is defined as the stationary probability distribution of a suitably modified Markov chain defined on the graph; this allows us to greatly alleviate the problem of spamming nodes.

However, the great majority of studies on social network analysis, properties, and behaviors are focused on static graphs and ignore the dynamics of real mobile networks. For example, in a static graph, an epidemic cannot break out if the initial infected node is in a disconnected component of the network; conversely, in a mobile network, node movements can ensure the temporal connectivity of the underlying dynamic graph. Moreover, an epidemic can take off or die out depending not only on the network structure and the initial carrier but also on the time when the disease begins to spread. These aspects cannot be captured by a static social network model. Because static graphs treat all links as appearing at the same time, they do not capture key temporal characteristics such as duration of contacts, inter contact time, recurrent contacts, and time order of contacts along a path. For this reason, they give us an overestimate of the potential paths connecting pairs of nodes, and they cannot provide any information about the delay associated with the information-spreading process.

2.3 Characterizing Time-Varying MSNs

2.3.1 Classifying Temporal Information

In introduction, a range of empirical data sets used in the literature was presented. Some of these have been used for static complex network analysis—where the temporal information is accessible from the original source (e.g., time stamps in the actor, coauthor data sets, etc.), where the temporal information is inherent but would be nontrivial to collect, and where the information was collected with time stamps present.

From these data set examples, four distinct sources of time information can be isolated, namely:

- Time stamps can be associated with nodes (new users or users leaving an OSN) and edges (a friendship being added or removed, a message being sent, a meeting between two people, etc.).
- Some form of duration in these time stamps is implicit—for example, the length of time a friendship lasts, the time it takes for a message to be sent and delivered, or how long two people meet.
- Frequency can be analyzed once we have a list of time stamps for an edge or node; this can uncover patterns in edge or node occurrences. Furthermore, periodicity is present in certain data sets such as transport traffic (e.g., during the morning and evening, before and after work) and human contact networks (e.g., daily meetings with colleagues or family members).
- Time order was highlighted in several data sets—for example, the timetable in public transport systems and a message or virus passing through a network. This is an important piece of information that is missed in static graph analysis. More generally, time order can be described as time dependency between events; for example, changing the order of time-stamped events would have an effect on metrics defined upon it.

In addition, we can also categorize two types of dynamic graph behavior. First, topological changes over time occur with fluctuations of the edges between nodes as meetings between people begin and end or as traffic moves along to congest and free up roads. Second, process changes are driven by some form of information exchange (i.e., a message or a virus).

2.3.2 Static versus Temporal Analysis

We should stress that the aim of this chapter is not to reject the use of static network analysis but merely to offer an alternative view whereby the incorporation of temporal information can potentially lead to more accurate analysis of networks where temporal information is inherent. Static graph analysis simplifies the analysis of real networks by ignoring time information, but it is still useful for many types of analysis where, for example, time information is not required or only a single snapshot is needed for analysis. In other cases, temporal analysis does not make sense because the changes would be minute; for example, the topology of the power grid does not change very frequently. However, the analysis of the traffic demands on the cables' carrying power would fluctuate frequently and, in this case, could potentially benefit from temporal analysis.

2.3.3 *Evolving versus Temporal Networks*

We should also distinguish between the concepts of the well-studied evolving (growth) network and the proposed temporal network analysis. Evolving networks are generative models—such as preferential attachment [19], which describes the accumulation of nodes and edges over time. The preferential attachment model was devised to understand how scale-free (where the degree distribution can be described by a power-law function) network topologies are formed over time as new nodes join a network. Basically, the recipe captures a snowball effect where new nodes have a higher probability of forming a link to popular nodes. This is used to explain the scale-free structure of the World Wide Web (WWW)—as new Web pages are added, they hyperlink to existing well-known Web pages; new researchers are more likely to coauthor a paper with well-known, respected peers, and so forth. Such evolving networks relate well to static analysis because the most current cumulative topology is of interest and has given rise to insightful results such as shrinking diameters (maximal shortest path length) and densification over time as nodes and edges are added to the graph. We are concerned here with a different time-dependent scenario where the population of nodes remains fixed from the outset, and the graph evolves through the appearance (birth) or the deletion (death) of edges.

In brief, the social network topology varies with time. Furthermore, the rate and/or degree of the changes is generally too high to be reasonably modeled in terms of network faults or failures. In these systems, changes are not anomalies but are rather an integral part of the nature of the system. We argue that static metrics such as path length, clustering coefficient, and centrality, to name a few, are sufficient where temporal information is not inherent in the network but give a too coarse-grained view in networks where the temporal dynamics is an essential component of the phenomenon under observation (such as human interactions over time).

Temporal concerns are an integral part of recent research efforts in complex systems. It is also apparent that the emerging concepts are in essence the same as those from the field of communication networks, again involving temporal definitions of the notions of paths, distance, and connectivity, as well as many higher concepts introduced in this chapter. Because the notion of (static) graph is the natural means for representing a static network, the notion of dynamic graph is the natural means to represent these highly dynamic networks. All the concepts and definitions in this chapter (and in existing works) are based on or imply such a notion, as is even expressed by the choices of names. For example, Flocchini et al. [20] and Tang et al. [21] independently employ the term time-varying graphs; Kostakos uses the term temporal graph [22]; and so on.

The main aim of this chapter is to integrate the existing models, concepts, and results proposed in the literature into a unified framework, which is called time-varying graphs (TVGs) [23]. Using it, it is possible to directly express not only the concepts common to all these different areas but also those specific to each. This, in turn, should enable the transfer of results from one application area to another.

2.3.4 Formalism in TVGs

Consider a set of entities V (or nodes), a set of relations E between these entities (edges), and an alphabet L accounting for any property such a relation could have (labels), that is $E \subseteq V \times V \times L$. The definition of L is domain-specific and therefore left open—a label could represent, for instance, the intensity of relation in a social network, a type of carrier in a transportation network, or a particular medium in a communication network; in some contexts, L could be empty (and thus possibly omitted). For generality, we assume L to possibly contain multivalued elements. The set E enables multiple relations between a pair of entities as long as these relations have a distinct label (e.g., Wi-Fi; range, bandwidth; energy consumption, encryption available).

Because this chapter addresses dynamic systems, the relations between entities are assumed to take place over a time span $T \subseteq \mathbb{T}$, which is called the lifetime of the system. The temporal domain $\mathbb{T}$ is generally assumed to be $\mathbb{N}$ for discrete-time systems or $\mathbb{R}^+$ for continuous-time systems. The dynamics of the system can be subsequently described by a TVG, $\mathcal{G} = (V, E, T, \rho, \xi)$, where $\rho: E \times T \rightarrow \{0,1\}$ is called the presence function and indicates whether a given edge is available at a given time; $\xi: E \times T \rightarrow T$ is called latency function and indicates the time it takes to cross a given edge if starting at a given date (the latency of an edge could vary in time).

The model can be naturally extended by adding a node presence function $\psi: V \times T \rightarrow \{0,1\}$ (i.e., the presence of a node is conditional upon time) and a node latency function $\varphi: V \times T \rightarrow T$ (accounting, for example, for local processing times).

Arguably, the TVG formalism can describe a multitude of different scenarios from transportation networks to communication networks, complex systems, or social networks.

Such formalism can describe a multitude of different scenarios, including:

■ Transportation networks (e.g., aviation, where nodes are the cities, directed edges are regular flights, the departure dates of which are given by punctual presences, and flight duration by non-negligible latencies)

- Communication networks (e.g., wireless mobile networks, where an edge is present whenever its two endpoints are within range, the latency corresponding here to the time to propagate a message)
- Complex systems, among which social networks (such scientific networks, where the nodes are scientists) and the edges (possibly both directed and undirected) account for citations or collaborations, for example.

These examples illustrate the spectrum of models over which the TVG formalism can stretch. As observed, some contexts are intrinsically simpler than others and call for restrictions (e.g., directed vs. undirected edges, single vs. multiple edges, punctual vs. lasting relations). Further restrictions may apply. For example, the latency function could be constant over time, over the edges, over both, or simply ignored. In fact, a vast majority of work in social networks does not require such information (e.g., the propagation time of an e-mail is of little interest to the understanding of a community behavior).

A number of analytical works on dynamic networks simply ignore ξ, or they assume a discrete-time scenario where every time step implicitly corresponds to a constant z. This value is also neglected, in general, when the graph represents dated interactions over a social network (the edges in this context are generally assumed to be punctual in terms of instantaneous presence and null latency). The formalism introduced in this chapter aims to address the general case, where $\mathcal{G} = (V, E, \mathcal{T}, \rho, \xi)$.

In particular, the temporal networks we consider in this review can be divided into two (rough and overlapping) classes corresponding to the two types of representations—contact sequences and interval graphs—illustrated in Figure 2.1 [24].

In the first representation (Figure 2.1a), there is a set of N vertices V interacting with each other at certain times, and the durations of the interactions are negligible. In detail, based on TVG, one can represent the system by V, a set of edges (pairs of vertices) E, and, for $e \in E$, a nonempty set of times of contact $T_e = \{t_1, \ldots, t_n\}$. Typical systems suitable to be represented as a contact sequence include communication data

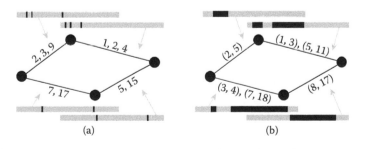

(a) (b)

Figure 2.1 (a) Contact sequences; (b) interval graphs. (Reprinted from *Physics Reports*, **519**, Holme, P., and J. Saramaki, Temporal networks, 97–125, Copyright (2012), with permission from Elsevier.)

(sets of e-mails, phone calls, text messages, etc.) and physical proximity data where the duration of the contact is less important (e.g., sexual networks). Typical edges in this scenario are available on a punctual basis, that is, the presence function ρ for these edges returns 1 only at particular date(s).

It is common to group the contacts happening at the same discrete-time step into one graph (or "graphlet" in the terminology of Ref. [25]) and present the temporal network as a time sequence of graphs. Because this representation makes it tempting to think of the temporal network structure as an evolving static network structure (which misses many of the unique points of temporal networks), in this book, the term "contact sequences" is preferred.

As illustrated in Figure 2.1b, in the second class of temporal networks (interval graphs), the edges are not active over a set of times but rather over a set of intervals $T_e = \{(t_1, t_1'), \cdots, (t_n, t_n')\}$, where the parentheses indicate the periods of activity—the unprimed times mark the beginning of the interval and the primed quantities mark the end. A typical presence function for this type of edge returns once for some intervals of time because the nodes are generally in a range for a non-punctual period of time. Note that the effective delivery of a message sent at time t on an edge e could be subjected to further constraints regarding the latency function, such as the condition that $\rho(e)$ returns once for the whole interval $(t, t + \xi(e, t))$.

Examples of systems that are natural to model as interval graphs include proximity networks (where a contact can represent that two individuals have been close to each other for some extent of time), seasonal food webs (where a time interval represents that one species is the main food source of another at some time of the year), and infrastructural systems such as the Internet.

Figure 2.1 illustrates the two fundamental temporal network representations in this chapter—contact sequences (a) and interval graphs (b). The times of the contacts are stated next to the edges. We also visualize the contact timelines (gray bars). In these, the contacts are marked by black bars or fields, and the timelines range from $t = 0$ to $t = 20$ (with $t = 0$ to the left). In the former, contacts occur at points in time, whereas the contacts are extended in time in the latter.

Note that the timeline plot shown in Figure 2.2b is another suitable depiction of contact sequences; the contact graph shown in Figure 2.1a is not as readable.

In Figure 2.2a, the times of the contacts between vertices A–D are indicated on the edges. Assume that, for example, a disease starts spreading at vertex A and spreads further as soon as a contact occurs. The dashed lines and vertices show this spreading process for four different times. The spreading will not continue further than what is indicated in the $t = \infty$ picture (i.e., D cannot become infected). However, if the spreading started at vertex D, the entire set of vertices would eventually be infected. Aggregating the edges into one static graph cannot capture this effect that arises from the time ordering of contacts. Panel (b) visualizes the same situation by showing the temporal dimension explicitly. The grayscales of the lines in (b) match the vertex grayscale in (a). The following

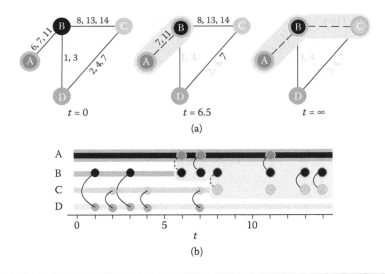

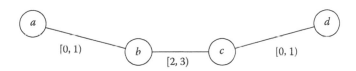

Figure 2.2 Illustration of the reachability issue and the intransitivity of temporal networks (more specifically, presentation for a contact sequence). (a) Contacts times indicated on the edges; (b) Explicit illustration of temporal dimension. (Reprinted from *Physics Reports*, 519, Holme, P., and J. Saramaki, Temporal networks, 97–125, Copyright (2012), with permission from Elsevier.)

Figure 2.3 An example of TVG that is not "connected over time."

will transpose and generalize a number of dynamic network concepts into the framework of a TVG.

Given a TVG $\mathcal{G} = (V,E,\mathcal{T},\rho,\xi)$, the graph $G = (V, E)$ is called the underlying graph of $\mathcal{G}$. This static graph should be seen as a sort of footprint of $\mathcal{G}$, which flattens the time dimension and indicates only the pairs of nodes that have relations at some time in $\mathcal{T}$. It is a central concept that is used recurrently. In most studies and applications, G is assumed to be connected; in general, this is not necessarily the case. Let us stress that the connectivity of $G = (V, E)$ does not imply that $\mathcal{G}$ is connected at a given time instant; in fact, $\mathcal{G}$ could be disconnected at all times. The lack of relationship, with regard to connectivity, between $\mathcal{G}$ and its footprint G is even stronger: the fact that $G = (V, E)$ is connected does not even imply that $\mathcal{G}$ is "connected over time." As illustrated in Figure 2.3, although its underlying graph G is connected, the nodes a and d have no means to reach each other through a chain of interaction.

2.3.5 Journeys and Related Temporal Concepts

A crucial concept in TVGs is that of journey, which is the temporal extension of the notion of path; it forms the basis of recently introduced temporal concepts. A sequence of couples $\mathcal{J} = \{(e_1,t_1),(e_2,t_2),...,(e_k,t_k)\}$, such that $\{e_1, e_2,...,e_k\}$ is a walk in G, is a journal in $\mathcal{G}$ if and only if $\forall i$, $1 \le i < k$, $\rho(e_i,t_i) = 1$ and $t_{i+1} \ge t_i + \xi(e_i,t_i)$. Additional constraint may be required in specific domains of application, such as the condition $\rho(e_i, [t_i, t_i] + \xi(e_i,t_i)) = 1$ in communication networks (the edge remains present until the message is delivered). The starting date t_1 and the last date $t_k + \xi(e_k,t_k)$ of a journey $\mathcal{J}$, respectively, are denoted by departure $(\mathcal{J})$ and arrival $(\mathcal{J})$. Journeys can be thought of as paths over time from a source to a destination; therefore, they have both a topological and temporal length. The topological length of $\mathcal{J}$ is the number $|\mathcal{J}| = k$ of hops in $\mathcal{J}$; its temporal length is its end-to-end duration: $\mathcal{J} = \text{arrival}(\mathcal{J}) - \text{departure}(\mathcal{J})$.

Using $\mathcal{J}^*$, let us denote the set of all possible journeys in a time-varying graph $\mathcal{G}$, and using $\mathcal{J}^*(u,v) \subseteq \mathcal{J}^*$, those journeys starting at node u and ending at node v. As observed, the length of a journey can be measured in terms of hops and time. This gives rise to two distinct definitions of distance in TVG $\mathcal{G}$.

■ The topological distance from a node u to a node v at time t is defined as $\text{Min}\{|\mathcal{J}| : \mathcal{J} \in \mathcal{J}^*(u,v) \cap \text{departure}(\mathcal{J}) \ge t\}$. For a given date t, a journey with a departure of $t' \ge t$ and a topological length that is equal to $d_t(u, v)$ is qualified as shortest.

From a temporal viewpoint, the following two distance definitions are given as follows:

■ The foremost distance from a node u to a node v at time t, note $\delta_t(u, v)$ is defined as $\text{Min}\{\text{arrival}(\mathcal{J}) - t : \mathcal{J} \in \mathcal{J}^*(u,v) \cap \text{departure}(\mathcal{J}) \ge t\}$.

■ For the fastest distance from a node u to a node v at time t, note $\hat{\delta}_t(u,v)$ is defined as $\text{Min}\{\mathcal{J} : \mathcal{J} \in \mathcal{J}^*(u,v) \cap \text{departure}(\mathcal{J}) \ge t\}$.

The problem of computing shortest, foremost, and fastest journeys in delay-tolerant networks (DTNs) was introduced in Ref. [26], and an algorithm for each of the three metrics was provided for the centralized version of the problem (assuming complete knowledge of $\mathcal{G}$). Temporal distance and related concepts have been used in various fields including social network analysis.

Recently, Tang et al. [21] defined several metrics for TVGs, including temporal path length, temporal clustering coefficient, and temporal efficiency. They showed that these metrics are useful for capturing temporal characteristics of dynamic networks that cannot be captured by traditional static graph metrics.

Most temporal concepts—including all those mentioned here—are based on replacing the notion of path by that of journey. Furthermore, the number of definitions built on top of temporal concepts could grow endlessly, and our aim is certainly not to enumerate all of them. Yet here is a short list of additional concepts that we believe are general enough to be possibly useful in several analytical contexts.

2.3.6 *Temporal Betweenness Centrality*

The betweenness of a node in a static graph measures the occurrences of that node within the shortest paths of other nodes. A temporal version of the betweenness is given as follows:

$$B(q) = \sum_{u \neq v \neq q \in V} \frac{|d_t(u,v,q)|}{|d_t(u,v)|} \tag{2.12}$$

where $|d_t(u, v)|$ is the number of shortest journeys between u and v in the time-varying graph $\mathcal{G}$, and $|d_t(u, v, q)|$ is the number of shortest journeys, among them, that pass through q. We can analogously define the temporal betweenness in terms of foremost or fastest distance by substituting $|d_t(u, v)|$ with $\delta_t(u, v)$ or $\hat{\delta}_t(u,v)$.

2.3.7 *Temporal Closeness Centrality*

In a static context, the closeness measures the mean of the shortest paths between a node and all the other reachable nodes. In a temporal dynamic network, it can be formally defined as:

$$TC(u) = \sum_{v \in V \setminus u} \frac{d(u,v)}{|\{w \in V : \exists \mathcal{J} \in \mathcal{J}^*(u,w)\}|} \tag{2.13}$$

and can again possibly define the foremost $\delta_t(u, v)$ or fastest $\hat{\delta}_t(u,v)$ versions.

2.3.8 *Temporal Eigenvector Centrality*

Another class of centrality measures takes its starting point in the assumption that something diffuses randomly around the network, instead of traveling from source to target along the shortest paths, such as for closeness and betweenness. For static graphs, this approach yields matrix-based centrality measures such as the eigenvector centrality and PageRank, which were introduced in Section 2.2. In this section, we describe the generalization of the eigenvector centrality in temporal networks [27].

Suppose we have a time-ordered sequence of unweighted graphs defined over a set of N nodes. Given the time points $t_0 < t_1 < \ldots < t_M$, we let $A^{[k]}$ denote the adjacency matrix for the network at time t_k. So the i, j entry of $A^{[k]}$ equals one if there is a link from node i to j at time t_k, and $A^{[k]}$ equals zero otherwise. For undirected links, where $A_{ij}^{[k]} \neq A_{ji}^{[k]}$ are allowed, no self loops exist, so $A_{ii}^{[k]} \equiv 0$.

Let $\Delta t_i := t_i - t_{i-1}$ denote the spacing between successive time points. We do not assume that the time points are equally spaced. Nonuniform spacing is natural; for example, if we have time-stamped e-mails or text messages with each $A^{[k]}$ recording one event.

One key observation that generalizes a simple result from graph theory is that the matrix product $A^{[0]} A^{[1]} \ldots A^{[w]}$ has i, j element that counts the number of dynamic walks of length w from node i to node j on which the mth step of the walk takes place at time t_m. Now, suppose that we were to quantify the propensity for node i to communicate or interact with node j. For each length $w = 1,2,\ldots$, we may count the number of dynamic walks from i to j, and this information may then be combined into a single, cumulative total overall w. Allowing for the fact that shorter walks are generally more important (because, for example, the noise or cost of transmission may increase with length), it makes sense to scale the counts according to the walk length. A particularly attractive choice is to downweight the walk of length w with a factor a^w, where $0 < a < 1$. Using the matrix multiplication setting, this leads to the task of summing all products of the form: $a^w A^{[0]} A^{[1]} \ldots A^{[w]}$. Letting I denote the $N \times N$ identity matrix, and noting that the resolvent $\left(I - a \cdot A^{[p]}\right)^{-1} = I + a \cdot A^{[p]} + a^2 \cdot \left(A^{[p]}\right)^2 + \cdots$, these arguments motivate the matrix product, the so-called dynamic communicability matrix:

$$\mathfrak{J}^k = \left(I - a \cdot A^{[0]}\right)^{-1} \left(I - a \cdot A^{[1]}\right)^{-1} \cdots \left(I - a \cdot A^{[k]}\right)^{-1} \qquad (2.14)$$

The parameter a is assumed to satisfy $a < 1/\max_k \mu\left(A^{[k]}\right)$, where $\mu(\cdot)$ denotes the spectral radius, that is, the largest eigenvalue in modulus. This ensures that the resolvent in the aforementioned equation exists and may be expanded according to $\left(I - a \cdot A^{[k]}\right)^{-1} = I + a \cdot A^{[k]} + a^2 \cdot \left(A^{[k]}\right)^2 + \cdots$. It is then easy to see that $\mathfrak{J}^k_{ij}$ is a weighted sum of the number of dynamic walks from i to j using the ordered sequence $\left\{A^{[0]}, A^{[1]}, \cdots, A^{[k]}\right\}$, where the number of walks of length w is scaled by a^w. The key idea here is that each possible walk around the network from node i to node j adds to the communicability measure, but longer walks are less influential than shorter walks. In the static case, where $k = 0$, this leads to the classical Katz centrality [28]; we note that Katz also offered the interpretation that a represents the independent probability that a message successfully traverses an edge.

In a sense, $\mathfrak{J}^k_{ij}$ can measure how well information can be passed from node i to node j. The ith row and column sums are given as follows:

$$C_i^{\text{broadcast}} = \sum_{n=1}^{N} \mathfrak{J}_{ni} \quad \text{and} \quad C_i^{\text{receive}} = \sum_{n=1}^{N} \mathfrak{J}_{ni} \qquad (2.15)$$

They represent the centrality measures that quantify how effective node i can broadcast and receive messages, respectively.

Because we are interested in the relative values of the centrality measures across all nodes, rather than their absolute sizes, it makes sense to avoid under- or overflow in the computation of $\mathfrak{J}$ using iteration such as:

$$\mathfrak{J}^{k+1} = \frac{\mathfrak{J}^k \left(I - aA^{[k+1]} \right)^{-1}}{\left\| \mathfrak{J}^k \left(I - aA^{[k+1]} \right)^{-1} \right\|}$$

(2.16)

where $\|\cdot\|$ denotes any convenient matrix norm. Note that $C_i^{\text{broadcast}}$ and C_i^{receive} reduce to Katz's centrality measure when there is a single time point with undirected edges.

It is important to note that the use of $\mathfrak{J}^k$ is strongly tied to the idea of a start point, t_0, and an end point, t_k. Any walk that took place in the time period $t_0, \ldots, t_k$ has equal influence. Also, by construction, the elements in $\mathfrak{J}^k$ are nonnegative and nondecreasing with k, so that pairs of nodes cannot become less communicative over time. These features are appropriate in some applications—for example, if the networks represent functional connectivity among brain regions in the course of a well-defined task [29]. However, there are many applications where we are interested in the current and recent activity but not in the activity that took place a long time ago—messages go out of date, rumors lose their timeliness, some viruses become less infectious.

In this chapter, we take the view that it is of interest to know whether node i recently had the opportunity to get a message to node j using short walks. At one extreme, the matrix $A^{[k]}$ gives us the most localized picture, telling us what is possible using a single step with only today's connectivity. At the other extreme, the matrix $\mathfrak{J}$ gives us the most historical view, telling us what is possible using all the connections that ever existed up to the current time. In the following, a matrix iteration that interpolates between these two extremes is presented [30].

Suppose we have several months' worth of hourly e-mail or phone activity, starting from some arbitrary day zero. It would be of interest to compute a time-dependent "running summary" of communicability between pairs of nodes. Here, at each point in time, we are interested in the capability of node i to pass messages to node j, where (i) as we discussed in the preceding paragraph (and has been noted in the derivation of many centrality measures for static networks), long walks are less important than short walks; but also, in this time-dependent setting, (ii) walks that started recently are more important than walks that started a long time ago.

These requirements motivate the idea of a running dynamic communicability matrix, $\mathcal{S}^k$, based on two parameters, $a \in (0,1)$ and $b > 0$. Here, as in the earlier equation, a is used to downweight walks of length w by the factor a^w. To explain the new parameter b, we refer to the current age, t, of a dynamic walk as the time that has elapsed since the walk began. The parameter b is then used to further downweight by the age-dependent factor e^{-bt}. So b is used to filter out "old" activity. We therefore propose the following iteration where, for convenience, $S^{[-1]} = 0$:

$$S^k = \left(I + e^{-b\Delta t_k} S^{[k-1]}\right)\left(I - aA^{[k]}\right)^{-1} - I, \quad k = 0, 1, 2, \cdots \tag{2.17}$$

To understand how this works, we can expand the right-hand side of this equation as:

$$a \cdot A^{[k]} + a^2 \cdot \left(A^{[k]}\right)^2 + \cdots + a^r \cdot \left(A^{[k]}\right)^r + \cdots + e^{-b\Delta t_k} S^{[k-1]}$$

$$+ e^{-b\Delta t_k} S^{[k-1]} a A^{[k]} + e^{-b\Delta t_k} S^{[k-1]} a^2 A^{[k]2} + \cdots + e^{-b\Delta t_k} S^{[k-1]} a^r A^{[k]r} + \cdots$$

This leads to the following interpretation:

- The first terms give a length-weighted count of all walks that start and finish at the current time, t_k.
- The second term deals with all "old" walks that do not involve time t_k. These get downweighted by the time factor $e^{-b\Delta t_k}$ because the age of each such walk has increased by Δt_k.
- The third term deals with all walks that began at an earlier time, but it makes use of one or more edges at the current time, t_k. The factor $e^{-b\Delta t_k}$ is used again because the age of each such walk has increased by Δt_k. Then, we have a length-downweighting factor a^r if r new edges are used.

This formulation has the following features: The i, j element of the matrix S^k records a scaled count of the number of dynamic walks from i to j that can be taken with the time-ordered sequence $A^{[0]}, A^{[1]}, \ldots, A^{[w]}$. The scaling is the product of (a) a factor a^w for walks of length w and (b) a factor e^{-bt} for walks that began t time units ago.

The factor $e^{-b\Delta t_k}$ may be interpreted as the independent probability that a message does not become "irrelevant" (or that a virus does not mutate into a harmless form) over a time length Δt_k. We also note that the iteration automatically incorporates the case of nonuniform time spacing.

When $b = 0$ (no down-scaling in time), we essentially recover the original iteration. Specifically, we have $\mathfrak{J}^k = I + S^k$. On the contrary, for $b = \infty$, that is, $e^{-b\Delta t_k} \equiv 0$ (complete down-scaling in time), we revert to Katz static centrality with $S^k = \left(I - aA^{[k]}\right)^{-1} - I$.

There is also a simple variation when we wish to restrict attention to walks that use at most one edge per time point. (For example, the time taken to pass a message along an edge may be comparable with a typical Δt.) In this case, we just replace the resolvent in Equation 2.17 by its first two terms: $S^k = \left(I + e^{-b\Delta t_k} S^{k-1}\right)\left(I + aA^{[k]}\right) - I$.

In brief, we can take the conventional approach of downweighting for length (messages become corrupted as they are passed along) and add the novel feature of downweighting for age (messages go out of date). This allows us to generalize widely used centrality measures that have proved popular in network science to the case of dynamic networks sampled at nonuniform points in time.

2.3.9 Small-World Properties in Temporal Social Network

A small-world network is one where the distance between two randomly chosen nodes (in terms of hops) grows logarithmically with the number of nodes in the network. TVG concepts, such as those of journeys, connectivity over time, and temporal distance, have been used to characterize the small-world behavior of real-world networks in temporal terms (i.e., the fact that there is always a journey of short duration between any two nodes). As the topological meaning (i.e., in terms of hops) of the small-world property in a dynamic context (e.g., the fact that mobile networks have a diameter of 7 [31]), it could be formalized as follows:

$$\forall u, v \in V, \ \forall t \in T, \ \exists \mathcal{J} \in \mathcal{J}^*(u,v) : \text{departe}(\mathcal{J}) \geq t, |\mathcal{J}| \leq 7$$

2.3.10 An Incarnation of TVG Framework

Specifically, in Ref. [32], a temporal graph can be represented by means of a sequence of time windows, where for each window we consider a snapshot of the network state at that time interval.

Given a real network interaction data set starting at t_{mim} and ending at t_{max}, the (undirected) temporal graph $\mathcal{G}^w(t_{\min}, t_{\max})$ is defined as an ordered sequence of undirected graphs $(G_0, G_1,\ldots, G_{\tau-1})$, where $G_t = (V_t, E_t)$ is a two-tuple consisting of a set of nodes V_t and edges E_t in the window t; $\tau - 1 = ((t_{\max} - t_{\min})/W) = |\mathcal{G}^w(t_{min}, t_{max})|$ is the number of graphs in the sequence; w is the duration of each time window expressed in some time units (e.g., seconds or hours).

Temporal path $p_{ij}^h = \left(n_0^{W_0},\ldots, n_\eta^{W_\eta} \right)$ starting at node $i = n_0$ and finishing at node $j = n_\eta$ can be defined over $\mathcal{G}^w(t_{min}, t_{max})$ as a sequence of η hops via a distinct node $n_\alpha^{W_\alpha}$ at time window W_α, where node n_α is passed a message if and only if there is an edge between $n_{\alpha-1}$ and n_α at time window $W_{\alpha-1} \leq W_\alpha$; and $0 \leq W_\alpha < \tau$.

To allow generality in this simplified temporal graph model and distance metric, the horizon parameter h is introduced, which is the maximum number of hops through which a message is replicated within the same window.

The discretization using fixed window size fits with many of the empirical data sets, such as annual friendship questionnaires, monthly snapshots of an MSN, and a constant scanning rate of Bluetooth sighting; however, it fails to model continuous time.

An important choice is that of the window size W, which is specific to application. However, some general guidelines for the selection of this parameter can be offered. Past work has made simplifying assumptions about the window size through arbitrary selection or has ignored this parameter completely due to the use of artificial simulation that also relied on known time steps. Also, the computational complexity of the calculation of temporal path length is $O(\tau \cdot (|V| + |E|))$, where τ is the window count; this means that although we could use a very fine window size, say for example seconds or milliseconds, for large networks that also extend a long observation time unnecessarily, small window sizes should be avoided.

Based on these observations and the experience of handling several empirical data sets, the following three guidelines could be helpful for selecting the size of the window.

- First, the data set collection timescale might provide a clear granularity to use—for example, the Bluetooth scanning rate in the REALITY Dataset, in which 100 participants were given a Bluetooth-enabled smartphone to carry on campus over the course of nine months with the goal of data mining human social behavior [9], such as predictability, in five minutes; hence, this provides a natural window size. The Gowalla friendship networks were collected in monthly intervals, and again no finer granularity is available.
- Second, the application timescale might motivate an appropriate window size, for example, daily interactions between people in an office or the seasonal effects of predator–prey relationships in food webs.
- Third, because the complexity of computing temporal shortest paths and path lengths is defined in terms of the number of windows w, computational power might limit the tractable window size. This is a limitation of the temporal graph model, but we shall see that any additional time information provides a better approximation to the real answer compared to a static graph (because increasing the window size eventually reduces down to a single window, which is the definition of a static graph).

Generally, the selection of an appropriate window size W plays an important part in the accurate analysis of temporal graphs, and with an appropriately fine window size, the horizon h parameter plays a small part in the calculation of the average temporal shortest path length. However, it should be noted that any increase in the number of windows τ from single windows (equivalent to an aggregated static graph) improves the accuracy of temporal analysis; hence, selecting a window size close to the collection interval gives a very good approximation of the true temporal path length. With this in mind, the following assumption is feasible: the typical time for a message to pass from a node to one of its neighbors is of the same order as the typical time at which the graph changes.

2.4 Spatiosocial Characteristics of MSNs

Online LBSNs have recently attracted millions of users and have experienced a huge increase in popularity over a short period of time. Thanks to the widespread adoption of location-sensing mobile devices, users can share information about their location with their friends. Among the biggest providers are Foursquare[*] and Gowalla,[†] although other hugely popular social networking services such as Facebook and Twitter have also introduced location-based features.

Location is increasingly becoming a crucial facet of many online services: people appear more willing to share information about their geographic position with friends, while companies can customize their services by taking into account where the user is located. As a consequence, service providers have access to a valuable source of data on the geographic location of users, as well as to online friendship connections among them. The combination of these two factors offers a groundbreaking opportunity to understand and exploit the spatial properties of the social networks arising among online users but also to open a potential window on real human sociospatial behavior. Specifically, these new features open novel research directions that are largely unexplored, such as the design of new social applications and the improvement of existing large-scale systems. Hence, it becomes important to investigate how geographic distance between individuals affects OSNs in order to deepen our understanding of these networks.

Several data sets exist from several popular LBN services such as Foursquare [5,33] and Gowalla [6]. With this user location information, researchers can construct a graph of user colocations (i.e., users who report that they are at the same location at the same time). Such graphs have been used in link prediction problems. User check-ins are time-stamped and so we have spatial–temporal information.

An analytical framework was described in Ref. [5], where network nodes are embedded in a metric space to study the relationship between social connections and geographic distance. Two new geosocial measures are defined: a node locality metric, which quantifies how much a node is engaged with a local rather than global set of individuals, and a geographic clustering coefficient, which extends the standard notion of clustering by taking into account how much clusters of people are connected by short-range ties.

In particular, a geographic social network can be represented as a graph G with N nodes and K links: nodes represent users and a link among two nodes exists if there is a social tie between them (e.g., a person lists another user as one of his or her friends). A link may be undirected or directed: in the latter case, the existence of a link from node i to node j does not imply the existence of the reverse link from j to i. Given a fixed location on the Earth for each user, nodes are embedded in a two-dimensional metric space where the distance between two nodes i and j is given by the geographic distance D_{ij} between their locations on the planet. This distance is used as the length of the link l_{ij} between nodes i and j.

---

[*] https://foursquare.com
[†] http://en.wikipedia.org/wiki/Gowalla

2.4.1 Node Locality

A metric of node locality is defined to quantify the geographic closeness (i.e., the locality) of the neighbors of a certain node to the node itself. Consider an undirected geographic social network, a node i with a particular geographic position, and the set Γ_i of its neighbors node degree k_i is the number of these neighbors, that is $k_i = |\Gamma_i|$. Then, the node locality of i can be defined as a measure of how geographically close are its neighbors, and it is computed as follows:

$$NL_i = \frac{1}{k_i} \sum_{j \in \Gamma_i} e^{-l_{ij}/\beta} \qquad (2.18)$$

where β is a scaling factor that avoids extremely small values of node locality when links have large lengths. By definition, NL_i is always normalized between 0 and 1. The value of β does not impact the relative values of node locality: a node with a higher locality than another one will always have a higher value, regardless of the value of β.

At the same time, β can be chosen so that networks with different geographic sizes can still be compared with each other, as we will discuss later in this book. Finally, we adopt an exponential decay for node locality to highlight social ties that span over short geographic distances and to reduce, at the same time, the impact of longer ties.

In a similar fashion, in the case of directed graphs, the node in-locality can be defined considering only the incoming connections of a node; the node out-locality is defined considering only outgoing links. A node without in-connections will have, by definition, a node in-locality equal to 0; the same applies to out-locality.

2.4.2 Geographic Clustering Coefficient

Although node locality captures how close the neighbors of a node are, another measure is needed to quantify how tightly connected the neighborhood of a node is. Thus, the geographic clustering coefficient can be defined as an extension of the clustering coefficient used for complex networks. The clustering coefficient measures the proportion of triangles among the neighbors of a given node: this geographic adaptation attempts to weigh triangles formed by nodes that are close to each other differently from triangles where nodes are at longer distance. The geographic clustering coefficient of node i is thus defined in the same way as the clustering coefficient, but each existing triangle between nodes i, j and k is assigned a weight ω_{ijk} defined as: $\omega_{ijk} = e^{\frac{-\Delta_{ijk}}{\beta}}$, where Δ_{ijk} is the maximum length among the three links, that is, $\Delta_{ijk} = \max(l_{ij}, l_{ik}, l_{jk})$. We define $\omega_{ijk} = 0$ if there is no link between j and k. Because this measure uses the maximum weight among all the links of a triangle, it focuses on nodes that are all close to each other: when just one of the three nodes is not close to the other two, the weight will immediately decrease. This emphasizes social triangles where users are extremely close to each other. Again, the parameter β is used to scale the values of the measure.

In the case of directed graphs, as in the case of the standard clustering coefficient, we consider triangles containing undirected links joining node i to its neighbors and directed links for the remaining side. Thus, if we consider as Γ_i the set of all the neighbors of node i (considering both incoming and outgoing links), with $k_i = |\Gamma_i|$, the geographic clustering coefficient is defined as:

$$GC_i = \frac{1}{k_i(k_i - 1)} \sum_{j,k \in \Gamma_i} \omega_{ijk} \qquad (2.19)$$

where the sum is extended only to existing triangles. Because there are exactly $k_i(k_i - 1)$ different ordered couples of neighbors in Γ_i, GC_i is normalized between 0 and 1 by definition.

Based on these metrics, from a geosocial point of view, it is possible to characterize some differences across various MSNs (OSNs). For instance, Scellato et al. [5] described and analyzed the social, geographic, and geosocial properties of four different data sets of real OSNs with geographic information—Brightkite, Foursquare, LiveJournal, and Twitter—and illustrated how different OSNs present contrasting characteristics. Whereas purely location-based social networking services such as Foursquare and Gowalla mainly focus on the geographic dimension of social interaction and have high node locality and geographic clustering close to standard clustering, OSNs based more on the idea of sharing information and content (such as Twitter) result in users with lower node locality and geographic clustering coefficient values. The aforementioned contrasting phenomenon could be explained by varying attitudes of their users toward the social and geographic aspects of online friendship: location-based MSNs engage their users in short-range social connections more than sharing-based services, which exhibit social ties on a wider scale.

In brief, by taking into account geographic location, we can understand which role distance plays in social phenomena such as the creation of friendship ties, the development of personal tastes, and the spreading of information. And furthermore, augmenting social structure with geographic information adds a new dimension to social network analysis, and a large number of theoretical investigations and practical applications can be pursued on sociogeographic systems. For example, applications such as social search, social recommendation, and advertising would greatly benefit from geographic information about users: Search queries about local content could be directed to nearby users with many social links in the area of interest, while advertising and recommender systems could better profile users by knowing how their social ties stretch over space. Moreover, information about social links and geographic placement can tell us a great deal about how culture and taste disseminate on an OSN.

2.5 Conclusion

This chapter thoroughly summarizes the temporal–spatial–social structure of dynamic MSNs. The research about dynamic social networks under the umbrella of network science is still new; there are still a few insights about how these social networks evolve at the spatial–temporal structural level. The understanding of these characteristics of dynamic networks, such as for the static networks' counterparts, is necessary for designing and implementing efficient network protocols, especially for solving the routing problem.

The study of temporal, spatial, and social networks is very much an interdisciplinary field, where much of the development has been taking place in parallel, seemingly without much communication among the different disciplines. This is reflected in a tremendous amount of overlapping terminology—one concept can easily have four or five different names in the literature. One of the goals of this chapter is to give an overview of this research area in different fields. We will not try to gather the theory into one unified framework. Instead, we hope that this chapter can help readers from one discipline read and understand papers in others by being aware of the confusing terminologies in characterizing the dynamic multidimensional mobile social networks.

References

1. Traud, A. L., P. J. Mucha, and M. A. Porter. Social structure of Facebook networks. *Physica A: Statistical Mechanics and its Applications*, Elsevier, 2012; 391(16): 4165–4180.
2. Wilson, C., B. Boe, A. Sala, K. P. N. Puttaswamy, and B. Y. Zhao. User interactions in social networks and their implications. In: Proceedings of the 4th ACM European Conference on Computer Systems (EuroSys), Nuremberg, Germany, April 1–3, 2009.
3. Kwak, H., C. Lee, H. Park, and S. Moon. What is Twitter, a social network or a news media. In: Proceedings of the 19th Conference on the World Wide Web (WWW), North Carolina, USA, April 26–30, 2010.
4. Scellato, S., C. Mascolo, M. Musolesi, and J. Crowcroft. Track globally, deliver locally: Improving content delivery networks by tracking geographic social cascades. In: Proceedings of the 20th International Conference on the World Wide Web (WWW), Hyderabad, India, March 28–April 1, 2011.
5. Scellato, S., C. Mascolo, M. Musolesi, and V. Latora. Distance matters: Geo-social metrics for online social networks. In: Proceedings of the 3rd Conference on Online Social Networks (WOSN), Boston, MA, June 22, 2010.
6. Scellato, S., A. Noulas, and C. Mascolo. Exploiting place features in link prediction on location-based social networks. In: Proceedings of the 17th ACM SIGKDD Conference on Knowledge Discovery and Data Mining (KDD), San Diego, CA, August 21–24, 2011.
7. Isella, L., J. Stehlé, A. Barrat, C. Cattuto, J.-F. Pinton, and W. Van den Broeck. What's in a crowd? Analysis of face-to-face behavioral networks. *Journal of Theoretical Biology* 2011; 271(1): 166–180.

8. Hui, P., J. Crowcroft, and E. Yoneki. Bubble rap: Social-based forwarding in delay tolerant networks. In: Proceedings of the 9th ACM International Symposium on Mobile Ad Hoc Networking and Computing (MobiHoc), Hong Kong SAR, China, May 27–30, 2008, pp. 241–250.

9. Eagle, N., A. Pentland, and D. Lazer. Inferring friendship network structure by using mobile phone data. *Proceedings of the National Academy of Sciences* 2009; 106(36): 15274–15278.

10. Rachuri, K. K., M. Musolesi, C. Mascolo, P. J. Rentfrow, C. Longworth, and A. Aucinas. EmotionSense: A mobile phones based adaptive platform for experimental social psychology research. In: Proceedings of the 12th ACM International Conference on Ubiquitous Computing (UbiComp), Copenhagen, Denmark, September 26–29, 2010.

11. Jackson, M. O. An overview of social networks and economic applications. In: *Handbook of Social Economics*, Edited by J. Benhabib, A. Bisin, and M. O. Jackson. Elsevier Press; 2011; 1: 511–585.

12. Jackson, M. O. and Rogers, B. W. Meeting strangers and friends of friends: how random are social networks? *American Economic Review* 2007; 97(3): 890–915.

13. Newman, M. E. J. The structure and function of complex networks. *SIAM Review* 2003; 45(2): 167–256.

14. Latora, V. and M. Marchiori. Economic small-world behavior in weighted networks. *The European Physical Journal B* 2003; 32: 249–263.

15. Watts, D. and S. Strogatz. Collective dynamics of small-world networks. *Nature* 1998; 393(6684): 440–442.

16. Kleinberg, J. Navigation in a small world. *Nature* 2000; 406(6798): 845.

17. Kleinberg, J. Authoritative sources in a hyperlinked environment. *Journal of the ACM* 1999; 46(5): 604–632.

18. Page, L., S. Brin, R. Motwani, and T. Winograd. *The PageRank Citation Ranking: Bringing Order to the Web*. Stanford Digital Library Technologies Project, ODU, Norfolk, 1998.

19. Barabási, A. and R. Albert. Emergence of scaling in random networks. *Science* 1999; 286: 509–512.

20. Flocchini, P., B. Mans, and N. Santoro. Exploration of periodically varying graphs. In: Proceedings of the 20th International Symposium on Algorithms and Computation (ISAAC), Honolulu, Hawaii, December 16–18, 2009.

21. Tang, J., M. Musolesi, C. Mascolo, and V. Latora. Characterising temporal distance and reachability in mobile and online social networks. *ACM Computer Communication Review* 2010; 40(1): 118–124.

22. Kostakos, V. Temporal graphs. *Physica A* 2009; 388(6): 1007–1023.

23. Casteigts, A., P. Flocchini, W. Quattrociocchi, and N. Santoro. Time-varying graphs and dynamic networks. *International Journal of Parallel, Emergent and Distributed Systems* (Taylor & Francis) 2012; 27(5): 387–408.

24. Holme, P., and J. Saramaki. Temporal networks. *Physics Reports* 2012 519 (3): 97–125.

25. Basu, P., A. Bar-Noy, R. Ramanathan, and M. P. Johnson. Modeling and analysis of time-varying graphs. e-print arXiv:1012.0260.

26. Bui-Xuan, B., A. Ferreira, and A. Jarry. Computing shortest, fastest, and foremost journeys in dynamic networks. *International Journal of Foundations of Computer Science* 2003; 14(2): 267–285.

27. Grindrod, P., M. C. Parsons, D. J. Higham, and E. Estrada. Communicability across evolving networks. *Physical Review E* 2011; 81: 046120.

28. Katz, L. A new index derived from sociometric data analysis. *Psychometrika* 1953; 18: 39–43.
29. Bassett, D. S., N. F. Wymbs, M. A. Porter, P. J. Mucha, J. M. Carlson, and S. T. Grafton. Dynamic reconfiguration of human brain networks during learning. *Proceedings of the National Academy of Sciences of the United States of America* 2011; 108(18): 7641–7646.
30. Grindrod, P. and D. J. Higham. A matrix iteration for dynamic network summaries. *SIAM Review* 2013; 55(1): 118–128.
31. Ramanathan, R., P. Basu, and R. Krishnan. Towards a formalism for routing in challenged networks. In: Proceedings of the 2nd ACM Workshop on Challenged Networks (CHANTS), Montréal, Québec, Canada, September 14, 2007.
32. Tang, J., M. Musolesi, C. Mascolo, and V. Latora. Temporal distance metrics for social network analysis. In: Proceedings of the 2nd ACM Workshop on Online Social Networks (WOSN), Barcelona, Spain, August 17, 2009.
33. Williams, M. J., R. M. Whitaker, and S. M. Allen. Decentralised detection of periodic encounter communities in opportunistic networks. *Ad Hoc Networks* 2012; 10(8): 1544–1556.

Chapter 3

User Behaviors and Interaction in MSNs

3.1 Introduction

Online social networks (OSNs) change the way humans connect and get in touch with each other. Novel OSNs are created almost every day, but only a few of them become popular worldwide; most of them vanish. The success of a social network, both short term and long term, depends on the behavior of its users. Basically, the users' activity has an important effect on the services; in particular, the activity of users and time spent on the mobile social network (MSN) are important aspects if the value of the social network has to be expressed. In addition, developers of new social networks can build their systems with more incentives if they merge the properties of successful OSNs. Based on these arguments, understanding user behavior is crucial for the success of OSNs.

However, users' behavior in social networks has received little attention so far. One of the reasons for lack of research and the application of users' behavior is the difficulty in gathering large-scale user behavioral data. Only the operators of the OSNs and access network operators have these data, although investors, seeding firms, and advertisers would appreciate and use this knowledge.

Actually, there does exist in literature some work on exploring the properties of user behaviors in OSNs. For example, although most social networking sites allow only a binary state of friendship, it has been unsurprisingly observed that not all links are created equal. It is demonstrated that the "strength of ties" varies widely, ranging from pairs of users who are best friends to pairs of users who even wished they were not friends [1]. In order to distinguish between these strong and weak links, researchers have suggested examining the activity network—the network that is formed by users who actually interact using one or many of the methods provided by the social networking site. Although the initial studies on activity networks have brought great insights as to how an activity network is structurally different from the social network and how system designers can use such information, little attention has been paid to a rather natural and important aspect of user interaction: the fact that the level of interaction between two individuals can vary over time. In particular, to capture this notion, Viswanath et al. [2] studied the evolution of activity between users in the Facebook social network through examining more than 60,000 users and more than 800,000 logged interactions between those users over a period of two years. The data set is now available to the research community in an anonymized form (http://socialnetworks.mpi-sws.org). It has been found that links in the activity network tend to come and go rapidly over time, and the strength of ties exhibits a general decreasing trend of activity as the social network link ages. For example, only 30% of Facebook user pairs interact consistently from one month to the next. Interestingly, it is also shown that even though the links of the activity network change rapidly over time, many graph-theoretic properties of the activity network remain unchanged.

In parallel, nowadays, due to the wide use of mobile devices, more and more Web applications have been expanded to mobile platforms, as have OSN services. We believe that it is the right time to highlight the importance of MSNs. In MSNs, mobile users can publish and share information based on the social connections among them. On one hand, most major OSN platforms such as Facebook, Twitter, and LinkedIn allow users to access their services through mobile devices. On the other hand, more mobile-centric functions have been integrated into OSNs, such as location-based services and mobile communication. Understanding the user behavior in MSNs is very helpful for the design and implementation of MSN systems, for improving the system efficiency in mobile environments, or for supporting better mobile-centric functions [3].

Finally, one important aspect that was neglected in most MSNs and OSNs is that almost all related work only focuses on collecting real MSN (or OSN) interaction data and on inferring some phenomenal results, but those studies do not try to formally model user interaction and evolutionary behaviors or go further to design some rules to guide users' behaviors to fulfill the designers' specific goals. We argue that a game theory (GT)–based model would especially help to formally model user interaction behaviors in MSNs and would help mechanism design (MD) to focus on the design of the "game rule" to achieve the global goals when facing rational and autonomous participants.

This chapter is organized as follows. Section 3.2 offers the measurement methodology used to collect large-scale user behavior data that naturally varies with time and space and summarizes the properties of user behaviors in typical OSNs and MSNs. In Section 3.3, several theoretical models for user behaviors are preliminarily summarized, especially for MSNs. The brief discussion on incentivizing users' behavior is introduced in Section 3.4. (More related content will be provided in Chapter 4.) Finally, we briefly conclude this chapter.

3.2 Measuring and Characterizing User Interaction in MSNs

Generally, understanding MSN (OSN) user behavior is important to different social networking–related entities in several aspects:

■ Because MSN (OSN) traffic is growing quickly and becoming significant, Internet service providers (ISPs) want to learn the evolution of the traffic pattern of MSNs (OSNs). This can guide them to some infrastructural activities (e.g., adding traffic optimization in network middle-boxes).

■ For MSN (OSN) service providers, studying user behavior helps them understand their customers' attitudes toward different functions, especially for some that are experimental. Moreover, from the perspective of infrastructure investment, such as which locations are most cost-effective for building data centers or which content delivery network (CDN) cluster could be leveraged to deliver frequently accessed data, understanding users' geographic distribution and traffic activity is vital.

■ For MSN (OSN) users, behavior study is important to enhance user experience, such as personalized service, location-based alerts and recommendations, and so on.

Dynamics are important for deeply understanding an OSN's user behavior. Much of the existing work has tried to investigate an OSN in a relatively static way, by collecting or studying a snapshot data set. However, the growth of OSNs is extremely rapid. Every day new users join OSNs, while existing users make new friends or end social connections, join or leave groups, and so on. Considering this dynamic can extract more inherent information than studying static data, not only by revealing the situation at a certain time but also by predicting some future activities. Also, studying different time intervals and time granularities would lead to more interesting findings. There are several challenges for performing dynamic analysis. One is fast data collection and timely processing, where an unbiased and efficient graph sampling algorithm can play an important role. Also, collecting dynamic data raises challenges for information storage.

3.2.1 Methodology of Measuring Users' Behaviors and Interactions

3.2.1.1 Crowdsourcing-Based Measurement Architecture

A rapid increase in the number of users makes the size of social graphs larger and larger, which presents researchers with a big challenge when performing any analysis with limited computation and storage capability. Graph sampling techniques are used to get a smaller but representative snapshot of social graphs, which preserves properties such as degree distribution. As shown in Ref. [4], the sampling results of breadth-first sampling (BFS) and random walk (RW) are biased toward high degree vertices, although they have been widely used in social graph analysis. The Metropolis-Hasting RW (MHRW) and a re-weighted RW (RWRW) have been proposed and have proven to perform uniformly in sampling Facebook. The article also introduces online convergence diagnostics to assess sample quality during the sampling process. Frontier sampling (FS), which leverages multidimensional RW, is proposed for achieving lower estimation errors than RW, especially in the presence of disconnected or loosely connected graphs [5]. Moreover, it is shown that FS is more suitable for estimating the tail of degree distribution than random vertex sampling. In addition, FS can be fully distributed without any coordination costs.

Basically, the monitoring and crawling of thousands of users (Internet-scale measurements of some popular OSNs) cannot be carried out from a single stand-alone machine. The volume of network traffic is the most important limitation factor of the methodology.

Actually, in current literature, most of the data sets used in OSN research were crawled by individual groups using their own network resources and computing clusters. However, because most of the OSN sites have very strict request rate limiting on Internet protocol (IP) addresses and user accounts, it becomes challenging for an individual research group to crawl sufficient data covering a significant portion of an entire OSN site.

A main reason is that the number of IP addresses a research group owns is typically limited, especially for universities outside the United States. It is difficult for a single group to compete against the increasing crawling requirement and unavoidable limitations of request rate. For example, Sina Weibo allows up to 1000 Application Programming Interface (API) requests per hour per IP address and 150 requests per hour per user account. Moreover, different research groups might be interested in the same OSN site, while they crawl data independently nowadays. Such uncoordinated crawling may have the following drawbacks: (1) wastes computing and network resources as well as researchers' programming efforts; (2) introduces unnecessary overhead (and extra cost) for OSN service providers; (3) leads to even stricter request rate–limiting policies from OSN service providers; and (4) leads to inability to provide universal/standard data sets for the research community to develop and evaluate different research methods on social network analysis.

Therefore, crowdsourcing-based crawling architecture is used in the literature, which allows multiple research groups to efficiently crawl data in a collaborative way. As shown in Figure 3.1, there are three components in the system: the initiator (result collection module), the task assignment module (TAM), and the partners. For every data-crawling project, the initiator can deploy the detailed crawling task to a coordinator that is trustworthy and responsible for assigning tasks to partners. Every partner is required to contribute some computers as crawlers and to provide

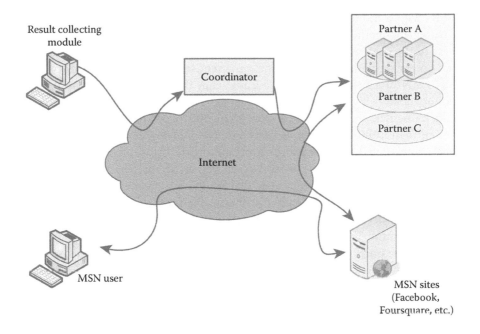

Figure 3.1   Crowdsourcing-based crawling architecture.

Internet connectivity for these crawlers. The initiator can collect crawled results from all partners and prepares an aggregate data set. The TAM is maintained by the coordinator and can be built based on a distributed hash table (DHT) to ensure scalability; crawlers are owned by partners, and they crawl OSNs collaboratively. The role of result collection module can be taken on by the initiator. However, for scalability, the result collection module could be installed in a cluster of distributed servers to ensure scalability.

Recently, crowdsourcing-based measurement methodology has been popular. For example, Gyarmati and Trinh [6] proposed the distributed measurement framework through adopting the PlanetLab* platform. First, the monitoring scripts are deployed to the PlanetLab nodes with the appropriate user identifiers. These scripts iteratively check whether the monitored users are online. If a monitored OSN user signs in to the OSN server from its own machine, the OSN server displays the user's online status on the profile page. Therefore, the monitoring script will notice in the next minute that the user is online and will save the processed usage information. Finally, at the end of the measurement, the usage data sets are downloaded from the PlanetLab nodes and analyzed on a local machine. The data are obtained by processing publicly accessible user

---

* PlanetLab; http://www.planet-lab.org/

profiles with a minute as sampling time over more than six weeks. The applied measurement methodology has a crucial property: it does not require having access to the databases of the OSN operators or being able to dump IP packets in an access network. Thus, those measurement settings can be named the poor man's OSN user behavior monitor because they do not require the cooperation of an operator.

In particular, the article used more than 500 nodes of PlanetLab, a cluster of more than 1000 machines, to carry out the measurement. The machines of PlanetLab executed the processing scripts automatically at one-minute intervals. Because PlanetLab is a global network, each PlanetLab machine saved the results in a GMT format for consistency. Any PlanetLab node can be out of order at any time; therefore, the authors evenly distributed the users to be monitored among the nodes in order to minimize the damage of an out-of-order machine. Therefore, they could continually monitor the vast majority of the selected users of the OSNs during the measurement period. To reduce the size of the data sets, the data of a user were only stored if the user was online. The main findings of the article include that the users' time spent online can be modeled with Weibull distributions; that soon after subscribing, a fraction of users tended to lose interest surprisingly fast; and that the duration of OSN users' online sessions shows power-law distribution characteristics.

In addition, a crowd crawling prototype on PlanetLab was implemented and deployed [7]. Specifically, the evaluations on PlanetLab have demonstrated the high efficiency of OSN data collection; that is, we can finish the crawling of 2.22 million Weibo users in 24 hours, including their profiles, social connections, and posted tweets.

3.2.1.2 Experience Sampling Method for MSNs

Understanding the behavior of users as they share information with mobile social applications is important for enhancing their experiences and for improving the services provided. Formal interviews and questionnaires allow us to collect self-reported information about users' behaviors when using mobile social applications, but users may forget some details about their experiences or may report inaccurate information when answering questionnaires. The behavior of mobile social application users can also be studied by analyzing the information shared on OSNs, but this only allows the examination of the information that has been shared (rather than the information that has not been shared or the contexts of which users do not wish to share). A third way to study users' behavior, that addresses some of these drawbacks, is the experience sampling method (ESM) [8,9]. ESM is a diary method that consists of asking participants to stop at certain times, either on a predetermined basis (signal contingent) or when a particular event happens (event contingent), and report about their experiences in real time. ESM allows collecting experiences in situ, which is more accurate than when data are collected later through a survey.

Compared to OSN analysis or traditional surveys, implementing ESM to study the behavior of mobile social application users is more complicated and time consuming. This method requires designing, implementing, and deploying an appropriate test-bed composed of smartphones to collect data and a server to monitor and store these data. But although it would be difficult for this special method to be as simple as a traditional survey or an analysis of participants' behaviors on OSN, there are a few main challenges that have to be addressed to improve the method and to avoid its potential shortcomings.

An initial challenge is to reduce the energy consumed by the smartphones. Using a single device to collect data, ask questions, and collect answers necessitates the use of more energy than the normal use of such a device to answer calls. In particular, continuously monitoring users' behavior may involve multiple sensors being triggered frequently, which may quickly deplete the battery. Hence, efficiently managing the sensors to save energy is an important challenge when collecting data on participants' behavior in their everyday lives. For instance, to save energy, the accelerometer embedded in most smartphones is commonly used to detect motion and to switch off the Global Positioning System (GPS) when the participant is not moving.

Another challenge is to make sure the experiment is not being too intrusive. Polling participants in their everyday lives may disturb them, and answering ESM questions may sometimes be inappropriate. A partial solution is to ask participants for the times they do not want to be asked ESM questions. Answering the questions may also take time, especially when they are received frequently. Furthermore, it is better to detect an activity or a context instead of asking the participant; detection not only provides other data than self-reported information but also helps in understanding the ESM answers given by the participants. For instance, the location can be detected instead of asking the participant.

Finally, in the case of an experiment involving human beings as participants, ethical considerations must be carefully taken into account—especially when the experiment is running during their everyday lives—because personal information may be collected. In particular, participants may experience privacy issues and, although unlikely, potential psychological harm, discomfort, or stress. For the latter, the risk is difficult to quantify or anticipate in full prior to the start of the experiment, but the participants always have the option to withdraw from the experiment at any time without any justification. As for privacy issues, what, how, and when data are collected must be made clear to the participant before they consent to participate, as well as where information is stored and who has access to it. Anonymization of personal data allowing participants' identification must be guaranteed.

In summary, because of the diversity and complexity of human social behavior, no one technique (crawling, traffic data collection, EMS, etc.) will detect every attribute that arises when humans engage in social behaviors.

3.2.2 Various Features of Interactions in MSNs

3.2.2.1 Connectivity and Interaction in Social Network

Usually, social networks can be modeled as undirected graphs (e.g., friendship graph, interaction graph) or directed graphs (e.g., latent graph, following graph) according to the properties of OSNs. Table 3.1 lists four different categories of social graphs. Based on these graph types, we discuss the connectivity and interaction among OSN users. Moreover, the huge size of the social graph challenges the effectiveness of analysis. Thus, the aforementioned distributed sampling and crawling techniques can be adopted to deal with this problem.

For a friendship graph, every user is denoted as a node, and the friendship between any user pair is represented by an edge. Wilson et al. [10] tried to find out whether social links are valid indicators of user interactions. They define wall posts and photo comments as interactions. Based on the crawled data from Facebook, they have found that users tend to interact mostly with only a small subset of their friends, while often having no interaction with up to half of their friends. Therefore, friendship in OSNs can hardly be viewed in the same way as friendship in the real world. Correspondingly, a new interaction graph is proposed to reflect the real user interactions in social networks, where only visible interaction between two users can create an edge in the graph, instead of just being friends.

Latent interactions are passive actions of OSN users (e.g., profile browsing) that cannot be observed by traditional measurement techniques. Latent interactions have been studied in Ref. [11] based on the crawled data of Renren, the largest OSN provider in China. Renren tracks the most recent nine visitors to every user's profile, thereby making the measurement of latent interactions possible. In a directed latent

Table 3.1 Four Typical Social Network Categories and Meanings of Their Edges

Category	Meanings of Edge in Social Graph
Friendship graph	Friendship between users
Interaction graph	Visible interaction, such as posting on blogs
Latent graph	Latent interaction, such as browsing profile
Follow graph	Subscribe to receive all messages (tweets)

graph, a directed edge from A to B indicates A has visited B's profile. Therefore, the in-degree of a node shows the number of visitors to that user's profile, while the out-degree reveals the number of profiles that user has visited. A comparison between latent interactions and visible interactions is conducted based on Renren's crawled data, which contains 42 million users and 1.66 billion social links. There are three major findings. First, latent interactions are significantly more prevalent and frequent than visible interactions. Second, latent interactions are nonreciprocal in nature. And, last but not least, the profile popularity is uncorrelated with the frequency of content updates or with the number of friends for very popular users. Characteristics of latent graphs are shown to fall between visible interaction graphs and classical friendship graphs.

Extensive measurements were performed on Twitter, the world's largest microblogging service, which revealed its power in information spreading on the news media level. In Twitter's following graph, a directed edge from A to B indicates A has subscribed to receive B's latest messages. The collected data are crawled over 24 days, with 41.7 million user profiles, 1.47 billion relations, 4262 trending topics, and 106 million tweets. It introduces a directed graph model to give a basic informative overview of Twitter, studies the distribution of followers/followees, and analyzes how the number of followers or followees affects the number of tweets. In addition, in order to show how Twitter acts as a social medium and how top users influence other users, this article tries to rank the users by number of followers, page rank, and retweets. The rankings by number of followers and page rank are almost the same, and the top users in the rankings are either celebrities or news media accounts. This article also analyzes the trending topics in Twitter and compares it with other media. It was found that the majority (more than 85%) of trending topics in Twitter are headlines or persistent news, which reveals Twitter's live broadcasting nature and confirms Twitter's role as a news medium.

3.2.2.2 Traffic Activities in Social Networks

Besides crawling, people can also study OSNs by monitoring the corresponding network traffic through so-called social network aggregators.

Commonly, social network aggregators pull content from multiple social networking sites to a single location, thereby helping users who belong to multiple networks manage diverse profiles more easily. Upon logging into a social network aggregator, users can access their social network accounts through a common interface, without having to login to each OSN site separately. This is done by a two-level real-time HTTP connection: the first level is between a user and a social network aggregator site and the second is between the social network aggregator site and the OSN sites. Social network aggregators typically communicate with OSN sites using open APIs that OSN sites provide. Figure 3.2 depicts the scheme interaction among users, a social network aggregator site, and OSN sites. Through the interface of the social network aggregator, a user can enjoy all features that

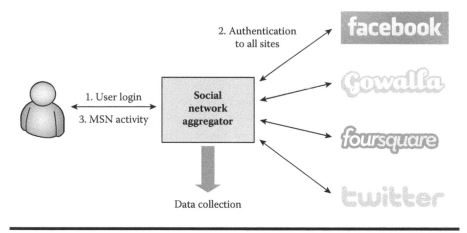

Figure 3.2 Illustration of data collection through a social network aggregator.

are provided by OSN sites—for instance, checking updates from friends, sending messages, and sharing photos.

Benevenuto et al. [12] analyze the user behavior of OSNs based on detailed clickstream data obtained from a social network aggregator. The clickstream data were collected over 12 days from HTTP sessions of 37,024 users who accessed popular social networks. This article defines and analyzes the OSN session characteristics:

- The frequency of accessing OSNs
- Total time spent on OSNs
- Session duration of OSNs

Through the clickstream data, user activities are also identified. Forty-one types of user activities are classified into nine groups, and the popularity of different activities and the traffic bytes are latent interactions such as browsing account for more than 90% of user activities. Also, they show how users have different activities in different OSNs. They also characterize how users transit from one activity to another using a first-order Markov chain.

Schneider et al. [13] also study clickstream data, but their focus is on feature popularity, session characteristics, and the dynamics within OSN sessions. The distribution of HTTP request–response pairs reveals the popularity of different features. The popularity of features can be different among users from different areas and among different OSNs. It can also differ in the time spent by the users. In addition, the distribution of transmission bytes per OSN session is given, which helps the ISPs learn the traffic pattern of different OSNs. Photo features account for most traffic bytes of OSNs. It also shows the duration of sessions and number of subsessions within a session. Moreover, the article reveals the dynamics within OSN sessions. It found that most users access Web sites other than OSNs during OSN sessions of more than 1 min. That is, users can be inactive when accessing the OSNs.

Clickstream data contributes a lot to the user behavior study of OSNs. However, it can be incomplete, which restricts its usage and performance. First, clickstream data are limited by the collection duration, and the behavior of inactive users in the collection duration is not monitored. Moreover, the data are restricted by the monitoring locations. That is, only the behavior of users using certain monitored ISPs is captured.

3.2.2.3 Locality of Interest and Navigation Characteristics

Facebook is heavily dependent on centralized U.S. data centers to provide consistent service to users all over the world. Therefore, users outside the United States experience slow response time. Also, a lot of unnecessary traffic is generated on the Internet backbone. Wittie et al. [14] investigate the detailed causes of these two problems and identify mitigation opportunities. They found that an OSN state is amenable to partitioning, and its fine-grained distribution and processing can significantly improve performance without a loss in service consistency. Based on simulations of reconstructed Facebook traffic over measured Internet paths, they showed that user requests can be processed 79% faster with 91% less bandwidth. Therefore, the partitioning of the OSN state is an attractive scaling strategy for OSN service providers.

Nowadays, OSNs represent a significant portion of Web traffic, comparable with search engines. Dunn et al. [15] try to understand the similarities and differences in the Web sites users visit through OSNs versus through search engines. Using Web traffic logs from 17,000 digital subscriber line (DSL) subscribers of a Tier 1 ISP in the United States, they found that OSN visitors are less likely to navigate to external Web sites. But when they visit external Web sites, OSN users will spend more time at those Web sites compared to search engine users. Also, OSNs direct visitors to a narrower subset of the Web than do search engines. Although Web sites related to games and videos are more commonly visited from OSNs, shopping and reference sites are common for search engines. Finally, OSNs send users to less popular domains more often than search engines. These findings can be useful to ISPs in network provisioning and traffic engineering.

The drawbacks of the aforementioned studies lie in the fact that most existing measurement and analysis projects are led by either academic groups or ISPs, without the active involvement of MSN/OSN service providers. Such a situation limits the insight of the study. On one hand, academic researchers always use extensive crawling to obtain the data, which encounters many restrictions from the MSN/OSN providers, such as traffic control (how many messages per IP and/or per account can be fetched in one hour). Also, some users may use privacy options to make their data unavailable. Last but not least, the huge number of users makes it almost impossible to obtain a timely snapshot, so data consistency cannot be guaranteed. On the other hand, although an ISP is able to capture and analyze all its traffic to/from an MSN/OSN site through traffic monitoring, it can only obtain a partial view of the whole site; that is, only users who access MSNs/OSNs through a specific ISP's infrastructure can be observed. As we have discussed, studies of user behavior can be beneficial for MSN/OSN providers themselves. MSN/OSN providers could collaborate with academia and industrial researchers in order to understand user behavior in an insightful way. This can enhance the user experience interactively and quickly. Also, this will save operational costs for MSN/OSN providers. On the contrary, there is still a lack of data for real deployment mobile social application on data analysis of human interaction and social behaviors in MSNs.

Zhu et al. [16] introduced the experimental methodology of deploying Goose software in two campuses located in Germany and China. Goose is an MSN application that allows microblogging and message sending. With the help of volunteers, the authors collected user interaction data for 15 days. Based on the collected data, the following aspects of user interactions and their influences were observed. First, overall user activities approximately match their daily life work patterns, with a slightly longer time duration and periodical appearance. Second, user encounters in an MSN follow the heavy tail distribution in small social communities, and user interactions follow the Pareto principle, where about 20% of users make close connections to the other users. Third, the communication path between a pair of mobile nodes is usually within six hops, and information diffusion using an epidemic strategy demonstrates that the informed population reaches 50% in the short term and approaches 80% in the long term.

3.2.3 Prediction of User Behavior in MSNs

Basically, we can obtain activity data from these online services. However, we cannot generally obtain data that directly describes which activity has had an impact on the following activities. The user interactions in some online communities have been referred to as "ballot box communications" [17] in which users do not directly interact with each other by exchanging messages, but they implicitly interact with one another by expressing their preferences on the shared items. Thus, predicting a user's future activity, preference, or location is extremely challenging.

3.2.3.1 Prediction of User's Future Activity Level

The study of users' social behaviors has gained much research attention since the advent of various social media such as Facebook, Renren, and Twitter. A major type of application is to predict a user's future activities based on his or her historical social behaviors. Zhu et al. [18] focused on a fundamental task—to predict a user's future activity levels in a social network (e.g., weekly activeness, active or inactive). The existing three properties of social networks—user diversity, social influence, and dynamic nature—raise new challenges and opportunities for accurately predicting users' future activity levels. First, the user diversity property implies that a global predictive model may not be precise for all users. On the contrary, historical data of individual users are too sparse to build precisely personalized models. Second, the social influence property suggests that relationships among users can be embedded to further boost prediction results for individual users. Finally, the dynamic nature of social networks means that users' behaviors may keep changing over time. Note that these three challenges are not specific to the activity level prediction problem. One or two of them are common to various prediction tasks. For example, the tasks of e-mail spam detection and e-mail importance ranking share the same problem of personalization; typical time series modeling needs to deal with the dynamics in the temporal space, and recently social recommender systems and node classification in social networks usually require encoding social structure into model learning.

However, dealing with those three challenges simultaneously makes the problem of predicting user future activity level unique. The proposed solution starts with a simple model based on logistic regression. Then the base model is extended by equipping different terms to capture the three properties of the user activity level prediction task, resulting in a unified learning framework. Specifically, to address the user diversity issue, the model is explicitly decomposed into two parts: the common part, which is for global optimization over all users, and the user-specific part, which is for personalized optimization over nonspecific users. However, as described earlier, historical data for each individual user are extremely sparse for optimizing personalized models independently. Therefore, the authors propose jointly learning personalized models for individual users by looking at the common part of the models as a bridge. To model the dynamics in user behavior, a time-decay term is introduced to penalize out-of-date training data. To model social influence, a social regularization term is deployed for smooth predictions of close friends and groups whose activity levels are also close to each other. Experiments on the social media Renren validate the effectiveness of the proposed model when compared with some baselines, including traditional supervised learning methods and node classification methods in social networks. In summary, prediction of these activities not only reduces the cost for user maintenance but also avoids disturbing normal users (e.g., by only sending notification and update e-mails to users who have the tendency to drop their activity level in social networks).

A significant majority of today's Internet users rely on Facebook and Twitter for their online social interactions. In June 2011, Google launched a new OSN called Google+ (or G+ for short) in order to claim a fraction of the social media market and its associated profit. G+ offers a combination of Facebook- and Twitter-like services in order to attract users from both rivals. There have been several official reports about the rapid growth of G+'s user population, although some observers and users dismissed these claims and called G+ a "ghost town."

This raises the following important question: Can a new OSN such as G+ really attract a significant number of engaged users and be a relevant player in the social media market? A major Internet company such as Google, with many popular services, is perfectly positioned to implicitly or explicitly require (or motivate) its current users to join its MSN/OSN. It is also interesting to assess to what extent and how Google might have leveraged its position to make users join G+. Nevertheless, any growth in the number of users in an MSN/OSN is really meaningful only if the new users adequately connect to the rest of the network (i.e., become connected), and if they become active by using some of the services offered by the MSN/OSN on a regular basis. Furthermore, today's Internet users are much more savvy about using MSN/OSN services and connecting to other users than were users a decade ago when Facebook and Twitter first became popular. This raises other related questions such as: How has the connectivity and activity of G+ users evolved over time as users have become significantly more experienced about using OSNs? And will these evolution patterns exhibit different characteristics compared to earlier major OSNs? These evolution patterns could also offer an insight on whether users willingly join G+ or are added to the system by Google.

These questions were tackled in Ref. [19] by presenting a detailed characterization of G+ based on large-scale measurements. The authors identify the main components of the G+ structure in detail and characterize the key features of their users and their evolution over time. Then, detailed analysis is conducted on the evolution of connectivity and activity among users in the largest connected component (LCC) of the G+ structure; these data are compared with other major OSNs. The authors infer that, despite the dramatic growth in the size of G+, the relative size of the LCC has decreased with time. Those investigations reveal that a significant fraction of new G+ users appear to be implicitly added by Google while they register for other Google services, regardless of the user's interest. Furthermore, the main connectivity features of the LCC have become relatively stable, which suggests that the G+ network has reached a steady state. Finally, it is shown that these stable connectivity features of the LCC component of G+ have a striking similarity with Twitter but are very different from Facebook. This similarity indicates that users use G+ for message propagation similar to Twitter rather than for pairwise user interaction, such as Facebook. In terms of user activity, even LCC users are not actively engaged in the G+ network. The contribution of user activity in terms of posting is skewed among LCC users (i.e., 10% of users are responsible for 80% of posts) and user reactions to activities is an order of magnitude more skewed (i.e., 1% of users generate 80% of reactions to all posts). Those findings collectively demonstrate that in the current OSN marketplace with two dominant players, namely Facebook and Twitter, a new OSN such as G+ might be able to attract a rather significant number of users to become part of the network (i.e., to connect to its LCC). However, it is much more challenging to get these users meaningfully engaged in the system.

3.2.3.2 Geographical Prediction in MSN/OSN

Geography and social relationships are inextricably intertwined. As people spend more time online, data regarding these two dimensions are becoming increasingly precise, allowing the building of reliable models to describe their interaction. The study of user-contributed address and association data from Facebook shows that the addition of social information produces an improvement in accuracy of predicting physical location [20]. First, friendship as a function of distance and rank is analyzed. It is found that at medium- to long-range distances, the probability of friendship is roughly proportional to the inverse of distance. However, at shorter ranges, distance does not have much influence. Next, the maximum likelihood approach is presented to predict the physical location of a user, given the known location of his or her friends. This method predicts the physical location of 69.1% of the users with 16 or more located friends to within 25 meters, compared to only 57.2% using IP-based methods.

Although human movement and mobility patterns have a high degree of freedom and variation, they also exhibit structural patterns due to geographic and social constraints. Using cell phone location data, as well as data from two online location-based social networks (LBSNs), Gowalla and Brightkite, Cho et al. [21] hoped to understand the basic laws that govern human motion and dynamics. They found that humans experience a combination of strong short-range spatially and temporally periodic movement that is not impacted by the social network structure, although social network ties have more of an influence on long-distance travel. Furthermore, social relationships can explain about 10%–30% of all human movement, while periodic behavior explains 50%–70%. Based on these findings, Cho et al. proposed a model of human mobility that combines periodic short-range movements with travel due to the social network structure and gives an order of magnitude better performance than previous models.

3.3 Modeling User Interactions in MSNs

3.3.1 Multidimensional Characterizing of Human Mobility in MSNs

In an MSN, especially in an MSNP (MSN in proximity, data exchange through direct P2P paradigm without centralized infrastructure), the movements of the communicating devices mirror those of their owners; finding a route between two disconnected devices implies uncovering habits in human movements and patterns in their connectivity (frequencies of meetings, average duration of a contact, etc.) and exploiting them to predict future encounters. Therefore, there is a challenge in studying human mobility, specifically in its application to MSNP research.

In this section, we review the field of human mobility analysis and present a survey of mobility models. We start by reviewing the most notable findings regarding the nature of human movements, which can be classified along the spatial, temporal, and social dimensions of mobility: geographic movement (where do we move?), temporal dynamics (how often do we move?), and the social network (how do social ties interact with movement?).

Basically, the way the agents are distributed on the plane and the rules that govern their behavior define a human mobility model. Depending on the principle used to define these rules, two classes of models can be distinguished. In trace-based approaches, the mobility model is defined by the set of distributions that fit some statistics extracted from the traces considered. Although the agreement between the statistical properties of the traces and those obtained from the model is usually extremely good, these solutions do not propose a general mobility model that describes users' movements. Thus, their applicability outside the environment from which they have been derived is not clear. On the contrary, synthetic models aim to reproduce driving forces of individual mobility such as social attitude, location preferences, and regular schedules. Traces in this case might be explored for validation.

Even though the aforementioned are some of the most fundamental questions and hypotheses about the dynamics of human mobility, answers to them remain largely unknown. This is mostly due to the fact that reliable large-scale human mobility data have been hard to obtain. Recently, however, location-based online social networking applications have emerged where users share their current location by checking in on Web sites such as Foursquare, Facebook, Gowalla, and so forth. Although records of calls made by cell phones traditionally have been used to track the location of the cell phone towers associated with the calls, LBSNs provide an important new dimension in understanding human mobility. In particular, although cell phone data provide coarse location accuracy, LBSNs provide location-specific data, such as how one can distinguish between a check-in at the office on the second floor and a check-in at a coffee shop on the first floor of the same building. On the contrary, check-ins to LBSNs are usually sporadic, although cell phone data provide better temporal resolution because a user "checks in" whenever he or she makes or receives a call. Both types of data also contain network information. LBSNs maintain explicit friendship networks, while a mobile phone network can be inferred from the communication network.

As described earlier, in recent years, studying human mobility has been one of the major focuses of different disciplines. The main findings can be classified along the three axes of spatial, temporal, and social connectivity properties (shown as Figure 3.3). Spatial properties pertain to the behavior of users in the physical space (e.g., the distance they travel), temporal properties pertain to the time-varying features of human mobility (e.g., the time users spend at specific locations), and connectivity properties pertain to the interactions between users.

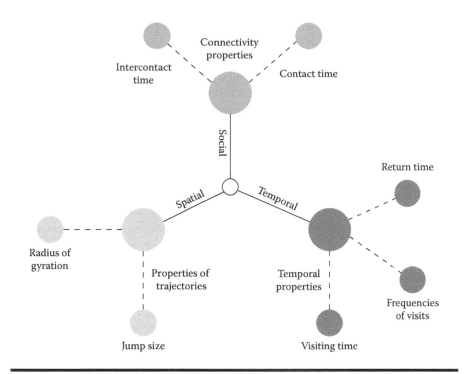

Figure 3.3   Properties of human mobility in mobile social networks.

It was observed that most individuals travel only a short distance, but a few regularly travel more than a hundred kilometers; furthermore, individual travel patterns collapse into a single spatial probability distribution, indicating that, despite the diversity of their travel history, humans follow simple reproducible patterns. More precisely, a "trajectory" can be defined as a spatiotemporal sequence of triples (x_i, y_i, t_i), where t_i is a time stamp and $\vec{r}_i^a (x_i, y_i)$ are points in R_2. Having the sequence of consecutive traces, jump size distribution is an important measure that many have been investigating with regard to human mobility. Brockmann et al. [22] analyzed a huge data set of bank note circulation records, interpreting them as a proxy for human movements. They showed that travel distances Δr (frequently called jump size) of individuals follow a power-law distribution $P(\Delta r) \sim (\Delta r)^{-(1+\beta)}$, where the exponent β is smaller than 2. This fits the intuition that people usually move over short distances, whereas occasionally they take rather long trips. The distribution known as Lévy flight was previously observed as an approximation of migration trajectories among different animal species. Studying data collected by tracing mobile phone users, Gonzalez et al. [23] complemented the previous finding with an exponential cutoff $P(\Delta r) = (\Delta r + \Delta r_0)^{-(1+\beta)} exp\left(\dfrac{\Delta r}{k} \right)$, with $\beta = 1.75 \pm 0.15$, $\Delta r_0 = 1.5$ km and k a cutoff value varying in different experiments, and showed that individual truncated

Lévy trajectories coexist with population-based heterogeneity. This heterogeneity was measured in terms of the radius of gyration (r_g), which depicts the characteristic distance traveled by a user. It was shown that the distribution of the radius of gyration can be approximated by a truncated power law, $P(r_g) = (r_g + r_g^0)^{\beta_r} exp(r_g / k)$, where $\beta_r = 1.65 \pm 0.15$, $r_g^0 = 5.8$ km, and $k = 350$ km. In other words, most people usually travel in a close vicinity to their home location, while a few frequently make long journeys.

As for the temporal properties of human movements, Gonzalez et al. [23] detected the tendency of people to return to a previously visited location with a frequency proportional to the ranking in popularity of the location with respect to other locations. The authors also computed the return time probability distribution (probability of returning at time t to a selected place) and concluded that prominent peaks (at 24, 48, 72, ... hours) capture the tendency of humans to return regularly to the location they visited before. Song et al. [24] extended the experiment to a larger data set and measured the distribution of the visiting time (i.e., the time interval Δt a user spends at one location). The resulting curve is well approximated by a truncated power law with an exponent $\beta = 0.8 \pm 0.1$ and a cutoff of $\Delta t = 17$ hours, which can be connected with the typical awake period of humans. In addition, it was found by Gonzalez et al. [23] that f_k, the frequency at which a user visits its kth most visited location, follows a Zipf's law ($f_k \sim k^{-\xi}$) with a parameter $\xi \approx 1.2 \pm 0.1$. This also suggests that the probability of a user visiting a given location a number of times is equal to f (i.e., visitation frequency) follows $P(f) \sim f^{-(1+1/\xi)}$.

Connectivity properties have been extensively studied in the context of opportunistic network (ON) research. The reason comes from an engineering concern—given that messages are forwarded from node to node when they get in touch with each other, the time between two consecutive contacts of two devices contributes to the overall delay, although the duration of the contact bounds the size of the data that can be exchanged at each encounter. Because the behavior of human-carried mobile devices can be considered as a proxy of human movements, a contact between two devices implies that the corresponding users are close to each other. Thus, by extension, we take the intercontact time and contact time as measures of how frequently and for how long two users spend time together. More specifically, we define the contact time between two mobile devices as the time intervals during which two devices are in radio range of one another, while the intercontact time is the length of the time interval from the end of the contact to the beginning of the next one. Chaintreau et al. [25] showed that the distribution of intercontact time has a power-law nature over a wide range of values, from a few minutes to half a day. Later, Karagiannis et al. [26] extended this result, suggesting that the power-law decay should be complemented with an exponential cutoff. Although less attention was dedicated to studying the duration of contact times, it was also shown by Hui et al. [27] that the duration of contact time follows an approximate power-law distribution.

We have observed several metrics that depict regularities in people's movements. These metrics, however, cannot capture some aspects of human mobility such as the distinction between periodic and frequent (but not periodic) trips. In order to capture periodical events in agents' moves, recent works have explored the concept of time-varying social graphs. The idea comes from the following reasoning. During each part of the day, the social communities to which we belong influence our behavior differently. For example, during working hours, a regular person tends to interact mainly with his or her colleagues at work, thus resulting in the person leaving home in the morning and spending part of the day in the office. On the contrary, our evening activities are usually connected with our family or friends; in order to meet them, we go back home or to a pub. In this scenario, the social strength of relationships among users (i.e., the weight of the edges connecting users in the social graph) changes with time, and the associated graph can thus be modeled as a time-varying graph (TVG), previously summarized in Chapter 2. The movement decision process at a specific time t for individual i consists of choosing one community among those communities that are relevant at time t and moving toward the location associated with the chosen community. Not only do they already account for two out of three mobility dimensions, the temporal and social ones, but they are also being extensively studied in the complex network research community, the results of which could be readily exploited. Future research efforts should then concentrate on how to incorporate the spatial dimension into a model based on time-varying social graphs. As a prelude to this, the relation between movements and user sociality should be better investigated using traces of real human mobility. In fact, although it has been shown in the literature that the social structure of the network does influence user movements, the dynamics of this causal relationship—both qualitative and quantitative—are not yet clear.

In summary, literature focuses on spatial, temporal, and social aspects in studying human mobility traces. As a result, three corresponding dominating techniques are explored in the models: maps of preferred locations, personal agendas, and social graphs. Unfortunately, each of these approaches is designed to reproduce only a subset of patterns and properties of movements; therefore, a combination of different techniques is required to construct more realistic models.

3.3.2 An Integrated Behavior Model in MSNs

Typically, connections established by users of OSNs are influenced by mechanisms such as preferential attachment and triadic closure. Yet recent research has found that geographic factors also constrain users: spatial proximity fosters the creation of online social ties. Although the effect of space might need to be incorporated into these social mechanisms, it is not clear to what extent this is true and in which way this is best achieved. To address these questions, Allamanis et al. [28] studied the temporal evolution of a LBSN: over a period of

16 weeks, daily snapshots were collected of a LBSN with hundreds of thousands of users (Gowalla), including the places visited by users and their social connections. Created in 2009, Gowalla is a location-based social networking service that allows users to add friends and share their location with them. It allows users to "check-in" at places through a dedicated mobile application, publicly disclosing their location on the service. These check-ins can then be pushed to friends. As a consequence, friends can see where a user is or has been. Users can create mutual friendship relationships requiring each user to accept friendship requests. Gowalla was discontinued at the end of 2011 because the company was acquired by Facebook.

Thanks to this fine-grained temporal information about network evolution and users' mobility, different edge attachment models were tested and compared that can explain the observed data by adopting an approach based on likelihood estimation. These core facets of temporal network evolution are analyzed in more detail.

- How edges are created: The authors test different edge attachment models based on the social and spatial properties of nodes and show that node degree and spatial distance are simultaneously influencing edge creation, demonstrating that a gravitational attachment model captures real network evolution better than purely social or spatial models.
- How social triangles are created: Because social networks tend to have a dominant fraction of new edges closing triangles, the authors test several different models of triadic closure, some of them also involve spatial distance; the authors find that social factors are more important than spatial constraints when an edge closes a triangle.
- How users' mobility affects new edges: Because social connections might arise among users visiting the same place, the authors study models of edge creation that exploit the properties of shared places to connect users; they found that both the popularity of a place and the popularity of users visiting that place help to predict which social connections are established.
- In addition, the temporal patterns of user behavior were investigated in which there was a special focus on the lifetime of a node (i.e., the amount of time a user is actively creating new edges) and on the inter-edge waiting time (which governs the amount of time elapsed before a node will create a new edge).

Based on those findings, the authors describe a new family of models of network growth that are able to reproduce the social and spatial properties observed in the real data. Such models combine a global gravitational attachment process with a local triadic closure mechanism based on shared friends and shared places, which is an evolutionary random process that grows a spatial network edge by edge. It is demonstrated that the resulting synthetic networks exhibit social and spatial properties similar to the real network, although a similar model that considers preferential rather than gravitational attachment—effectively ignoring the effect of geographic distance—fails to reproduce real properties. This work sheds light on the effect of geographic constraints on the evolution of OSNs. The obtained results offer useful insights for researchers and practitioners, with promising implications for the wide range of applications that already take advantage of the spatial properties of online social services. The integrated model is composed of the processes outlined in the following sections.

3.3.2.1 Edge Creation

In this section, we provide an overview of how the creation of individual edges is influenced by social and spatial properties of the nodes, exploring the effect of node degree, node age, and spatial distance on the edge attachment process.

The effect of node degree on network evolution is well captured by the preferential attachment model, where the probability of connection among nodes i and j, P_{ij}, is proportional to the degree of node j, $P_{ij} \propto k_j$. This model generates networks with a degree of distribution exhibiting a heavy tail because there are a few nodes (the so-called hubs) that accumulate an extremely high number of connections. Real-world examples such as transportation and communication networks can be described by a preferential attachment model, but geographic distance is an important parameter as well. In fact, long-range connections tend to exist mainly between well-connected hubs [29].

The effect of geographic distance can be included in the attachment probability, $P_{ij} \propto k_i k_j f(D_{ij})$, where f is a decreasing deterrence function of the geographic distance D_{ij} between the nodes. Thus, long distances tend to be covered only to connect to important hubs, although nodes with fewer connections become attractive when they can be reached over a short distance. When the deterrence function has a simple functional form such as $f(d) \sim d^{-\alpha}$, then the probability of a connection between two nodes becomes similar to the gravitational attraction between celestial bodies, $P_{ij} \propto \dfrac{k_i k_j}{d_{ij}^{\alpha}}$. Hence, this family of attachment models has been known as the gravity models. The main driving factors in edge attachment are node degree and geographic distance, and a gravity model that combines them is the most suitable option.

3.3.2.2 Social Triadic Closure

The edge attachment mechanisms previously investigated only take into account the influence of global network properties on new edge creation. However, local network properties can be equally or more important. For instance, new links tend to connect users that already share friends, creating social triangles that are extremely common on social networks. This mechanism, where a node just copies a connection from a node it is already connected to, has turned out to be essential for reproducing the structure observed in many networks. Here, we briefly summarize the extent to which new links generate social triangles and whether different models based on local network properties can reproduce the patterns observed in the data.

3.3.2.3 Triangle-Closing Models

Because a vast majority of new edges close social triangles, our aim is now to understand what factors influence which node to choose when an edge is closing a triangle. The maximum likelihood principle is explored to test and compare whether different triangle-closing models would be able to generate the triangles created during the real network evolution.

We consider the case of when a source node s has to choose another target node t that is two hops away in order to create a new link. A simple model would be for nodes to choose t uniformly at random from all the nodes at a distance of two hops, which will be our baseline model. We then take into account more complex models where a source node s first chooses an intermediate node i among its neighbors, according to a given strategy, and then picks a target s among i's neighbors with, potentially, a different strategy. The edge (s, t) is then created, closing the triangle $(s; i; t)$. Because every strategy involves only choosing a node among the neighbors of a given node, we consider five different strategies to choose a neighbor v of a given node u:

■ Random: Uniformly at random
■ Shared: Proportional to the number of shared friends between u and v
■ Degree: Proportional to the degree of the neighbor v
■ Distance: Inversely proportional to the geographic distance between u and v
■ Gravity: Proportional to the degree of v and inversely proportional to the geographic distance between u and v

Because there are five different triangle-closing models, there are 25 combinations. We compute the log-likelihood of each combination, and we measure the percentage improvement over the log-likelihood of the baseline model. The results are presented in Table 3.2. The general trend is that random and shared offer the largest improvements over the baseline, with a maximum improvement of 14.54% in the combination shared–random and 12.34% for random–random. Instead, models based on degree or on distance have performance much lower than the baseline, with degradation up to 40% when the gravity model is adopted. In particular, the random–random model works surprisingly well because it favors connections between nodes that have multiple two-hop paths between them and that have higher degrees, while being extremely simple and computationally fast. These results show that triadic closure is mainly driven by social processes. Nonetheless, it only

Table 3.2   Performance of Different Triangle-Closing Models: on Each Row There Is the Model to Pick the Intermediate Node and on Each Column the Model to Then Pick the Target Node

	Random	*Shared*	*Degree*	*Distance*	*Gravity*
Random	12.34	9.48	−3.47	−28.17	−35.26
Shared	14.54	11.47	−0.95	−24.74	−35.46
Degree	7.33	5.16	−6.79	−25.17	−41.98
Distance	−0.92	−3.70	−16.94	−39.32	−41.53
Gravity	2.71	0.25	−12.11	−33.01	−43.18

Source: Allamanis, M. et al., Evolution of a location-based online social network: Analysis and models, In: Internet Measurement Conference, 145–158, 2012. With permission.

Note: The value in each cell gives the percentage improvement over the baseline, which is the log-likelihood of choosing a random node two hops away from the source.

reproduces some aspects of network evolution because edges that do not close triangles are also arising in the network. As we will see, we can exploit users' mobility information to understand how other online social connections are created, adopting closure models based on the places visited by online users.

3.3.2.4 Mobility-Driven Closure

The edge attachment mechanisms discussed so far do not include any information on mobility of the users. However, the places where users check-in could help explain how new social ties are created. According to the common focus theory [30], individuals who visit the same places tend to establish new social connections. In this section, we measure the impact that users' mobility has on network evolution. In agreement with the common focus theory, we study edge attachment mechanisms that connect users that visit the same places. We consider mobility-driven closure models to be two-step processes. A source node s first selects a place $p \grave{o} P_S$, where P_S is the set of all places where node s has checked in; then, given place p, a target note $t \grave{o} Q_p$ is selected, where Q_p is the set of all nodes that have checked in at place p. We consider different strategies that a node u adopts to select a place p from the set P_u:

- ■ Random: Uniformly at random
- ■ Friends: Proportional to the number of user u's friends that have visited p
- ■ User-check-ins: Proportional to the number of check-ins made by user u at p

■ Tot-check-ins: Proportional to the total number of check-ins made at p by all users
■ Tot-users: Proportional to the total number of users who have checked in at p
■ Place-distance: Inversely proportional to the distance between user u's home location and p
■ Place-gravity: Proportional to the total number of check-ins made by all users at place p and inversely proportional to the distance between user u's home location and p

Given a selected place p, we then consider another set of strategies to select a target user t from Q_p:

■ Random: Uniformly at random
■ Degree: Proportional to user t's degree
■ Deg-diffusion: Proportional to user t's degree and inversely proportional to the logarithm of user t's total number of visited places
■ User-check-ins: Proportional to user t's number of check-ins at p
■ Tot-check-ins: Proportional to user t's total number of check-ins
■ Inv-tot-check-ins: Inversely proportional to user t's total number of check-ins
■ Distance: Inversely proportional to the distance between user t's home location and p
■ Gravity: Proportional to user t's degree and inversely proportional to the distance between user t's home location and p

To test and evaluate mobility-driven models, we again use the maximum likelihood principle; we only evaluate the likelihood that a model has to reproduce real edge attachments where the source and target nodes share at least one place. We adopt a baseline model that selects at random target users from the set of all users that share places with the source. Table 3.3 presents the results for all the possible combinations.

Generally, in the first step, the best improvement is produced by selecting a popular place that has already been visited by many users, friends or not. For the second step, node degree plays an important role, akin to a local preferential attachment. The greatest improvement over the baseline is provided by first selecting a place that has been visited many times (tot-check-ins) and then choosing a node proportional to the degree it is "diffused" over the number of visited places (deg-diffusion). This mobility measure corrects for the fact that popular users that visit only a few places might be more related to that place, thus enticing other visitors to connect. The tot-check-in-degree model has a similar but slightly inferior performance, yet it is simpler and computationally faster. In addition to the models presented in Table 3.3, we experimented with variations of tot-users and tot-check-ins where we use a probability of attachment inversely proportional to the total number of users or check-ins. All these models provided inferior performance compared to the baseline.

Table 3.3 Performance of Mobility-Driven Closure Models: on Each Row There Is a Model to Pick the Intermediate Place and on Each Column a Model to Then Pick the Target Node

	Random	Degree	Dig-diffusion	User-check-ins	Tot-check-ins	Inv-tot-check-ins	Distance	Gravity
Random	0.28	6.88	9.24	0.16	-17.02	-4.51	-19.36	-7.04
Friends	4.70	11.60	13.63	4.74	-10.63	-1.56	-14.88	-1.71
User-check-ins	0.05	6.59	8.94	-0.03	-17.27	-4.80	-19.69	-7.41
Tot-check-ins	6.09	13.13	15.18	6.14	9.29	0.04	-13.15	-0.02
Tot-users	5.10	12.33	14.33	5.16	-9.96	-1.08	-14.19	-0.84
Place-distance	-23.41	-15.57	-13.21	-23.56	-40.82	-28.27	-43.67	-30.17
Place-gravity	0.37	7.22	9.46	0.32	-16.26	-5.29	-19.60	-6.81

Source: Allamanis, M. et al., Evolution of a location-based online social network: Analysis and models, In: Internet Measurement Conference, 145–158, 2012. With permission.

Note: The value in each cell gives the percentage improvement over the baseline, which is the log-likelihood of choosing a node at random among all the nodes that share at least one place with the source.

3.3.2.5 Temporal Evolution

In this section, we introduce how users create new connections as they spend more time on the network. We study the amount of time users remain active for their lifespan; then, we investigate the inter-edge temporal gap between the creation of consecutive edges. We consider only users that joined the service after our measurement process started, in order to observe their behavior from the very first moment.

- Node lifespan: The life span of a node is defined as the difference between the time the node created the last and the first edge. It is shown that an approximately exponential behavior, with a deviation only at longer life spans for few users who were early adopters and started using the service from the very first days. The fit is reasonably accurate for a wide range of life span values.
- Inter-edge temporal gap: Different users can show significant differences in the pace at which they add new edges: users with a higher degree create new ties at a faster rate. Thus, the term of $\delta_i(k)$ is investigated, which represents the temporal gap between the kth and $(k+1)$-th edges of user i, for different values of k. The probability distribution of $\delta(1)$, the amount of time between the first and the second edges created by a user, can be reproduced by different functions and has the following form: an exponentially truncated power law $\delta(1)^{-\alpha_1} exp(-\delta(1)/\beta_1)$ yields a slightly higher log-likelihood than a pure power law, a shifted exponential and an exponential. This result also holds for different values of k. Furthermore, users with higher degrees tend to wait, on average, for a shorter amount of time. In fact, users wait 20 days on average before adding their second edge but wait only seven days when they have about 100 friends. Although α_k tends to be unrelated to k, the exponential cutoff β_k becomes smaller as k grows larger. The final effect is that nodes with higher degrees are more likely to wait for a shorter time span because the truncated tail of the power law $P(\delta(k))$ increasingly constrains larger gap values.

It is not surprising that nodes with higher degrees add links at a higher pace. Given a fixed temporal period, such as in our measurement, higher degree nodes add more links than lower degree ones, so their activity has to be faster in the same temporal period. Nonetheless, this heterogeneous temporal behavior is crucial to fostering the heterogeneity observed in the degree distribution of social systems.

3.3.2.6 Putting It All Together: New Models Emerging

It has been shown that a gravity-based attachment, combining spatial distance and node degree, influences how new edges are created. At the same time, we have discussed that triadic and mobility-based closure are mainly shaped by social factors rather than by geographic ones. These two mechanisms seem to be complementary. Although the gravity attachment is responsible for edges connecting different parts of the network together, the closure mechanisms seem to be involved in the creation of local edges among nodes that already share either a friend or a place. Finally, nodes tend to become faster and faster in creating new edges as they obtain more connections.

Building on all these results, one aim is to now define network growth models that are able to reproduce the spatial and social properties observed in the real network. We stress that the goal of our models is not to accurately reproduce the network or to predict edge creation events but to describe the fundamental mechanisms affecting user behavior.

Following the methodology presented in Ref. [1], we describe the network-growing model as a simple algorithm to grow a network one node, and one edge, at a time. Our model combines global attachment mechanisms and local closure mechanisms:

1. A new node u joins the network according to a certain arrival discipline and positions itself over the space.
2. A new node u samples its lifetime from an exponential distribution.
3. Node u adds its first edge to node v according to a global connection model (preferential or gravity-based attachment).
4. A node with degree k samples a time gap δ from a distribution $P(\delta(k)) \propto \delta(k)^{-\alpha_k} exp(-\delta(k))/\beta_k$ and then goes to sleep for δ time steps.
5. When a node wakes up, if its lifetime has not expired yet, it creates a new edge. With probability $q = 1/3$, it creates an edge according to the global attachment mechanisms. Otherwise, the model proceeds as follows: with probability p, the node uses the random–random social triangle-closing model. If not, it uses the tot-check-ins-degree mobility-based closure.
6. The node repeats Step 4.

It is demonstrated that this model could reproduce the social and spatial properties observed in real social network traces.

In summary, although preferential attachment and triadic closure together are already able to reproduce the global social properties observed in real social networks, namely the degree distribution and the level of clustering, neglecting spatial information about where users are located fails to account for the effect of distance. In real systems, users preferentially connect over short distances, resulting in a considerable fraction of short-range ties; ignoring spatial constraints would predict an unlikely majority of long-range connections. This goes against empirical evidence, in offline and online social systems. In detail, it is observed that a gravity-based attachment, combining spatial distance and node degree, influences how new edges are created [28]. At the same time, Allamanis et al. [28] reported that triadic and mobility-based closures are mainly shaped by social factors rather than by geographic ones. These two mechanisms seem to be complementary. Although the gravity attachment is responsible for edges connecting different parts of the network together, the closure mechanisms seem to be involved in the creation of local edges between nodes that already share either a friend or a place.

The overall picture is that proximity over physical and social dimensions fosters the creation of new social links. In other words, the result is that the likelihood of a new connection increases when two individuals share many other connections or when two individuals are close to each other. This dual role of proximity has promising applications in a wide range of systems (such as friend recommendation services).

One interesting question should be discussed, however. Researchers recently have begun to use geosocial mobility traces (e.g., Foursquare check-in traces) as the representatives of our behaviors in MSNs because of their availability and scale. But are those data accurate enough to reflect our real behaviors? Zhang et al. [31] took the initial steps toward quantifying the value of geosocial data sets using a large ground truth data set gathered from a user study. By comparing GPS traces against Foursquare check-ins, the authors found that a large portion of visited locations is missing from check-ins, and most check-in events are either forged or superfluous events. In particular, 75% of events in Foursquare check-in traces are extraneous check-ins generated by users to achieve in-system rewards, and check-in events only capture 10% of actual visited locations from real physical mobility traces. These discrepancies would translate to significant deviations in application results relying on these traces. Looking forward, two major challenges should be addressed for those geosocial mobility traces of MSN. Identifying extraneous check-ins is the first step toward a trace that more accurately captures real mobility patterns; a more difficult challenge is to fill in the missing locations visited, but not reported, by users.

3.4 Motivating User Interaction in MSNs

One of the greatest features of MSNs is that the locus of control in creation and configuration of resource/content/service has been shifting to the grassroots: stemming from people, relying on people, and providing service for people. This principle also implies that, for the huge user-generated contents, a key aspect to modeling, analyzing, and finally designing mechanisms for MSN/OSN is to recognize that participation in all these systems is voluntary. Contributors have a choice whether to participate in the system at all. Second, even after having decided to participate, contributors can decide how much effort to put into their contributions, which affects the quality of the output they produce.

The question of user participation and contribution in online domains is a broad topic to which a number of active lines of research have contributed, including social psychological studies of user engagement [32], incentives for effort and high-quality contribution in social media [33], and mechanisms for distributed online recruitment [34,35]. Different systems have been used successfully in different settings.

For example, today's LBSNs allow users to embed current locations into social activities (e.g., checking in their nearby point-of-interest [POI] via mobile devices).

Foursquare is one of the largest LBSNs, with more than 30 million registered users and 3 billion "check-ins" (as of January 2013). Other popular sites include United States–based Yelp and Gowalla (now part of Facebook) and China-based Sina Weibo and JiePang. LBSNs incentivize user check-ins using (virtual) rewards. In Foursquare, for example, the user who checks in to a location most frequently in the last 60 days is awarded with the "mayor" designation. In addition, "badges" are given to users for achieving certain check-in requirements (e.g., visiting five different coffee shops). Other LBSNs have similar incentives.

The incentives in Foursquare, badges and mayorships, also motivate individuals to game the mechanisms. For example, these superfluous and remote check-ins, which are intentionally conducted by rational individuals to earn badges, account for a large fraction of gathered mobility traces. As shown in Ref. [31], in a check-in trace containing 14,297 check-ins (correspondingly, the GPS trace included 30,835 visits), 10,772 check-in events (75% of total check-ins) do not match up with any matching visit in the GPS trace (so-called extraneous check-ins), and these 27,310 visits in the GPS trace (89% of all visits) do not match any Foursquare check-in events (so-called missing check-ins or unmatched visits).

Finally, this chapter's section relates to the more general question of incentives for contribution in social media. This is a very broad area, encompassing a number of approaches beyond just the use of badges for expressing incentives. As noted in the introduction, two methodologies in the computing literature that have been brought to bear on this question are (i) social-psychological perspectives on the notion of engagement and social motivators and (ii) algorithmic game-theoretic and economic approaches, including incentives for recruitment, contribution quality, and crowdsourced effort. More detailed contents related to incentive mechanisms will be offered in the next chapter (Chapter 4).

3.5 Conclusion

Although advances in understanding the structure and dynamics of MSNs have been enormous, a conceptual framework to study this issue from the perspective of interaction design is still lacking. Generally, the global level phenomenon in social networks is affected by local behaviors and local properties of substructures. Thus, one important research field includes various approaches to modeling the social network evolution and formation, which serves as the main goal of this chapter. Specifically, several measurement methodologies used to collect large-scale user behavior data are provided, including crowdsourced crawling architecture, social network aggregator, and ESM. Then, various features of interactions in MSNs are offered. Following, we summarize the multidimensional characterizing of human mobility in MSNs, which is classified along the three axes of spatial, temporal, and social connectivity. Moreover, a new family of models of network growth is introduced that combines a global gravitational attachment process with a local triadic closure mechanism based on shared friends and shared places, which could reproduce the social and spatial properties observed in the real data. Finally, we briefly discuss how to incentivize users' behavior.

References

1. Gilbert, E. and K. Karahalios. Predicting tie strength with social media. In: Proceedings of CHI, Boston, MA, April 4–9, 2009.
2. Viswanath, B., A. Mislove, M. Cha, and K. P. Gummadi. On the evolution of user interaction in Facebook. In: Proceedings of WOSN, Barcelona, Spain, August 17, 2009.
3. Jin, L., Y. Chen, T. Wang, P. Hui, and A. V. Vasilakos. Understanding user behavior in online. *IEEE Communications Magazine* 2013; 51(9): 144–150.
4. Gjoka, M. et al. Practical recommendations on crawling online social networks. *IEEE Journal of Selected Areas in Communications* 2011; 29(9): 1872–1892.
5. Ribeiro, B. and D. Towsley. Estimating and sampling graphs with multidimensional random walks. In: Proceedings of IMC, Melbourne, Australia, November 1–3, 2010.
6. Gyarmati, L. and T. A. Trinh. Measuring user behavior in online social networks. *IEEE Network* 2010; 24(5): 26–31.
7. Ding, C., Y. Chen, and X. Fu. Crowd crawling: Towards collaborative data collection for large-scale online social networks. In: Proceedings of the COSN, Boston, MA, October 7–8, 2013.
8. Larson, R. and M. Csikszentmihalyi. The experience sampling method. *New Directions for Methodology of Social and Behavioral Science* 1983; 15: 41–56.
9. Abdesslem, F. B. and T. Henderson. Understanding mobile social behaviour using smartphones. In: Proceedings of the First Workshop on Observing the Mobile User Experience, Reykjavik, Iceland, October 16–20, 2010.
10. Wilson, C., B. Boe, A. Sala, K. Puttaswamy, and B. Y. Zhao. User Interactions in social networks and their implications. In: Proceedings of EuroSys, Nuremberg, Germany, April 1–3, 2009.
11. Jiang, J., C. Wilson, X. Wang, P. Huang, W. P. Sha, Y. F. Dai, et al. Understanding latent interactions in online social networks. In: Proceedings of IMC, Melbourne, Australia, November 1–3, 2010.
12. Benevenuto, F., T. Rodrigues, M. Cha, and V. Almeida. Characterizing user behavior in online social networks. In: Proceedings of IMC, Chicago, IL, November 4–6, 2009.
13. Schneider, F., A. Feldmann, B. Krishnamurthy, and W. Willinger. Understanding online social network usage from a network perspective. In: Proceedings of IMC, Chicago, IL, November 4–6, 2009.
14. Wittie, M. P., V. Pejovic, L. Deek, K. C. Almeroth, and B. Y. Zhao. Exploiting locality of interest in online social networks. In: Proceedings of CoNext, Philadelphia, PA, November 30–December 3, 2010.
15. Dunn, C. W., M. Gupta, A. Gerber, and O. Spatscheck. Navigation characteristics of online social networks and search engines users. In: Proceedings of WOSN, Helsinki, Finland, August 17, 2012.
16. Zhu, K., P. Hui, Y. Chen, X. Fu, and W. Li. Exploring user social behaviors in mobile social applications. In: Proceedings of the 4th Workshop on Social Network Systems (SNS), Salzburg, Austria, April 10–13, 2011.

17. Xia, M., Y. Huang, W. Duan, and A. B. Whinston. Ballot box communication in online communities. *Communication of the ACM* 2009; 52(9): 138–142.
18. Zhu, Y., E. Zhong, S. J. J. Pan, X. Wang, M. Zhou. and Q. Yang. Predicting user activity level in social networks. In: Proceedings of CIKM, Maui, Hawaii, October 29–November 02, 2013.
19. Gonzalez, R., R. Cuevas, R. Motamedi, and R. Rejaie. Google+ or Google-? Dissecting the evolution of the new OSN in its first year. In: Proceedings of WWW, Rio de Janeiro, Brazil, May 13–17, 2013.
20. Backstrom, L., E. Sun, and C. Marlow. Find me if you can: Improving geographical prediction with social and spatial proximity. In: Proceedings of WWW, Raleigh, NC, April 26–30, 2010.
21. Cho, E., S. Myers, and J. Leskovec. Friendship and mobility: User movement in location-based social networks. In: Proceedings of KDD, San Diego, CA, August 21–24, 2011.
22. Brockmann, D., L. Hufnagel, and T. Geisel. The scaling laws of human travel. *Nature* 2006; 439: 462–465.
23. Gonzalez, M. C., C. A. Hidalgo, and A.-L. Barabasi. Understanding individual human mobility patterns. *Nature* 2008; 453: 779–782.
24. Song, C., Koren, T., Wang, P., and Barabasi, A. L. Modelling the scaling properties of human mobility. *Nature Physics*, 2010; 6: 818–823.
25. Chaintreau, A. et al. Impact of human mobility on opportunistic forwarding algorithms. *IEEE Transactions Mobile Computing* 2007; 6: 606–620.
26. Karagiannis, T. et al. Power law and exponential decay of intercontact times between mobile devices. *IEEE Transactions Mobile Computing* 2010; 9: 1377–1390.
27. Hui, P., A. Chaintreau, J. Scott, R. Gass, J. Crowcroft, C. Doit, et al. Pocket switched networks and human mobility in conference environments. In: Proceedings of ACM SIGCOMM Workshop on Delay-Tolerant Net, Philadelphia, PA, August 26, 2005; pp. 244–251.
28. Allamanis, M., S. Scellato, and C. Mascolo. Evolution of a location-based online social network: Analysis and models. In: Internet Measurement Conference, Taipei, Taiwan, July 23–28, 2012, pp. 145–158.
29. Barthelemy, M. Spatial networks. *Physics Reports* 2011; 499: 1–101.
30. Feld, S. L. The focused organization of social ties. *American Journal of Sociology* 1981; 86(5): 1015–1035.
31. Zhang, Z., L. Zhou, X. Zhang, G. Wang, Y. Su, M. Metzger, et al. On the validity of geosocial mobility traces. In: Proceedings of ACM Hotnets, College Park, MD, November 21–22, 2013.
32. Burke, M. and B. Settles. Plugged in to the community: Social motivators in online goal-setting groups. In: Proceedings of the International Conference on Communities and Technologies, Brisbane, Australia, 29 June–2 July, 2011, pp. 1–10.
33. Ghosh, A. and R. P. McAfee. Incentivizing high-quality user-generated content. In: Proceedings of the International World Wide Web Conference (WWW), Hyderabad, India, March 28–April 1, 2011, pp. 137–146.
34. Babaioff, M., S. Dobzinski, S. Oren, and A. Zohar. On Bitcoin and red balloons. In: ACM Conference on Electronic Commerce, Valencia, Spain, June 4–8, 2012, pp. 56–73.
35. Cebrian, M., L. Coviello, A. Vattani, and P. Voulgaris. Finding red balloons with split contracts: Robustness to individuals' selfishness. In: ACM Symposium on Theory of Computing, New York, USA, May 19–22, 2012, pp. 775–788.

Chapter 4

Incentive Mechanisms in Mobile Social Networks

4.1 Introduction

Mobile phones have become ubiquitous in our daily lives. With the proliferation of smartphones, a vast majority of individuals carry small devices that not only provide a means of communication but are also equipped with storage and computational resources. Such developments in mobile computing have helped us to envision a future with human-centric mobile applications and the development of mobile social applications in a distributed environment [1]. Naturally, the success of mobile-based social applications largely depends on resource sharing and selfless/altruistic human contributions.

Meanwhile, mobile social networks (MSNs) and online social networks (OSNs) are increasingly centered on contributions by their users. User-generated content (UGC) such as Amazon and Yelp reviews, Wikipedia articles, blogs, or YouTube videos now constitute a large fraction of the relevant, easily accessible content that makes MSNs/OSNs extremely useful; crowdsourcing tasks to the online public is increasingly common—ranging from systems based on unpaid contributions such as games with a purpose or online question and answer (Q&A) forums (Y! Answers, Quora, and Stack Overflow to name a few) to platforms for paid crowdsourcing such as Amazon's Mechanical Turk and TopCoder. But although some Web sites consistently attract high-quality contributions, other seemingly similar sites are overwhelmed by junk, and still others fail due to too little participation. There is a growing body of work in the social psychology literature on what factors motivate, or constitute rewards for, participants in these social computing systems, and on user experience and interface design to exploit these factors.

For instance, some researchers have explored the factors affecting answer quality on Q&A sites. Raban and Harper [2] point out that a mixture of intrinsic factors (e.g., perceived ownership of information, gratitude) and extrinsic factors (e.g., reputation systems, monetary payments) motivate Q&A site users to answer questions. Beenan et al. [3] confirmed that intrinsic motivations, such as visibility of expertise and the feeling of making a unique contribution, influence participation in such systems. Results regarding extrinsic motivators have been more mixed: Hsieh et al. [4] found market-based incentives did not increase answer speed or high-quality answers, but Harper et al. [5] found fee-based sites produced higher quality answers than free sites. The Defense Advanced Research Projects Agency's (DARPA) red balloon challenge was a highly publicized instance of human computation in the sense of a distributed network of human sensors that required incentivizing the rapid mobilization of a large number of participants on a social network. The challenge took place in December 2009 and consisted of locating 10 eight-foot high red balloons that had been moored at ten unknown locations throughout the United States; the first team to correctly identify the locations of all ten balloons would receive a cash prize of $40,000. For a team to win the challenge, it was necessary not only to recruit members who would look for and report sightings of the balloons themselves but also to incentivize recruits to further recruit team members because increasing the number of searchers increased a team's chance of quickly locating the balloons. That is, in addition to the problem of incentivizing participation, a team also had to incentivize further participation. The recursive incentive scheme that was used by the winning MIT team—to split the prize money among its participants—is described and analyzed in Ref. [6].

In brief, these approaches cover a broad spectrum of incentive mechanisms: from extrinsic, through social, to intrinsic. Some of them are based on different theories from the fields of social psychology and behavioral economics and involve economic rewards mechanisms, reputation, open group user modeling, and social visualization.

Irrespective of any incentives inspired by intrinsic, social, or extrinsic motivation theories, there generally are two components to the problem of incentive design for MSNs that rely on rational users' participation and contribution: (1) Identifying the costs and benefits of potential contributors to the system (the components that help formulate a model of agent behavior) and (2) deciding how to assign rewards, or benefits, as a function of contribution (analysis and design) [7] [8].

The first component—identifying costs and benefits—relates closely to the question of why people contribute, that is, what constitutes a benefit or a reward? The answer to this question, of course, varies depending on the particular system in question. Although some systems offer financial incentives for participation, a vast majority of human computation is driven by social-psychological rewards from participation; such rewards include, for example, intrinsic motivators such as fun, interest, or the satisfaction of benefiting a cause, as well as extrinsic social rewards such as attention, reputation or status, and so on. There is now a growing literature in social psychology addressing what motivates, or constitutes a reward for, users in such systems. But even after answering the question of why people contribute, there is a second question, which relates to how rewards are allocated. Given that users value rewards (financial, social, or psychological) and incur costs (in time and effort) associated with different actions in the system, how rewards are assigned will influence what actions users take. That is, when a system depends on self-interested agents with their own benefits from and costs of participation, the quality and quantity of contributions will depend on the incentives created by the reward allocation scheme being used by the system. Given the understanding from the social psychology literature on what constitutes a reward, how should the allocation of these rewards be designed to incentivize desirable outcomes?

A given design for an MSN corresponds to, or induces, some rules that specify the allocation of rewards or benefits given each set of possible actions by agents. Note that, in general, an individual's reward can depend not only on their output but also on the outputs (determined by the action choices) of other agents. Given a particular system design and the corresponding rules it induces, strategic agents will choose actions that maximize their utility (the difference between benefit and cost) from the system. Agents' choices of actions lead to outputs, which in turn define the benefit, or reward, that each agent receives from the system. A vector of action choices by agents, roughly speaking, constitutes an equilibrium if no agent can improve their payoff by choosing a different action. Naturally, a game-theoretic approach could help us understand why and could provide guidance on designing systems that incentivize high participation and effort from contributors.

A formal game-theoretic approach to incentive design, very broadly, proceeds by constructing an appropriate model where users (agents) make choices over actions, which are typically associated with costs (note that the term "cost" does refer not only to financial costs such as an entry fee but also to nonmonetary quantities such as the cost in time or effort). For example, action choices in MSN systems can consist of the following: (1) In most systems, participation is a voluntary action choice with an associated cost (e.g., the time required to create an account or to log in to the system to participate), and mechanisms must be designed to induce adequate participation when entry is an endogenous, strategic, choice. (2) In many systems, individuals can make a choice about how much effort to expend on any given task, potentially influencing the quality of their output and therefore its value to the system; mechanisms must be designed so as to induce agents to expend a high level of effort (which is more costly than lower effort). (3) Finally, in some systems, agents may hold information that they can potentially strategically misreport to their benefit, such as voting or rating; this leads to the problem of designing mechanisms that induce agents to truthfully reveal this information. Naturally, any real system might contain a combination of these choices as well as others unique to its function.

Generally, there are two aspects to a game-theoretic approach to incentives: analysis and design. Analyzing equilibrium behavior under the reward allocation rules of a given system leads to a prediction about the behavior of agents and therefore what kind of outcomes one might expect from that system. Choosing (or altering) the rules according to which rewards are allocated to induce individual behavior that achieves some particular outcome, or family of outcomes, constitutes design. Although a game-theoretic approach to the analysis and design of any system with strategic agents has the general structure described earlier, each setting or system comes with its own unique features, depending on the choices of available actions, the nature of the available rewards and differing constraints on how they can be allocated, and the observability of individuals' output.

In brief, in MSN applications, it is imperative to incorporate mechanisms and tools in the design of the social application that can motivate users to participate in and contribute to high-quality and high-quantity contents, and more generally, to change their behavior in a desirable way that is beneficial for the community. Because different people are motivated by different things, it can be expected that personalizing the incentives and the way the rewards are presented to the individual would be beneficial.

The chapter is organized as follows. Section 4.2 introduces basic concepts and theories on motivating rational individuals, including a classical economic view, a behavioral economic view, and related social psychology theories. Some typical incentive mechanisms inspired by gamification (especially badges) are summarized in Section 4.3. In Section 4.4, we discuss the concept of crowdsourced sensing in MSN and introduce some existing schemes. Finally, in Section 4.5, we briefly conclude this chapter and point out some future considerations.

4.2 Basic Concepts about Motivation Theories

Why people act in particular ways is a fundamental question that has been the focus of economists and psychologists since these disciplines came into existence. Vassileva [9] presented an overview of different approaches to motivate users to participate. These approaches are based on various theories from social psychology and behavioral economics and involve rewards mechanisms, reputation, open group user modeling, and social visualization.

4.2.1 Motivation Approaches

4.2.1.1 Economic View of Motivation: Example Design and Challenges

Classical economics approaches the issue of motivation by assuming that people are rational agents who act to maximize their utility (payoff) in a world where behaviors have certain payoffs (negative or positive). Thus, to make people behave in particular way, one needs to create an appropriate system of incentives (rewards) for the desirable behaviors. Incentive mechanism design (also simply called "mechanism design") is a very active area of research in mathematical economics and game theory. The goal is to design rules of encounter that, when followed by the participants, will ensure that the overall system fulfills a particular goal or fits a set of criteria (e.g., optimizes the joint welfare for all participants, ensures a fair chance for them to maximize their utilities, or simply maximizes the utility of the owner of the system). The payoffs for particular actions may be subjectively different; that is, each participant may have his or her own unique utility function. Most of the applications of mechanism design are in tightly constrained systems (e.g., auctions). Approaching the problem of motivating participation in an MSN community as an economic mechanism design emphasizes the benefit of the system or the community as a whole, rather than that of the individual users.

An example of an economic mechanism based on virtual currency can be found in a peer-help community called "I-Help" [10]. The mechanism regulated the demand and supply of help in the community. Students could be buyers and/or sellers of help on various questions/topics. Virtual currency was used to complete trades. The price depended on the scarcity of helpers competent in answering a question on a given topic at the moment of the request. At the end of the term, the currency accumulated by students was exchanged for something of real-world value.

4.2.1.2 Some Challenges in an Economic View of Motivation

4.2.1.2.1 Challenges in Creating an Appropriate Market Model for the Community

Although introducing currency and a market is fairly straightforward, challenges arise related to the specifics of the community in which the market is introduced. For example, I-Help was a learning community, where help is the main traded good. Help is quite different from tangible goods (such as those traded on eBay). In any group, there are weaker students who mostly need help and are rarely able to provide help to others. In a pure market-based system, these students are likely to become "bankrupt" (i.e., unable to buy help anymore) and thus are shut out of the system. In I-Help, of course, such an outcome was undesirable because the main goal of the system was to increase the knowledge of all students by creating incentives for students who had knowledge to provide help and by giving everyone a fair chance to buy help. Therefore, a "social welfare system" had to be introduced. However, providing a fresh supply of virtual money (e.g., a weekly allowance) complicated the economy significantly, leading to inflation. The total amount of currency was no longer fixed but could grow without limits, which made it hard to adequately match the virtual currency earned by the students with real-world rewards. An economic solution to this problem, such as introducing taxation on earnings, would have further complicated the mechanism and would likely have not been motivational for active helpers. Putting a cap on the earnings of active helpers would have also been a disincentive to continue helping after they had reached their cap. The pedagogical goal of the system was to encourage students to always help—even if they were the top helpers in the community—because one learns more by helping than by receiving help. This could not be achieved with the economic model of I-Help, where the currency was injected in the system from outside rather than generated from within the community; knowledge was a positive externality that was generated from within the community during help sessions but formally unaccounted for in the model.

4.2.1.2.2 Challenges in Designing the User View of the Mechanism

Another challenge arising in the system design is how much of the underlying economic model should be revealed to the user. This is a common problem with all incentive mechanisms whether they are based on a market model, a game-like system where users collect points or earn reputation (as discussed in the next section), or on a psychological theory of motivation. In I-Help, because the general purpose of the system was to facilitate learning, it was important to keep the students' attention focused on learning rather than on trading help and earning currency. For this reason, instead of having the users explicitly trading for help, the economic transactions (negotiating for the price and dealing with the payments) were delegated to the software infrastructure (the personal agents of the users), who maintained models of user preferences with respect to price, availability, topics of competence, and so forth.

4.2.1.2.3 Challenges in Adapting the Parameters of the Mechanism at Run Time

Another challenge is that often the mechanism is a part of a dynamic system and requires user input to dynamically adapt the rewards to the situation at hand. For example, in I-Help, the mechanism did not consider the quality of help exchanged in the price negotiation—only the help demand/supply ratio at the moment. It could happen that after a help session started, one of the parties discovered that the session was a waste of time. The system, however, allowed students to quit a session at any time to avoid being charged for useless chatter. A time-meter mode of payment was used (similar to a telephone call); that is, the price per minute of help was negotiated by the agents (instead of a total price for the session), thus allowing any partner to interrupt the chat session if they felt they were not getting value from it.

An alternative solution more typical of current online communities is to collect feedback (ratings) after the session from both partners and to compute reputation for each helper and helpee. This would allow the agents of users with a high reputation to charge higher prices for their services and would have also provided an incentive for users to give good help.

Two large-scale communities similar to I-Help, though not in an educational context, are Google Answers and Yahoo Answers. Google Answers operated between 2001 and 2006 and was based on a market model using real dollars. However, research into the user motivations in these communities [11] shows that the participation of experts is associated with a hybrid of economic and social motivators such as "star" ratings and user feedback on answers.

Monetary rewards were responsible for the demise of Google Answers because the community was riddled with gamers trying to exploit the system and make money, while not providing any valuable answers and causing a lot of user complaints. To avoid following Google Answers missteps, Yahoo Answers uses a modified currency mechanism that rewards active users with a range of honor badges ("power users," "top contributor," etc.) that are visible to other users and that represent their reputation in the community. This kind of mechanism is more in line with modern behavioral economics; incentive mechanism designs along these lines will be discussed in the next section.

4.2.1.3 Behavioral Economics View of Motivation

In contrast to classical economics, behavioral economics views people as irrational and investigates—usually experimentally—the social, cognitive, and emotional factors in understanding the economic decisions of individuals. Many behavioral economic theories relate to why people make certain choices and what drives or motivates people's behaviors, showing that many theoretically sound economic mechanisms are not psychologically valid and fail when tried with real users [12]. Some psychological phenomena have been studied, such as "fairness," "inequity aversion," and "reciprocal altruism," that question the classical economics assumption of "perfect selfishness." In the context of proliferating social networking sites, smartphones, and pads with sensors, the theories of behavioral economics gave rise to user engagement design approaches aiming to increase participation in MSNs. One major direction is the so called "gamification" of MSNs or the introduction of gaming elements into MSN user interaction design.

4.2.1.3.1 Gamification

"Gamification" is defined as: "the integration of game mechanics in nongame environments to increase audience engagement, loyalty, and fun" (www.gamification. org, for academic references see Ref. [13]). The related area of practical expertise called "game mechanics" has accumulated a number of patterns, rules, and feedback loops, that are motivational, create user engagement and loyalty, and can be applied to develop game-like elements in virtually any application or community. Examples of the most commonly used patterns are ownership (allowing the user/player to own things, such as points, tokens, and badges, because that creates loyalty to the application, game, or community), achievements (providing a virtual or physical representation of having accomplished something that can be easy, difficult, surprising, funny, and accomplished alone or as a group), status (computing and displaying a rank or level of a user), and community collaboration and quests (posing challenges to the users related to time limit or competition that can be resolved by working together). The use of badges can be viewed as part of the growing phenomenon of gamification. And moreover, badges are a growing trend in the design of online communities, social computing applications, and electronic commerce, and researchers have begun to study the role that badges play in these sites. Antin and Churchill [14] present a conceptual organization for different types of badges, considering among other things the motivation they provide from a social-psychological viewpoint.

As explained here, the idea of adding game elements to nongame applications and social sites has a lot of potential. Recently, however, there have been criticisms of the gamification trend, pointing out that most of the current gamified sites make their users collect points for trivial actions, thus devaluing the rewards. Some influential bloggers predicted that soon it will be "game over" for this type of applications. The reason is that ubiquitous point gathering is based on a simplistic economic and behavior model and leads to short-term motivation. The current hype of gamification will unavoidably disincentivize the most creative elite users, who are most valuable for any community or social application. Moreover, it is emphasized that there is a need for development of different types of games that foster a sense of achievement rather than rewarding users with points and badges, that create intrinsic motivations rather than replacing them with extrinsic rewards (points and badges) and that reintroduce genuine play and genuine delight. As Deterling explains [15]: "... we play games because we inherently enjoy the activity. If you look further at what makes an activity inherently enjoyable, then you see that games deliver on all three things in the current major theory of intrinsic motivation, self-determination theory: they give you experiences of competence, autonomy, and relatedness".

4.2.1.3.2 Reputation

For a long time, reputation has been used in online communities to motivate participation. Slashdot pioneered this approach by introducing the notion of "karma" in the mid-1990s to reward users (who gave good comments) with visibility and power in the community. Currently, most social sites provide ways for users to build their reputation based on the ratings received by their contributions. The most prominent examples are eBay's seller and buyer reputation ratings and Amazon's reviewer ratings. Yet designing successful reputation schemes can be quite challenging.

The difference between status and reputation is that although status can be earned by the user in isolation by performing certain actions, reputation is based on the opinions of others about the user or his/her contribution. Reputation can be developed, for example, by posting articles that earn very high ratings. Rock groups and celebrities on Twitter measure their reputation by the number of fans/followers. Users on Facebook keep track of their reputation by the number of friends they have. However, the term "reputation" has often been used interchangeably with status. For example, Amazon calculates what they call the "reputation" of book reviewers based on the number of reviews they have written (this would be their *status* using our definition) and the ratings these reviews have obtained (this would be their *reputation* according to our definition).

Selecting which user actions should be rewarded with status and/or reputation in a particular community is not a straightforward process, nor is determining what privileges should be granted for what levels of status and/or reputations.

Comtella is a small-scale, peer-to-peer online community developed at the MADMUC lab at the University of Saskatchewan that is used for sharing links to academic papers and class-related Web resources among students (http://madmuc. usask.ca/peer-motivation.htm). The first incentive mechanism applied in Comtella rewarded users with points for actions that were beneficial to the community (contributing new papers, downloading papers from others, and making them available for sharing with others, etc.). These were actions over which the user had full control and that did not reflect the opinion of others about the user's actions. Thus the reward was called "membership level/status" rather than "reputation." Each user was classified, depending on their accumulated points, into one of three different status levels (gold, silver, or bronze). Different status levels implied different privileges (e.g., interface appearance, number of ratings to give out). The results of the evaluation of this mechanism showed a significant but short-term increase in participation. There were attempts by some users to game the system by performing unreasonably high numbers of the rewarded actions. Because the quality of the contributions was not evaluated, the users' participation in the system deteriorated due to the overwhelming amount of low-quality contributions and the resulting cognitive overload.

If increasing participation was the only goal, the first Comtella incentive mechanism based on status was quite effective. However, because of the problem of gaming, increasing participation could not be tackled without introducing a measure of quality of contribution into the mechanism; therefore, reputation had to be introduced in the next version of Comtella.

Basically, the reputation of a user was calculated as a function of the ratings the user's contributions received from others. However, incentivizing users to rate the contributions of others is not easy; again, this is a problem of increasing participation (of a different kind) that is encountered by all systems that rely on user ratings (e.g., recommender systems). To encourage users to give ratings, a market-based model with virtual currency (c-points) was introduced. The user could earn c-points by rating a resource and could spend them to promote his or her own contributions (such as Google's sponsored links). The currency model was very successful in stimulating ratings, and it resulted in twice the amount of ratings generated by the experimental group versus the control group in a controlled experiment. With many ratings, the computation of user reputation became more accurate.

Unfortunately, in most real communities and applications, there are also general system goals similar to those existing in the area of market design. For example, it may be desirable that the user contributions follow a particular time pattern because early contributions usually are more important than late contributions; they set the tone of future contributions and provoke users to respond or to share their own contributions. Later on, as the volume of contributions increases, it becomes important to get users to rate the contributions of others so that good resources can be found more easily. Also, high-quality contributions should be rewarded any time. Therefore, a need arises for the creation of dynamic incentives that "orchestrate" the individual user behaviors to produce a harmonic overall behavior of the system. The patterns of game mechanics are insufficient for accomplishing this.

However, no general theories or guidelines exist for designing mechanisms with dynamic rewards. They are crafted according to the specific needs of the community. The second version of the Comtella incentive mechanism is presented here as an example.

The new incentive mechanism aimed to encourage contribution of links to high-quality articles, to discourage excessive contribution, and to encourage timely contributions. The rewards for each participative action (contributing papers and contributing ratings) were increased or decreased dynamically according to the individual's reputation for contributing high-quality papers and high-quality ratings. Because Comtella was deployed in an educational context, where students were sharing articles related to the weekly topics discussed in their class, one of the overall goals was to ensure that students shared their articles early in the week so that there was time for their colleagues to read, rate, and comment on them. The weight of each action depended on the day of the week and on the number of resources that already had been contributed by the community. To prevent over-contributions by students who might have tried to game the system to achieve high status, there was also a personal cap on the number of rewarded contributions that depended on the quality of the previous contributions by the user and the desired number of contributions for the week for the entire community, as set by the instructor. In this way, the status of the user was calculated based on dynamic, adaptive rewards that took into account a model of the community's needs and the model of the individual contributions of each user.

The results of a controlled study evaluation with 21 students showed that the mechanism was very effective and stimulated exactly the behavior that was desired. The conclusion was reached that a mechanism with rewards adaptive to the individual patterns of contribution and to the needs of the community could orchestrate/conduct the desired patterns of behavior in the individual users, leading to a sustainable level and to higher quality of contributions.

However, it is very possible that such mechanisms have been implemented in real systems but never revealed. Generally, most successful large-scale communities do not reveal details about the incentive mechanisms that are deployed; otherwise, they would be challenged by gamers.

4.2.2 Motivation Theory from Psychology

Motivation has been studied extensively in the field of psychology, where a wealth of theories of motivation has been developed over the last 100 years. The focus of these theories is the individual and his or her experiences with the environment and other individuals or with society as a whole. It is impossible to provide an overview of any depth here regarding the existing theories of motivation due to their sheer number. To navigate among the spectrum of theories, we will consider a distinction that may serve as a watershed between two clusters of influential psychological motivation theories (Figure 4.1). This is the distinction between extrinsic motivation (from the outside, driven by external rewards or by pressure from the environment and other individuals) and intrinsic motivation (from within, driven by interest or enjoyment that the individual experiences from the activity). This classification will help the reader navigate within the theories because the three positions (Intrinsic, Social and Extrinsic, as shown in Figure 4.1) of the spectrum can be found in existing design patterns. However, we do not claim that this classification has any larger validity. In fact, the distinction among the categories in Figure 4.1 is quite blurred, and there are researchers who question even the distinction between intrinsic and extrinsic motivation, emphasizing that it is all a matter of individual differences.

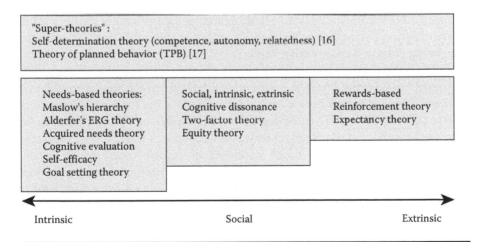

Figure 4.1 A spectrum of motivation theories. (With kind permission from Springer Science+Business Media: *User Model and User Adapted Interaction,* **Motivating participation in social computing applications: a user modeling perspective, 22, 2012, 177–201, Vassileva, J.)**

Reiss [15] proposes a theory of 16 basic desires that can exist simultaneously or at different strengths at different times in different individuals.

Extrinsic motivation (reward) is the focus of Skinner's reinforcement theory and of the expectancy theory. On the other side of the spectrum, intrinsic motivation is the focus of the need-based theories of Maslow; Alderfer's existence, relatedness, and growth (ERG) theory; acquired needs theory; as well as self-efficacy theory and goal setting theory. In the middle of the spectrum, Herzberg's two-factor theory, equity theory, and cognitive evaluation theory consider the interplay of intrinsic, extrinsic, and social motivators. Social comparison, status, and reputation can clearly provide a strong motivation for participation for many users.

Many of the aforementioned theories can explain the motivational effect of the game design patterns mentioned previously. Clearly, all of these patterns provide extrinsic rewards for the users that, according to extrinsic motivation theories, should provide motivation for the users to perform the actions or behaviors that lead to rewards (e.g., collect points, badges). Theories in the middle of the spectrum explain the motivational effect of reputation, which has meaning only in a social environment. On the other side, because different people consider different things as rewarding—depending on their intrinsic needs, values, and goals—the theories on the intrinsic end of the spectrum and those in the social category explain the different possible needs that people have. For example, the motivational effect of reputation and status can be explained by all need-based theories—such as Maslow's theory, Alderfer's ERG theory, and acquired needs theory, and social psychology—all of which point to the human need to socialize and seek out social recognition and status. It can be also explained by Bandura's self-efficacy theory because usually social status and reputation are a result of recognized mastery, which is one of the four major sources of self-efficacy. A visible reputation in a group sets conditions for another source of self-efficacy, social modeling, or witnessing people successfully completing tasks or demonstrating mastery.

For example, let us focus on the theory of social comparison, which states that people tend to compare themselves with others they perceive as similar to them in order to evaluate or enhance some aspect of the self. Whether the social comparison serves a self-enhancement function depends on whether the comparer assimilates or contrasts his or her self relative to others who are superior or inferior. Two processes can be observed: assimilation, which facilitated by the belief that one can obtain the same status as the target (the role model), and contrast, which is the comparison with dissimilar others to enhance or protect the subjective well-being and thereby satisfy the self-enhancement motive. The social comparison theory can explain the motivational effect of the leaderboard pattern in game mechanics and has been the inspiration for incentive mechanism design in several research projects. In Comtella, Julita Vassileva sought to encourage upward assimilation by visualizing the status and reputation of users by utilizing a star–sky metaphor [9]. Specifically, each user was shown as a star on a night sky with colors corresponding to status (gold, silver, bronze), with brightness corresponding to the reputation of the user, and with size corresponding to the number of shared papers. It was determined that many users checked their reputation status in the visualization, and that users who checked their status more frequently contributed more. Similar results have been reported by other authors proposing similar incentive mechanisms based on reputation status and social comparison. It seems that social comparison can be used as a powerful incentive and can effectively increase contributions to online communities.

More recently, newer areas of social science such as social psychology, educational psychology, organizational science, and sociology (media studies) have contributed more theories focusing on motivation in particular types of environments. These newer theories have more emphasis on intrinsic motivation and therefore hold a promise of inspiring newer motivational patterns and incentive mechanisms that emphasize achievement, altruism, and genuine delight in gaming (in contrast to the currently used gamification patterns).

In summary, this section has provided a broad overview of existing approaches, design patterns, and theories related to motivating participation in social applications. There exist simple approaches and design patterns that have been shown to successfully engage users and that are widely applied in the gamification of social applications. An approach using adaptive dynamic incentives guided by community needs models and individual user models was briefly presented that orchestrated a particular overall time and activity pattern beneficial for the entire community. Finally, some of the most commonly cited theories of motivation that have inspired research on design of motivational patterns and mechanisms (emphasizing those providing intrinsic and social motivation) were presented.

4.2.3 Future Trends

4.2.3.1 Exploring Designs of Mechanisms Inspired by Theories of Motivation

One clear ongoing trend is further exploration regarding which of the numerous contemporary motivation theories in the areas of social psychology and behavioral economics can be usefully applied in designing reward mechanisms for particular types of communities.

Incentive mechanisms that are applied in a real community are rarely grounded on a single theory; usually, they rely on motivations following two or three theories in combination. It is therefore hard to attribute success or failure to a particular theory. It is also very hard to control external factors that can unpredictably influence participation (e.g., certain external events that may fascinate the community and trigger unexpected bursts of participation and software or system failures that can kill participation if they happen in a critical phase at the start). Moreover, the success of a particular incentive mechanism design in one community does not guarantee that the same mechanism will be successful in another community, so there will be a great need for repeated studies under different conditions to confirm earlier findings.

4.2.3.2 Personalized Incentive Mechanism Design

The purpose of incentive mechanisms is to change the state of the user (their goals, beliefs, motivations)—to adapt the individual user to the benefit of the overall system or community. This is the opposite of the purpose of user-adaptive environments, which is to adapt the system to the needs of the individual user. Most work on incentive mechanism design can be viewed as orthogonal to personalization because it is based on the assumption that, to ensure fairness, a community needs common (not personalized) rules for rewarding user behavior. However, stepping up from the individual (micro) to the community (macro) level, an incentive mechanism can be viewed as an adaptation mechanism focused toward the behavior of a community of users. It monitors the actions of the community represented in a community model or in a collection of individual user models, and it makes adaptations to the interface, information layout, or functionality of the community in order to respond to the changes in the user model according to some predefined goal (e.g., maximizing participation). The parallels between an adaptation mechanism in a personalized environment and an incentive mechanism in an online community are summarized in Table 4.1.

It seems logical that incentive mechanisms need to be personalized because every person has different motivations depending on personality, gender, age, education, stage in life, cultural background, interests, priorities, and so forth. However, in current literature, most existing incentive mechanisms are not personalized. Even in the adaptive rewards version of Comtella introduced earlier in this chapter, where the weights of different activities (i.e., the rewards) depended on a model of the user's previous contributions, the mechanism as a whole was still the same for all users and was geared toward earning reputation, status, and power in the community.

Intuitively, there is a good reason for having just one incentive mechanism in a community—designing an incentive mechanism is like making up the rules of a game. Normally, all the players in a game are bound to follow the same rules. However, if the game is complex enough, it has many rules and some players may choose to follow some of the rules while not violating the others. Similarly, there may be several incentive mechanisms embedded in a community (e.g., one targeting the people who are motivated by reputation, another one for people primarily motivated by power, and a third one for people motivated by building balanced relationships with other users). Although the resulting system will not necessarily be "personalized," it will provide an opportunity for users to choose and to pursue their intrinsic personal motivations and to set their goals accordingly. Yet the introduction of different mechanisms in the same system is not straightforward. Interactions among different incentive mechanisms can lead to mutual cancelling out of their motivational effects. The investigation of the motivational effects of different incentives, their combinations, and side effects is currently an active area of research in behavioral economics.

Table 4.1   Parallels between Personalization Mechanism and Incentive Mechanism

Personalization Mechanism in User-Adaptive System (microlevel)	Incentive Mechanism in an Online Community (macrolevel)
User model: Individual user's preferences, interests, ratings, knowledge, goals, etc.	*Community model:* Community participation, represented according to a certain set of metrics *Individual models:* Individual participation represented according to certain metrics
Purpose of adaptation: Optimizing system behavior toward the individual user Recommending content of interest for the user Adapting interface to the preferences/level of knowledge/experience/current goal of the user Stimulating reflection in user	*Purpose of adaptation:* Optimizing system behavior with respect to all the users in the system Increasing the number and quality of user contributions Binding the users in social ties Enticing users to commit to a common goal Making the community self-sustainable, ensuring growth and stability
Adaptation interventions: Showing recommendations, sorting list of search results, reducing complexity of interface to text, visual signaling	*Adaptation interventions:* Providing rewards for particular actions (individually weighted) Visualizing the community adaptively to emphasize particular incentives

Finally, an important issue that has not yet been discussed in this chapter is the ethics of motivating people to stimulate particular behaviors. Although it was implicitly assumed that motivational and incentive mechanisms are designed for "good" purposes, nothing prevents their exploitation for commercial purposes (such as a very high interest by companies in gamification that has been observed) and for darker purposes.

In summary, there are interesting challenges lying ahead for designing proper incentive mechanisms in MSNs: how to adapt the motivational approach to the individual without disturbing the effects of the general incentive mechanism in the community, how to create models of groups and communities that can support adaptive incentive mechanisms, and how to design open group models and social visualizations with particular motivational purpose.

An interesting feature of MSNs that is related to incentive mechanism design is that an MSN is typically characterized as a two-sided market. (This will be thoroughly described in Chapter 9.) In particular, the most common form of MSN systems are developed and maintained by independent and specialized platform providers. A typical example is that merchants can post product information and advertisements on the platform so as to direct them to potential customers, while consumers may search and browse the platform for desired products. The two user groups/two sides (consumers and merchants) are interdependent with regard to platform value. In other words, a platform with more merchant users would be more attractive to consumers and vice versa. Interestingly, although two-sided market theories have been extensively discussed with regard to economics and information systems, related literature has primarily focused on pricing strategies in platform settings, such as payment systems or matching intermediaries. Few efforts have addressed the issue of user adoption behaviors under the two-sided market effects. An integrated conceptual model was proposed for analyzing user adoption behaviors toward mobile marketing platforms from a two-sided market perspective [18]. Both the consumer side and the merchant side are modeled based on extending classical theories with newly introduced factors reflecting cross network effects; the two sides are integrated in the overall model, which reveals the dynamic interaction between the evolutions of the two user groups through the platform. In brief, the cross network effect should be properly used in designing incentive mechanism in MSNs.

4.3 Typical Incentive Schemes in MSNs

As described earlier, UGC refers to a very wide spectrum of online content generated by end users, ranging all the way from collaborative information sites such as Wikipedia to Q&A forums and discussion boards to content on blogs and social media such as Facebook, as well as comments on blogs and news articles. On the one hand, UGC constitutes a large fraction of the high-quality, easily accessible content that makes the Web useful. On the other hand, because there is no barrier to entry (unlike traditional publishing), it also attracts junk and spam.

Therefore, it imperative to identify the economical, social, and psychological motives, also called incentives, of contributors in these systems and to design user interfaces and systems to better reward those motives to encourage participation. There is a vast literature in social psychology that employs this approach. A game-theoretic approach complements this approach and can also benefit from this literature in appropriately modeling cost and benefit functions for contributors.

In particular, Gosh and McAfee [19] addressed this question from a game-theoretic perspective, which uses a model with strategic contributors where the primary motivator is exposure or viewer attention. A contributor is motivated by the amount of exposure their content will receive. However, without some connection between quality of a contribution and amount of exposure, such exposure-motivated contributors will flood a site with low-quality contributions—as is indeed observed in practice. Is there a way to allocate the available attention from viewers among the contributions, a mechanism that encourages high-quality contributions while also maintaining a high level of participation?

In the model offered by Gosh and McAfee [19], both the quality of contributions as well as the extent of participation is determined endogenously in a free-entry Nash equilibrium. The model predicts, as observed in practice, that if exposure is independent of quality, there will be a flood of low-quality contributions in equilibrium. An ideal mechanism in this context would elicit both high quality and high participation in equilibrium, with near-optimal quality as the available attention diverges, and should be easily implementable in practice. A very simple elimination mechanism is designed that subjects each contribution to a rating by some number S of viewers and eliminates any contributions that are not uniformly rated positively. The authors construct and analyze free-entry Nash equilibriums for this mechanism, and they show that S can be chosen to achieve quality that tends to optimal, along with diverging participation, as the number of viewers diverges.

During the past few years, gamification has been a trending topic and a valuable means of supporting user engagement and enhancing contribution effort. These desired use patterns are considered to emerge as a result of positive, intrinsically motivating, "gameful" experiences brought about by game/motivational affordances implemented into a service. As a result, gamification is touted as a next generation method for marketing and for customer engagement in popular discussion. For instance, Gartner [20] estimates that more than 50% of organizations managing innovation processes will gamify aspects of their business by 2015. Furthermore, there is an increasing number of successful start-ups, the entire service of which is focused on adding a gamified layer to a core activity (e.g., Codecademy[*] a service that uses game-like elements to help teach users how to code) or assisting more traditional companies in gamifying their existing services (e.g., Badgeville[†]).

According to the conceptualization provided by Hamari et al. [21], gamification can be seen to have three main parts: (1) the implemented motivational affordances, (2) the resulting psychological outcomes, and (3) the further behavioral outcomes. The authors provide a thorough literature review of empirical studies on gamification, and they attempt to answer the question: Does gamification work? The finding is that, indeed, gamification does work, but some caveats exist. The majority of the reviewed studies did yield positive effects/results from gamification.

---

[*] www.codecademy.com
[†] http://badgeville.com

However, most of the quantitative studies concluded that positive effects exist only in part of the considered relationships between the gamification elements and studied outcomes. Further, the studies that investigated gamification qualitatively revealed that gamification as a phenomenon is more manifold than the studies often assumed. These observations suggest that some underlying confounding factors exist. Most prominently, the studies bring forth two main aspects: the role of the context being gamified and the qualities of the users.

For the first aspect, the present studies on gamification and motivational affordances suggest that the context of the service might be an essential antecedent for engaging in gamification. For example, Hamari [22] suggested that, for services oriented to strictly rational behavior (such as e-commerce sites), gamifying systems might prove challenging because the users could be geared toward optimizing economic exchanges.

Theory and real games both suggest that outside pressures (such as extrinsic rewards) undermine intrinsic motivations and hence would, in essence, undermine gamification. Understanding the contextual factors would benefit from considering the following perspectives: (1) the social environment—the voluntariness of carrying out a task is one of the main antecedents for attitude formation and behavior; (2) the nature of the system—whether the system in question is utilitarian or hedonic in nature; and (3) the involvement of the user—whether it is cognitive or affective in nature.

For the second aspect, user qualities were believed to have an effect on attitudes toward gamification, which could explain why in certain environments or with certain users, gamification had significant effects. Basically, people interact with game-like systems in different ways and for different reasons. Thus, the experiences created by the gamifying motivational affordances are also likely to vary. Eickhoff et al. [23] mention the emergence of distinct "worker types" in their service, which gamified crowdsourced relevance assessments. In addition, the series of studies done on gamifying IBM's Beehive social networking service [24] also note that the users fell into distinct behavioral patterns; for example, some users wanted to be at the very top of the leaderboard, while for others it was enough to simply appear on the leaderboard, regardless of ranking. Hamari [22] suggests that the sporadic nature of usage might not be compatible with persistent gameful affordances because the users might not spend enough time within the service in order to become interested.

A common theme in a growing number of online communities and social media sites that rely on user contributions is gamification via badges, leaderboards, and other such forms of (competition- or accomplishment-based) social-psychological rewards. These rewards, meant to provide an incentive for participation and effort on a given system or site, usually reflect various site-level accomplishments based on a user's cumulative "performance" over multiple contributions. Such badges or top contributor lists clearly appear to motivate users, who actively pursue and compete for them. For example, users on StackOverflow are observed to increase their effort levels when they get close to the contribution level required for a badge [25].

There are entire discussion communities on the Web that are centered around how to break into Amazon's top reviewer list or how to maintain a top contributor badge on Yahoo! Answers, although users who have just earned entry into top contributor lists often find an increased number of negative votes from other users attempting to displace them. Given that the rewards created by these virtual badges and leaderboards appear to be valued by users (a phenomenon that appears to be quite general, occurring across a range of online communities) and that participating and putting in the effort required to obtain them is costly, a particular way of allocating these rewards creates a corresponding set of incentives or, more formally, induces a mechanism in the presence of self-interested contributors. So gamification also involves reasoning about incentives in a game-theoretic sense. Given that there are several different ways to gamify a site, how should these rewards for overall contribution be designed to incentivize desired levels of contribution?

Generally, badges play multiple roles in all these settings. First, they function as a credentialing system, summarizing the skills and achievements of the individuals who receive them. But they also work powerfully as incentives; experience across many domains shows that people will direct a considerable amount of effort in pursuit of a badge. It is this incentive function of badges, and particularly the ways these badge-based incentives can be used in online applications, that is important. Badges are in several respects simpler than some of these other incentive mechanisms, lacking the direct competition of auctions and leaderboards and the exchangeability of currency-like systems. Despite their simplicity, however, many social sites have positioned badges as an important part of their incentive systems. Badges provide a rich language for expressing incentives but with little existing framework for reasoning about their effects.

Anderson et al. [25] addresses a set of questions that can help provide insight into badges and their use. In particular, a natural first question is: Do badges work? That is, can we find concrete evidence that badges increase site participation or steer users toward taking actions they might not have taken otherwise? If badges do have an effect on users, how can we model user behavior in the presence of badges? And to the extent that designers can indeed steer user behavior with badges, how should they define badges to achieve the outcomes they want? Particularly, the authors introduce a formal model for reasoning about user behavior in the presence of badges and, in particular, for analyzing the ways in which badges can steer users to change their behavior. To evaluate the main predictions of our model, the authors study the use of badges and their effects on the widely used Stack Overflow question answering site, and they find evidence that their badges steer behavior in ways closely consistent with the predictions of our model. Finally, the problem of how to optimally place badges in order to induce particular user behaviors is investigated. Several robust design principles emerge from our framework that could potentially aid in the design of incentives for a broad range of sites.

There are a number of exciting directions for further work to develop the game-theoretic foundations for MSN systems, both extending the existing models to relax simplifying assumptions as well as developing mechanism and market design questions requiring entirely new models. One dimension in particular that is unexplored by the current literature relates to the temporal aspect.

At the level of individual tasks, most works assume that agents make simultaneous choices about their contributions. However, many UGC environments are perhaps better suited to an alternative (although harder to analyze) sequential model in which potential contributors arrive at different times and make decisions about their own contributions after viewing the existing set of contributions. What mechanisms elicit adequate participation and quality in such dynamic sequential models? And furthermore, moving beyond individual tasks and users, an interesting direction relates to investigating an entire site's evolution over time in terms of attracting and sustaining adequate contributions and contributors, with models that explain both observed site growth and decline as well as that allow the designing of mechanisms for sustained participation and quality over time.

4.4 Mobile Crowdsourcing Sensing in MSNs

The proliferation of sensor-enabled smartphones is making it increasingly feasible to build crowdsourcing systems that gather large-scale mobile sensor data. Recently, numerous research prototypes have demonstrated this—for example, near real-time bus monitoring [26] and urban noise and pollution maps [27]. Further, early stage commercial examples are emerging from companies such as Gigwalk [28] and FieldAgent [29] that recruit users for tasks, such as gathering photos of store interiors (to assess product displays or stock levels) or real estate.

Chon et al. [30] reported on the systematic study of a large-scale urban deployment using a representative crowd-sensing system built on commodity smartphones with the goal of examining questions of core interest to mobile crowdsourcing including quantifying and scaling coverage, coping with and classifying noisy data, incentives, and privacy. Specifically, the findings are: (1) from even a relatively small number of contributors, it is feasible to collect a surprising amount of data and achieve impressive coverage levels across a large city; (2) users are clearly cautious of sensitive sensing data (e.g., audio and images). However, they still participate, and their concerns do not prevent the building up of a large multimodal data set.

With the increasing ubiquity of sensor-embedded mobile devices (e.g., smartphones), MSNs—which integrate sensor data collection techniques and services into many kinds of social networks—have been the focus of considerable research effort in recent years due to two changes. First, the terminal devices for social network applications have changed from PCs to mobile phones. Second, the interactive mode extends from the virtual space to the real physical world. MSNs provide a new opportunity for crowd sensing, which takes advantage of the pervasive mobile devices to solve complex sensing tasks. Different from existing systems for wireless sensing or crowdsourcing, crowd sensing exploits sensing and processing behavior abilities of mobile devices to provide sensor data to be used toward a specific goal or as part of a social or technical experiment.

Extensive user participation and submission quality are two crucial factors determining whether crowd-sensing applications in MSNs can achieve good service quality. Most of the current crowd-sourced sensing applications are based on a common hypothesis that all users voluntarily participate in submitting the sensing data. However, mobile devices are controlled by rational users in order to conserve energy, storage, and computing resources. Selfish users could be reluctant to participate in sensing data for crowd-sensing applications. Thus, it is indispensable to provide some incentive schemes to stimulate selfish participants to cooperate in MSNs.

In addition to the issue of extensive user participation, the participant submission quality issue is also challenging in crowd-sensing applications. If the submission quality of participants is not well guaranteed, even though the extensive user participation offers useful information, the service quality from participants is far from satisfactory for the requesters of crowd-sensing applications. For example, some protocols study the extensive user participation to improve the network coverage performance, but the extensive user participation may make the participants with high-quality sensing data drop out of a crowd-sensing application due to the limits of coverage constraints [31]. Other protocols use traditional incentive mechanisms such as the Vickrey–Clarke–Groves (VCG) mechanism and its variants to promote the adequate user participation; however, adequate user participation will also make the participants with higher true valuation frequently become starved to win and may make these participants drop out of crowd-sensing applications [32].

Although extensive user participation and submission quality issues have been identified as two crucial human factors for the wide acceptance of crowd-sensing applications, the tendency of recent research has been to study them separately in crowd-sensing applications. The reason is that, if the extensive user participation and submission quality problems are addressed at the same time as crowd-sensing applications, the issue would become more challenging. For example, some submission quality-enhanced techniques stimulate participants to generate high-quality sensing contents to achieve good service quality, but they could make it hard for some incentive strategies—especially the reputation-based incentive strategies under budget constraints—to implement extensive user participation coverage constraints for crowd-sensing applications because it is not practical to assume that the requester will always provide an unlimited budget to achieve good service quality. Therefore, how to simultaneously address both extensive user participation and submission quality issues becomes particularly challenging for crowd-sensing applications with budget constraints.

To address the aforementioned challenges, a behavior-based incentive mechanism was proposed by Sun [33] for practical crowd-sensing applications with budget constraints. Figure 4.2 illustrates the crowd-sensing system model and its

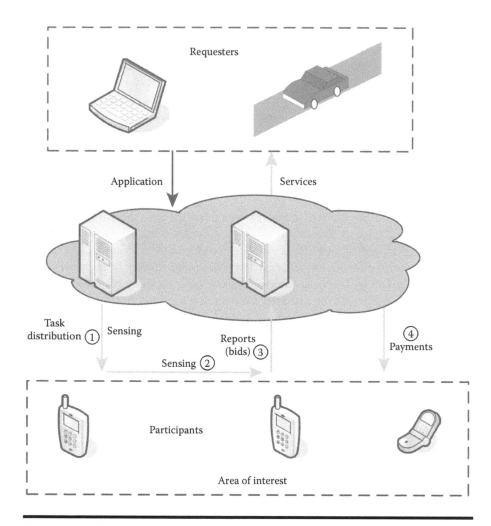

Figure 4.2 Illustration of crowd-sensing framework with all-pay auctions and its operational processes. (With kind permission from Springer Science+Business Media: *User Model and User Adapted Interaction,* **Motivating participation in social computing applications: a user modeling perspective, 22, 2012, 177–201, Vassileva, J.)**

operational processes. The system consists of a crowd-sensing application platform to which a requester with a budget posts a crowd-sensing application that resides in the cloud and consists of multiple sensing servers and many mobile device users, which are connected to the cloud by cellular networks (e.g., Global System for Mobile Communications [GSM]/3G/4G) or by Wi-Fi connections. The crowd-sensing application first publicizes a sequent sensing contest in an area of interest (AoI) at each period.

In order to avoid risk-aversion behavior from the participants, the authors apply the sequential all-pay auction theory to simultaneously generate extensive user participation and high-quality sensing content submission to achieve the better service for the requester of crowd-sensing applications with budget constraints in MSNs, where users arrive in a sequential order. Specifically speaking, the main results and contributions are summarized as follows:

- A behavior-based incentive mechanism was explored in MSNs for practical crowd-sensing applications with budget constraints. In order to simultaneously satisfy the requirements of extensive user participation and high-quality sensing content submission, the authors combine the all-pay auction theory and the greedy budgeted maximum coverage (GBMC) idea to stimulate the participants to generate high efforts and adequate coverage constraints to achieve the better service.
- A more specific crowd-sensing scenario was contemplated where participants arrive one by one online in a random order, which can be modeled as an online sequential all-pay auction. Each participant would exert the most effort for sensing data service quality to win more payments over time. Sensing data are submitted sequentially, and the users with the high-quality sensing contents are selected as the winners. Further, after observing previous submissions, every user's best response effort bidding function for sequential crowd-sensing applications with budget constraints is derived, which influences user participation levels and sensing content submission quality levels.

4.5 Conclusion and Future Directions

Frankly speaking, it still remains more of an art than a science to design and deploy successful mobile social applications with built-in incentive mechanisms that attract and sustain active contribution by their users. In this chapter, we thoroughly summarize incentive mechanisms in MSNs. Specifically, we first introduce basic concepts and theories on motivating rational individuals, including the classical economic view, behavioral economic view, and related social psychology theories. We offer some typical incentive mechanisms inspired by gamification, especially badge, and introduce some existing schemes about crowdsourced sensing in MSNs.

In the previous sections, we saw how a game-theoretic or, more broadly, an economic approach can help with analyzing strategic behavior and incentive design in MSNs. But there remain many challenges—unique to such online contribution domains—that need to be understood before we can fully develop the game-theoretic foundations for incentives in MSNs.

First, there are a number of immediate questions regarding theoretical modeling and analysis. There exists an interesting family of problems arising from the diversity of roles that participants play in many systems (for example, contribution versus moderation in an online community). How should incentives be designed to ensure that each participant is incentivized to properly contribute to their role(s) in the system given that different roles might require different incentives and that these incentives could potentially interact with each other? A principled framework that helps answer this question will need to begin with new models that appropriately capture such multirole participation as well as interactions among different sets of incentives. An issue relates, at least in spirit, to the question of what incentives are created by simultaneously using different forms of gamification on an MSN site. A further question along these lines, arising from the voluntary nature of participation, is how to structure incentives to also induce different potential participants to choose their socially optimal roles in the system.

In addition to problems related to modeling and theoretical analysis, there are also a number of cross-disciplinary questions. One family of problems lies at the interface of game theory and interaction design. By influencing usability and usage, the design of the user interface in an MSN also interacts with incentives in a game-theoretic sense. After all, any game-theoretic analysis involves modeling the behavior of the agents (i.e., users) in an MSN system, which is determined not only by its rules for reward allocation but also by its interface. As a very simple example, consider a system that rewards contributors based on the quality of their outputs as measured by the ratings, or votes, provided by users who view these contributions. An interface design that leads to very little rating by users (for example, a hard-to-find rating button or an overly complex menu of options) or one that leads to ambiguity in the meaning of a rating (such as a thumbs-up button that is interpreted by some users to mean "helpful" and others to mean "I agree") results in noisier ratings than an interface that elicits meaningful votes from a large number of users. A greater degree of noise, roughly speaking, means that reward depends on effort in a more uncertain way, which in turn affects the incentives for agents to put in effort in the system. It is easy to see that even in this specific example there is much more to consider at the interface of interaction design and incentives, such as the questions of which users are allowed to rate contributions and whether raters are offered a more or less expressive set of ratings to choose from. Another example of the connection between interaction design and game theory can be found in the context of badges and gamification. How much information about users' behavior and performance is revealed to other users can potentially affect users' valuations of badges and, consequently, their strategic choices. Generally, how users respond to a given mechanism in a strategic or game-theoretic sense, as well as the space of available mechanisms itself, can depend on the choice of interface in the interaction design phase. An ideal design paradigm would take into account the influence of the user interface and the reward allocation rules on user behavior to provide an integrated, complete approach to the design of incentives in human computation systems.

Finally, a very important family of questions relate to properly understanding contributor motivations and rewards in a more nuanced fashion.

One particularly interesting issue that is pertinent to most MSN systems is that of mixed incentives. Unlike in most traditional economic analysis, MSN systems typically involve a mixture of potential contributor rewards. How do people value these different kinds of rewards in combination, and also, how do they value them relative to each other? What happens when virtual points are used to create an economy with money-like properties (a currency for exchange of goods and services) versus using virtual points to create psychological rewards (such as status)? Second, how do social-psychological rewards, even individual ones, aggregate in terms of the perceived value to contributors? Although utility from money (both in terms of value as a function of total wealth and in the change in value of wealth with time) is a relatively well-studied subject in the economics literature, very little is known or understood about how social-psychological rewards aggregate and how they retain (or gain or lose) value over time. Also, unlike with financial rewards, this could be partially controlled by system design. Understanding how multiple rewards influence incentives when they occur simultaneously in a system, and how social-psychological rewards provide value (starting with understanding agent preferences from a behavioral economics perspective and then integrating this understanding into formal game-theoretic models) is an essential component to a strong foundation for incentive design for human computation and one of the most exciting directions for future work in this area.

References

1. Hameed, S., A. Wolf, K. Zhu, and X. Fu. Evaluation of human altruism using a DTN-based mobile social network application. In: Proceedings of the 5th Workshop on Digital Social Networks (DSN), 2012.
2. Raban, D. and F. Harper. Motivations for answering questions online. In: *New Media and Innovative Technologies*. Ben Gurion University of the Negev Press, 2008.
3. Beenan, G., K. Ling, X. Wang, K. Chang, D. Frankowski, P. Resnick, and R. E. Kraut. Using social psychology to motivate contributions to online communities. In: Proceedings of the 2004 ACM Conference on Computer Supported Cooperative Work, Chicago, IL, November 6–10, 2004, pp. 212–221.
4. Hsieh, G. and S. Counts. Mimir: A market-based real-time question and answer service. In: Proceedings of the SIGCHI Conference on Human Factors in Computing Systems, 2009, Boston, MA, April 4–9, pp. 769–778.
5. Harper, F. M., D. Raban, S. Rafaeli, and J. A. Konstan. Predictors of answer quality in online Q&A sites. In: Proceedings of CHI, Florence, Italy, April 5–10, 2008.
6. Pickard, G., W. Pan, I. Rahwan, M. Cebrian, R. Crane, A. Madan, and A. Pentland. Time-critical social mobilization. *Science* 2011; 334: 509–512.
7. Ghosh, A. Social computing and user-generated content: A game-theoretic approach. *ACM SIGecom Exchanges* 2012; 11(2): 16–21.
8. Ghosh, A. Game theory and incentives in human computation systems. In: *Handbook of Human Computation*. Springer, New York, USA, 2013, pp. 725–742.

9. Vassileva, J. Motivating participation in social computing applications: A user modeling perspective. *User Model and User Adapted Interaction* 2012; 22: 177–201.

10. Greer, J., G. McCalla, J. Vassileva, R. Deters, S. Bull, and L. Kettel. Lessons learned in deploying a multiagent learning support system: The I-Help experience. In: Proceedings of the 10th International Conference on Artificial Intelligence in Education (AI-ED 2001), San Antonio, TX, May 19–23, 2001.

11. Rafaeli, S., D. R. Raban, and G. Ravid. Social and economic incentives in Google answers. In: K. Sangeetha and P. Sivarajadhanavel, eds. *Google's Growth, a Success Story* (pp. 150–161). ICFAI University Press, Hyderabad, 2007.

12. Ariely, D. *Predictably Irrational: The Hidden Forces that Shape Our Decisions.* 2nd ed. HarperCollins, Noida, 2008.

13. Deterding, S., R. Khaled, L. Nacke, and D. Dixon. Gamification: Toward a definition. In: Proceedings of CHI Workshop on Gamification, Vancouver, BC, May 7, 2011.

14. Antin, J. and E. Churchill. Badges in social media: A social psychological perspective. In: Proceedings of CHI Workshop on Gamification, Vancouver, BC, May 7, 2011.

15. Bozarth, J. An interview with Sebastian Deterding. eLearn. ACM Press, New York, USA, 2011.

16. Deci, E. L. and R. M. Ryan. *Intrinsic Motivation and Self-Determination in Human Behavior.* Plenum, New York, 1985.

17. Ajzen, I. The theory of planned behavior. *Organizational Behavior and Human Decision Processes* 1991; 50: 179–211.

18. Guo, X., Y. Zhao, Y. Jin, and N. Zhang. Two-sided adoption of mobile marketing platforms: Towards an integrated conceptual model. In: Proceedings of the Ninth International Conference on Mobile Business, Athens, Greece, June 13–15, 2010.

19. Gosh, A. and P. McAfee. Incentivizing high-quality user-generated content. In: Proceedings of the 20th International Conference on World Wide Web, Hyderabad, India, March 28–April 1, 2011.

20. Gartner. Gartner says by 2015, more than 50 percent of organizations that manage innovation processes will gamify those processes. Available from: http://www.gartner.com/newsroom/id/1629214 [cited April 12, 2011].

21. Hamari, J., J. Koivisto, and H. Sarsa. Does gamification work?—A literature review of empirical studies on gamification. In: Proceedings of the 47th Hawaii International Conference on System Sciences, Hawaii, USA, January 6–9, 2014.

22. Hamari, J. Transforming homo economicus into homo ludens: A field experiment on gamification in a utilitarian peer-to-peer trading service. *Electronic Commerce Research and Applications* 2013; 12(4): 236–245.

23. Eickhoff, C., C. G. Harris, A. P. de Vries, and P. Srinivasan. Quality through flow and immersion: Gamifying crowdsourced relevance assessments. In: Proceedings of the 35th International ACM SIGIR Conference on Research and Development in Information Retrieval, Portland, OR, August 12–16, 2012.

24. Thom, J., D. Millen, and J. DiMicco. Removing gamification from an enterprise SNS. In: Proceedings of the ACM Conference on Computer Supported Cooperative Work, Seattle, WA, February 11–15, 2012.

25. Anderson, A., D. Huttenlocher, J. Kleinberg, and J. Leskovec. Steering user behavior with badges. In: Proceedings of the WWW, Rio de Janeiro, Brazil, May 13–17, 2013.

26. Zhou, P., Y. Zheng, and M. Li. How long to wait?: Predicting bus arrival time with mobile phone based participatory sensing, In: Proceedings of the MobiSys, Low Wood Bay, UK, June 25–29, 2012.

27. Schweizer, I., C. Meurisch, C. Gedeon, R. Bärtl, M. Mühlhäuser. Noisemap: Multi-tier incentive mechanisms for participative urban sensing. In: Proceedings of the Third International Workshop on Sensing Applications on Mobile Phones (PhoneSense), Toronto, CA, November 6, 2012.
28. Gigwalk. The official website of the company of Gigwalk, Available from: http://gigwalk.com/
29. FieldAgent. The official website of the company of FieldAgent, Available from: http://www.fieldagent.net/
30. Chon, Y., N. D. Lane, Y. J. Kim, F. Zhao, and H. Cha. A large-scale study of mobile crowdsourcing with smartphones for urban sensing applications. In: Proceedings of Ubicomp, Zurich, Switzerland, September 8–12, 2013.
31. Jaimes, L. G., I. Vergara-Laurens, and M. A. Labrador. A location-based incentive mechanism for participatory sensing systems with budget constraints. In: Proceedings of the IEEE International Conference on Pervasive Computing and Communications (PerCom), Lugano, Switzerland, March 19–23, 2012.
32. Lee, J.-S. and B. Hoh. Dynamic pricing incentive for participatory sensing. *Pervasive and Mobile Computing* 2010; 6(6): 693–708.
33. Sun, J. A behavior-based incentive mechanism for crowd sensing with budget constraints. arXiv:1310.5485.

Chapter 5

Information Diffusion in Mobile Social Networks

5.1 Introduction

Humans behave in a viral fashion and have a natural inclination to share information so as to gain reputation, trustworthiness, or money. This "word-of-mouth" (WOM) dissemination of information through social networks is of paramount importance in our everyday life. Basically, the flow of information or influence through a large-scale network can be thought of as unfolding with the dynamics of an epidemic. As individuals become aware of new ideas, technologies, fads, rumors, or gossip, they have the potential to pass them on to their friends and colleagues, causing the resulting behavior to cascade through the network. There is a wide range of situations where agents coordinate their decisions and form conventions (epidemiology, computer virus, marketing, political science, agriculture, etc.), and there are a great deal of applications. For example, traditional marketing is being replaced by a new strategy in which the product is turned into an "epidemic," where consumers do the marketing themselves. A common ingredient in all these situations is that a certain behavior spreads across the social network due to some "contagion process" (as a typical example, the more agents choosing a certain action, the more it becomes appealing for another neighbor in the social network to do so as well).

There is a large body of work that identifies the effects of social interactions on behavior diffusion. Seminal studies [1] examined the effects of social connections on the adoption of a new behavior, specifically the adoption of hybrid corn as a crop in the United States. By looking at aggregate adoption rates in different states, these authors illustrated that the diffusion of hybrid corn followed an S-shaped curve over time—starting out slowly and accelerating and then ultimately decelerating. Refer to Young [2] for a more recent and complementary analysis.

However, these studies did not explicitly incorporate social structure in examining behavior diffusion. Pastor-Satorras and Vespignani [3] modeled the network as random, adopted mean field theory to investigate the epidemic spreading in a scale-free network and found the absence of an epidemic threshold and its associated critical behavior. More recently, how a behavior spreads in a network of interaction agents was studied in Ref. [4], in which agents' actions are determined by the actions of their neighbors, according to a simple diffusion rule. In comparison [3], more general results were obtained, indicating that a threshold exists for the rate above which the behavior spreads and becomes persistent in the population. This threshold crucially depends on the connectivity distribution of the social network and on specific features of the diffusion rule.

Although these models provide some ideas about how social structure impacts diffusion, they are limited to settings where becoming infected or avoiding infection is not a choice, that is, contagion is nonstrategic; the impact on behavior is somewhat mechanical and not strategic (this mostly stems from the epidemiology literature). For example, in understanding the diffusion of a disease or information about jobs and so forth, network structure matters mainly as a conduit, and the transmission can be modeled probabilistically. For other situations, such as the trade of goods and services, the adoption of a technology, the provision of local public goods, and other decision making that is influenced by friends and acquaintances, network structure also matters but with the added features of strategic interactions among networked agents. Such interaction naturally calls on game theory as a tool for modeling these richer interactions.

The games on social networks were analyzed in Refs. [5,6], where agents select one of two actions (whether to adopt a new technology, to withdraw money from the bank, to become politically active, etc.). An agent's payoff from each of the two actions depends on how many neighbors an agent has, which actions the agent's neighbors choose, and some agent-specific cost and benefit parameters. The goal of this study is twofold. First, a general dynamic model in which agents' choices depend on the underlying social network of connections is proposed; the second goal is to show the usefulness of the model in determining when a given behavior expands within a population or when it disappears. The authors' works enriched the analysis of behavior diffusion in terms of moving from the mechanical spread of disease to a strategic interaction and provided general results on how social structure impacts diffusion. Spreading a particular behavior among agents located in a random social network was investigated in Ref. [7], in which neighboring agents interact strategically by playing a 2 × 2 coordination game. The authors proved that there exists a threshold for the degree of risk dominance of an action such that, below that threshold, contagion of the action occurs. This threshold depends on the connectivity distribution of the network. Lopez-Pintado [8] studied a model where agents, located in a social network, decide whether to exert effort in experimenting with a new technology (or acquiring a new skill, etc.)

in which agents are assumed to have a strong incentive to take a free ride on their neighbors' effort decisions. Based on the analysis of mean field dynamics, the authors show how the pattern of free riders in the network depends on properties of the connectivity distribution.

In this chapter, we pay special attention to the information (behavior) diffusion in autonomous and rational mobile social networking environments. By autonomous, it means all behaviors in a mobile social network (MSN) are voluntarily chosen and determined by independent and autonomous individuals; by rational, it means that each individual wants to maximize his or her own utility by intentionally selecting behaviors. Naturally, the diffusion of behavior in autonomous and rational networks often exhibits features that do not match well with those of the epidemic models. Unlike virus spreading, information diffusion depends on the voluntary nature of humans. It has a perceived transmission cost and is only passed by its host to individuals who may be interested in it. The general goal of this chapter is to systematically characterize the behavior diffusion pattern in rational and autonomous networks. Here, the diffusion pattern includes two components—the diffusion model and the diffusion process.

- Diffusion models describe how a rational individual determines whether to adopt behavior/spread information in a local interaction. These models are mainly inspired by game theory literature on social and economic networks, especially by random utility theory. Specifically, an individual's utilities are parameterized as a deterministic function derived from the diffusion model plus a random error term that can accommodate the uncertainty in the inference of utility. For completeness, we also briefly introduce the traditional threshold model and the cascade model.
- The diffusion process characterizes the unfolding of behavior dynamics adopted by the whole population in a social network, in which mean field theory is often used to approximately analyze the diffusion dynamics of adopted behaviors. Basically, at the outset, a small portion of the population is randomly set and selected to adopt the behavior. Then the diffusion process is defined so that at each period, the agent best responds to the actions taken by his or her neighbors in the previous period. Note that in most cases, the analytical results of the exact model are extremely complicated and thus will not be tackled. Nevertheless, to proceed, two complementary approaches can be considered. On one hand, the analysis of the model can be simplified using the mean field theory. On the other hand, the dynamics can be comprehensively simulated to obtain numerical approximations of the results for the exact model.

The chapter is organized as follows. Section 5.2 presents a detailed description of various diffusion models, including the general threshold model and the general cascade model. We argue that both of those models, in a sense, could be regarded as special cases of a game-theoretic model. In Section 5.3, the problem of influence maximization is introduced, and some existing algorithms and approaches are analyzed. Section 5.4 summarizes several extensions to diffusion modeling and influence maximization through incorporating various aspects, including competition, budget, and time criticality. Finally, we briefly conclude this chapter in Section 5.5.

5.2 Information Diffusion Models

Here, we consider a simple diffusion model in which each peer is classified as either an "active (behavior 1)" or a "potential (behavior 0)" consumer and is represented by a node in the network structure. Typically, behavior 1 and behavior 0 can represent whether users spread a rumor, adopt a specific technology or software, and so on. Without a loss of generality, we consider the behavior 0 to be the default behavior (e.g., the status quo technology). This section introduces two traditional diffusion models: the general threshold model and the cascade model [9]. A general game theory-based diffusion is proposed, and its equivalence to those two traditional diffusion models is also given.

5.2.1 General Threshold Model

In a general threshold model, each peer q has an arbitrary function $g_q(X)$ defined on subsets of its neighbor set $N(q)$: for any set of neighbors $X \subseteq N(q)$, there is a value $g_q(X)$ between 0 and 1. This function is always assumed to be monotone, in the sense that if $X \subseteq Y$, then $g_q(X) \leq g_q(Y)$. Specifically, each peer's threshold θ_q (the fraction of neighbors required for it to adopt the new behavior) is chosen uniformly at random in [0,1]; there is an initial set S of active nodes; and for time steps $t = 1,2,3,\ldots$, each q becomes active if its set of currently active neighbors satisfies $\rho(q, X) = \Pr(g_q(X) \geq \theta_q)$. This model is extremely general; it encodes essentially any threshold rule in which influence increases (or remains constant) as more friends adopt. Moreover, the assumption that the threshold is selected uniformly at random (rather than from some other distribution) is essentially without a loss of generality because other distributions can be represented by appropriately modifying the function g_v.

5.2.2 General Cascade Model

The previous subsection formulated models for the spread of a behavior strictly in terms of node thresholds—as some people adopt the behavior, the thresholds of others are exceeded, they too adopt, and the process spreads. It is natural, however, to ask whether we can pose a different model based more directly on the notion that new behaviors are contagious—a probabilistic model in which you "catch" the behavior from your friends. The resulting model is equivalent to the general threshold model.

Next, we define the cascade model to incorporate these ideas. Again, there is an initial active set S, but now the dynamics proceeds as follows: whenever there is an edge (p,q) such that p is active and q is not, the node p is given one chance to activate q. This activation succeeds with some probability that depends not just on p and q but also on the set of nodes that have already tried and failed to activate q. If p succeeds, then q may now, in turn, try to activate some of its (currently inactive) out-neighbors; if p fails, then p joins the set of nodes that have tried and failed to activate q.

This model thus captures the notion of contagion more directly and also allows us to incorporate the idea that a node's receptiveness to influence depends on its history of interactions with its neighbors. The concrete model is described as follows. In place of a function g_q, end node q now has an incremental function that takes the form $\rho(q, X; p)$, where p is a neighbor of q and X is a set of neighbors of q not containing p. The value $\rho(q, X; p)$ is the probability that p succeeds in activating q, given that the set X of neighbors has already tried and failed. We will only consider function $\rho(q, X; p)$ that is order-independent; if a set of neighbors $p_1, p_2,...,p_k$ all try to influence q, then the overall probability of success does not depend on the order in which they try.

Although the cascade model is syntactically different from the general threshold model, these two are, in fact, equivalent. One can translate from a set of incremental function $\rho(q, X; p)$ to a set of threshold function g_q, and vice versa, so that the resulting processes produce the same distribution on outcomes. The detailed proof can be found in Ref. [10]. Next, we give the schematic description. First, suppose we are given an instance of the general threshold model with function g_q; we define corresponding $\rho(q, X; p)$ as follows. If a set of peers X has already tried and failed to activate q, then we know that q's threshold θ_q lies in the interval $(g_q(X), 1)$; subject to this constraint, it is uniformly distributed. In order for p to succeed after all the nodes in X have tried and failed, we must also have $\theta_q \leq g_q(X \cup \{p\})$. Hence, we should define the incremental function as

$$\rho(q, X; p) = \frac{g_q(X \cup \{p\}) - g_q(X)}{1 - g_q(X)}. \tag{5.1}$$

Conversely, suppose we have incremental function $\rho(q, X; p)$. Then, the probability that peer q is not activated by a set of neighbors $X = \{p_1, p_2,...,p_k\}$ is $\prod_{i=1}^{k}(1 - \rho(q, X_{i-1}; p_i))$, where $X_{i-1} = \{p_1,...,p_{i-1}\}$. Note that order-independence is crucial here to ensure that this quality is independent of the way in which the elements of X are labeled. Hence, we can define a threshold function by g_q setting

$$g_q(X) = 1 - \prod_{i=1}^{k}(1 - \rho(q, X_{i-1}; p_i)). \tag{5.2}$$

This completes the translations in both directions and establishes the equivalence of the two models.

5.2.3 Game Theory–Based Diffusion Model

As described in Section 5.1, diffusion models specify how a rational individual determines whether to adopt behavior/spread information through maximizing the utility inferred from interactions with his or her neighbors. Note that those interactions can be specified by game-theoretic models. An agent's payoff from each action depends on his or her location within the network (specifically, the number of neighbors he or she has, his or her neighbors' choices, and a random cost determined at the outset). We consider the action 0 to be the default behavior, and an individual peer has a cost of choosing behavior 1, denoted η, that is randomly and independently distributed with probability density function $h_c(\eta)$.

Let $u(q, X; a)$ denote the accumulated utility of an arbitrary agent q obtained from interactions with his or her neighbors, when he or she takes the behavior $a \in \{0, 1\}$, where X denotes the subset of q's neighbors composed by peers adopting behavior 1. Peer q's added payoff from adopting behavior 1 over sticking to the action 0, called peer q's marginal utility, is then $v(q, X) - \eta$, where $v(q, X) = u(q, X; 1) - u(q, X; 0)$. Naturally, agents adopt the new behavior 1 only if it appears worthwhile for them to do so, which implies that $v(q, X) - \eta \geq 0$ [i.e., $\eta \leq v(q, X)$]. If we let H_c represent the cumulative distribution function of density function $h_c(\eta)$, then the probability that peer q chooses action 1, $\rho(q, X)$, can be given as

$$\rho(q, X) = H_c\big(v(q, X)\big). \tag{5.3}$$

Note that if we regarded $v(q, X)$ and cost distribution $h_c(\eta)$ as the threshold function $g_q(X)$ and the threshold distribution θ_q in the general threshold model, then this game theory–based model is equivalent in format to the general threshold model (and is also equivalent to the cascade model). And moreover, unlike the threshold function $g_q(X)$ that is always assumed to be monotone, the property of the utility difference function in our proposed game theory diffusion model can be arbitrary.

We next attempt to formulate the effect of a network's structural characteristics on the behavior diffusion, especially the effect of degree distribution. For slightly misusing notations, let $u_d(a, d_i)$ denote the accumulated utility of an arbitrary agent of degree d obtained from interactions with the agent's neighbors when he or she has d_i neighbors using behavior 1. Assume that each of his or her neighbors independently chooses the action 1 with probability x (mean field approximation). Thus, the probability that a peer with d links has exact d_i neighbors choosing behavior 1 follows a binomial distribution given by

$$\binom{d}{d_i} x^{d_i} (1-x)^{(d-d_i)}. \tag{5.4}$$

Then, that peer's average utility, when adopting behavior *a*, can be represented as

$$\bar{u}_d\left(a,x\right)=\sum_{d_i=0}^{d}u_d\left(a,d_i\right)\binom{d}{d_i}x^{d_i}\left(1-x\right)^{\left(d-d_i\right)}.\qquad(5.5)$$

The payoff of the agent of degree *d* is $v(d, x) - \eta$, where $v\left(d,x\right)=\bar{u}_d\left(1,x\right)-\bar{u}_d\left(0,x\right)$; thus, the probability $\rho(d, x)$ that a random agent of degree *d* chooses the action 1 when anticipating that each neighbor will choose 1 with an independent probability *x* can be denoted as

$$\rho(d, x) = H_c(v(d, x)).\qquad(5.6)$$

5.3 Influence Maximization Problem

5.3.1 Definition

WOM or viral marketing differentiates itself from other marketing strategies because it is based on trust among an individual's close social circle of families, friends, and coworkers. Research shows that people trust the information obtained from their close social circle far more than the information obtained from general advertisement channels such as TV, newspaper, and online advertisements. Thus, many people believe that WOM marketing is the most effective marketing strategy.

The increasing popularity of many social network sites such as Facebook, MySpace, and Twitter presents new opportunities for enabling large-scale and prevalent viral online marketing. Consider the following hypothetical scenario as a motivating example: A small company develops an online application and wants to market it through a social network. The company has a limited budget and can select only a small number of initial users in the network to use the application (by giving them gifts or payments). The company wishes that these initial users would love the application and would start encouraging their friends in the social network to use it and that their friends would then influence their friends' friends and so on. Thus, through the WOM effect, a large population in the social network would adopt the application. The problem is selecting who will be the initial users so that they eventually influence the largest number of people in the network. This is called influence maximization.

Influence maximization is first formulated as a discrete optimization problem: A social network is modeled as a graph with nodes representing individuals and edges representing connections or a relationship between two individuals.

Influence is propagated in the social network according to the diffusion models introduced in Section 5.2.

In brief, given a social network graph and a specific diffusion model, the problem of influence maximization, first posed by Domingos and Richardson [11], is stated as follows: If we can try to convince a subset of individuals to adopt a new product and the goal is to trigger a large cascade of further adoptions, which set of individuals *S* should we target in order to achieve a maximized influence?

5.3.2 Existing Algorithms in Influence Maximization

5.3.2.1 Greedy-Based Algorithms

We have shown that finding the influential set of initial nodes is an Non-deterministic Polynomial-time hard (NP-hard) problem. The greedy algorithm can approximately work only for submodular function of diffusion model, with bounded threshold. Specifically, Kempe et al. have proven that a simple greedy algorithm (choosing the nodes with maximal marginal gain) can approximate the optimal solution by a (1-1/*e*) (i.e., within 63% of optimal) [10]. The greedy algorithm works as follows. For an initial active (target) set *A*, let $\sigma(A)$ be the expected number of infected nodes at the end of the process. The simple greedy strategy chooses the number of k initial influential nodes with maximal marginal gain. The algorithmic procedure is

  1. Start with $S = \Phi$
  2. For $i = 0$ to k do
  3. Let v_i be a node that maximizes the marginal gain $\sigma(S \cup \{v\}) - \sigma(S)$
  4. Set $(S \cup \{v_i\}) \rightarrow S$
  5. End for

However, the simple greedy-based algorithm has a serious drawback, computation inefficiency. Rather than finding an exact algorithm, Monte-Carlo simulations of the influence cascade model are run for a sufficient number of times to obtain an accurate estimate of the influence spread. The more rounds the enumeration takes, the more accurate the result is. However, when the network size increases, the computational time will increase dramatically, which prevents the greedy algorithm from becoming a feasible solution for a large-scale influence maximization problem in the real world.

To solve the problem of efficiency, the research community has intensely studied the algorithmic aspects of maximizing influence in social networks, primarily from two directions: improving the greedy algorithm to further reduce its running time and proposing new heuristics.

An optimized greed algorithm, the "cost-effective lazy forward (CELF)," was presented in Ref. [12]. This algorithm uses the submodular property of the influence maximization objective to reduce the number of evaluations on the influence spread of nodes. The experimental results demonstrated that, compared with a simple greedy-based approach, CELF optimization could make seed selection 700 times faster. A CELF++ algorithm was proposed to further optimize CELF by exploiting submodularity. This algorithm improved CELF efficiency by 35%–55% [13].

Considering that the fundamental step of the greedy algorithm is to pick a node in each iteration from the remaining nodes and then attempt to make the maximum marginal contribution to the process of spread of information, Narayanam and Narahari [14] proposed the Shapley value-based influential node (SPIN) algorithm for computing the marginal contributions using the concept of Shapley (a well-known solution concept in cooperative game theory). The Shapley value of a coalitional game provides the marginal contributions of the individual players to the overall value that can be achieved by the grand coalition of all the players.

Even with these improvements, the running time of greedy algorithms is still great, and they may not be suitable for large social network graphs.

5.3.2.2 Heuristic Schemes

A possible alternative is to use heuristics. In sociology literature, degree and other centrality-based heuristics are commonly used to estimate the influence of nodes in social networks. To simplify, in a so-called degree centrality scheme, the remaining nodes with the highest degree are chosen. Surprisingly, if we choose all target users with high centrality, the resultant scheme only outperforms random selection for small target sets; this is due to the so-called overlapping effect. This occurs because the selection approach is based on degree centrality and does not take into account the neighborhood overlapping (the so-called overlapping effect). A given group of connected nodes may have a high degree, but if their adjacent nodes are overlapped, then the information may not propagate through the rest of the network.

A degree discount heuristic algorithm called DegreeDiscount was proposed to alleviate the effect of overlapping. It intentionally discounts the degree of each node by removing the neighbors that are already in the activation seed set [15,16].

The basic idea of new degree discount heuristics is given as follows. Let v be a neighbor of vertex u. If u has been selected as a seed, then when considering selecting v as a new seed based on its degree, the edge vu toward its degree should not count. Thus, v's degree should be discounted by one due to the presence of u in the seed set; the same discount be taken on v's degree for every neighbor of v that is already in the seed set. This is a basic degree discount heuristic that is applicable to all cascade models.

For the independent cascade (IC) model with a small propagation probability p, there is a more accurate degree discount heuristic. Because v is a neighbor of u that has been selected into the seed set, with probability at least p, v will be influenced by u, in which case v does not need to be selected into the seed set. This is the reason why a further discount is more accurate. When p is small, the indirect influence of v on multihop neighbors can be ignored, and the focus should be on the direct influence of v on its immediate neighbors, which makes degree discount calculation manageable. This forms the guideline for computing the degree discount amount.

The experimental results show that the influence spread of new degree discount heuristics is close to that of the greedy algorithm and that it always outperforms the centrality-based heuristics and the classic degree. The biggest advantage of new degree discount heuristics is the speed; they are many orders of magnitude faster than all greedy algorithms.

The results provide us with a new perspective on the study of the influence maximization problem. Instead of focusing our effort in further improving the running time of the greedy algorithm, fine-tuned heuristics may provide truly scalable solutions to the influence maximization problem with a satisfying influence spread and blazingly fast running time.

Basically, one's influence is not only embodied by the number of friends one has but also reflected by the kind of friends one has. In other words, the neighbors' influence should also be considered when measuring one's influential power. The spreading power of nodes in a mobile disconnected network was specifically characterized in Ref. [17], in which eigenvector centrality (EVC) was shown to be a meaningful measure of the ability of the nodes to spread an epidemic in the network. Specifically, the EVC is computed as the eigenvector relative to the spectral radius (i.e., the largest eigenvalue) of the network's adjacency matrix.

Information diffusion models and the top-k nodes problem are also appropriately considered from the view of the blogspace where a blogger may have a certain level of interest in a topic and is thus susceptible to talking about it. By discussing the topic, the blogger may influence other bloggers [18]. A mechanism for detecting contagious outbreaks in social networks was proposed in Ref. [19], which demonstrated that by monitoring only the friends of these randomly selected students, an early detection of flu at Harvard College can be obtained by up to 13.9 days. Masuda and Kori [20] proposed to relate the PageRank (originally designed for Web graphs) to the influence by reversing all the links of the original network (a so-called reverse PageRank) because in a Web graph, receiving links increases a page's ranking (which is opposite to the content of the influence). Specifically, considering a continuous-time simple random walk on the network generated by reversing the direction of all the links of the original network, such as in the definition of the PageRank, random global jumps were introduced to the continuous-time random walk on the link-reversed network in order to solve the known problem of trapping a random walker inside a local neighborhood when the graphs are disconnected or loosely connected. The destination of the random jump is chosen from all the nodes with equal probability $1/N$. The influence of each node is equal to the stationary density of the continuous-time random walk on the link-reversed network (the long-term probability that the walker visits node i.).

5.3.3 Distributed Realization of Influence Maximization in MSNs

Han and Srinivasan [21] proposed a lightweight and distributed protocol, iWander, with low message overhead, to identify influential users through fixed-length random walks. The approach is motivated by the "friendship paradox" [22] that "your friends have more friends than you do" and leverages random walks to identify critical users. The reason behind this paradox is that people with larger numbers of friends may have a high probability of being observed among one's friend circle.

5.3.3.1 Basic Concepts about Random Walk

Given a graph and a starting vertex, a random walk [23] is the mathematical formalization of a random path on the vertices of the graph. At the beginning of the walk, we select one of the neighbors of the starting vertex at random and move to this neighbor; then, we select a neighbor of this vertex at random; and so on. A random walk is simply the sequence of vertices selected by this randomized process. It is a very simple yet powerful concept.

Actually, a random walk can also be seen as a finite Markov chain (a randomized process defined by transitions from one state to another between a finite or countable number of possible states). And moreover, if the transition probability of the Markov chain is stochastic and irreducible, then the Markov chain is guaranteed to have a unique positive stationary distribution. Specifically, in the case of undirected graph $G = (V,E)$, with $|V| = n$, and $|E| = m$, the stationary distribution of a random walk on G is the probability distribution π such that: $\pi(u) = \dfrac{d(u)}{\sum_{v \in V} d(v)} = \dfrac{d(u)}{2m}$,

for every vertex $u \in V$.

This means that the probability of being at any given vertex v tends to be a well-defined limit independent of the starting vertex of the random walk (or Markov chain process), and this value depends on the degree of v.

We can extend the simple result from the stationary distribution of random walks on undirected graphs. Let $G_w = (V, E)$ be a graph with a weigh label e_{ij} associated to each edge $(i,j) \in E$. Then, the probability distribution π on the nodes of the graph such that: $\pi(u) = \dfrac{d(u)}{\sum_{v \in V} d(v)}$ is the stationary distribution on the graph, noting that in this case the degree of the vertex is defined as the sum of all edges incident to the vertex, that is, $d(u) = \sum_{(u,v) \in E} e_{uv}$ [24].

Several variants of the basic random-walk model have been integrated into centrality measurement in social science. For instance, Newman [25] proposes the random-walk betweenness centrality, a relaxation of the shortest-path betweenness. This measure defines how often a node in a graph is visited by a random walker between all possible node pairs. Noh and Rieger [26] introduce the random-walk closeness centrality metric, which measures how fast a node can receive a random-walk message from other nodes in the network.

Random walks have also been widely explored for various purposes in other fields, such as in computer science, economics, biology, and psychology. For example, Braginsky and Estrin [27] route queries on a random walk to sensor nodes around which a particular event occurs. Yu et al. [28] propose SybilGuard, which uses a special kind of random walk, where every node chooses the next hop based on a precomputed random permutation, to limit the bad effect of Sybil attacks on peer-to-peer systems.

5.3.3.2 The Process of Distributed iWander Protocol Inspired by Random Walk

Han and Srinivasan [21] proposed leveraging random walks for designing a distributed protocol, iWander, for identifying influential users in MSNs. The theory is that, if we periodically initialize random walks from a small group of smartphones, influential mobile users may be visited by these random walks more frequently than average.

The proposed iWander protocol works as follows. Every T hours, iWander generates a tiny probing message on each smartphone with a given probability q. The message only contains a preconfigured time-to-live (TTL) field L. During the smartphone's contacts with its peers, if it has a probing message in its local queue, it sends this message to another randomly selected peer. When a smartphone receives a probing message, it decreases L in the message by 1 and then stores it in its local queue, waiting for the opportunity to forward the message to other peers. A probing message with $T = 0$ eventually will be discarded. iWander maintains a random-walk counter initialized to zero on each smartphone, to record how many times it has received the probing messages (i.e., has been visited by these random walks).

After collecting the random-walk counters recorded by their smartphones from all users, the set of k critical users can be determined from the head of the user list sorted by these counters. The reason for this is that, based on the friendship paradox, influential users have high probabilities of being visited by random walks and thus own large random-walk counters.

Unlike the random-walk betweenness metric proposed by Newman [25], iWander applies fixed-length—instead of all-pairs—random walks; this is for two reasons. First, in practice, it is difficult for a mobile user to know every other user and thus specify random-walk destinations. Second, the message overhead of all-pairs random walks may be much higher than fixed-length random walks, which makes them unsuitable for battery-powered smartphones. The update and reset of random-walk counters are determined by the upper layer applications. In practice, they may reset these counters periodically—for example, everyday at midnight. They can also apply an exponential moving average to update these counters by assigning a higher weight to recent counters.

In summary, the performance of iWander relies on three parameters: q, the probability that a smartphone generates a probing message (i.e., the fraction of mobile users that initialize random walks); L, the length of random walks (i.e., the number of mobile users visited by a single random walk); and ΔT, the frequency of generating new random-walk probing messages. It is important to understand the impact of these parameters on the performance of iWander because they determine the quality of identified influential users and the number of probing messages spreading over the network.

To reduce energy consumption on smartphones, iWander prefers short random walks with only a few steps. "Static" versions of social-contact networks are often very dense and "expander-like" [29]. In such highly mixing networks, it is well-known that a random walk of length $O(\log N)$ or less, where N is the number of nodes in the network, suffices for coming very close to the stationary distribution of the random walk (in which each vertex has a probability proportional to its degree). MSNs are inherently mobile and thus not static, but their static snapshots will likely be expander-like. The mobile networks will also likely mix well. Thus, the short random walks adopted by iWander will likely come quite close to sampling vertices approximately according to their degrees.

5.3.3.3 The Evaluation of iWander

Han and Srinivasan developed a trace-driven simulator in C, using the Dartmouth data set [30], to evaluate the performance of random walk–based target set selection. In this simulator, it is assumed that the underlying wireless communication is reliable, and Bluetooth is used as the underlying communication protocol for iWander due to its low energy consumption.

5.3.3.3.1 Simulation Setup

The performance of the proposed random walk–based target set selection (abbreviated as RW-1) was compared with random selection (Random) and with degree-based selection (Degree).

The simulator first generates the contact trace of mobile users under the same assumption that they are in contact if their wireless devices are associated with the same access point. It then replays the contact events for the given information dissemination period, from 6:00 p.m. to 10:00 p.m. On the basis of the preconfigured information dissemination probability, the simulator randomly determines whether a user can receive information from peers after each device discovery. Usually information providers will send information to uninfected users at the end of dissemination period to guarantee that every user can finally receive the delivered information.

The interval of device discovery is 60 seconds, which means that smartphones have the chance to start the exchange of information every 60 seconds. Degree uses the number of other smartphones with which a smartphone has contact as the metric to select target users. For RW-1, smartphones generate one-step random-walk messages of iWander with a probability of 0.1 every hour. RW-1 and Degree choose target users based on the updated random-walk counters and the number of contacts of smartphones at the beginning of the information dissemination period.

Simulations are run 1000 times, and average values with standard deviations are reported. The information dissemination probability p is set as 0.01, 0.05, and 0.005. The target set size varies from 10 to 2000.

5.3.3.3.2 Simulation Results

Simulation results show that RW-1 and Random outperform Degree when the size of the target set is larger than 10. RW-1 performs better than Random for small target sets. For example, for a target set with 50 users, RW-1 can deliver information to 51% more users than Random when p is 0.005. The improvement is 37% when p is 0.01 and 14% when p is 0.05.

However, the performance of RW-1 becomes worse than Random when the size of the target set is larger than 1000. One possible reason is that noninfluential users (i.e., users with low centrality in social-contact networks) also play an important role in information dissemination. Zyba et al. [31] call these users vagabonds, which demonstrates that under certain circumstances the effectiveness of information dissemination in MSNs primarily depends on the number of vagabonds. When the target set size is large, Random has a higher probability of selecting more vagabonds into the target set that then may have very little chance to receive information before the delivery deadline. However, Degree and RW-1 select only mobile users with high centrality into the target set and ignore these vagabonds. To verify this, RW-1 was modified by selecting 90% of target users with low centrality from the end of the user list sorted by random-walk counters. This enhanced scheme is referred to as Mix-1, and it also uses one-step random walks. Simulation results clearly show that Mix-1 outperforms Random for large target sets. Zyba et al. also tried this with other percentages of noninfluential target users, and these variations perform similarly.

In summary, when information service providers can deliver information directly to only a small number of users, it is better to use the pure random walk–based target set selection policy. However, the enhanced scheme that mixes influential and noninfluential users into the target set is preferable when it is possible to deliver information directly to a large number of users.

5.4 Extensions to Influence Maximization

The basic influence maximization problem is stated as follows. Given that the constant number *k* seeds could be selected, how can those seeds be chosen to maximize information/behavior diffusion irrespective of various constraints such as competition, budget, time, and so on. In this section, we introduce several additional aspects for extending the influence maximization problem, including competition, budget, and time criticality, among many others.

5.4.1 Budget and Cost in Information Diffusion

Information propagation in delay-tolerant networks (DTNs) is difficult due to the lack of continuous connectivity. Earlier studies focused on information propagation in a static network. One simple way to a model time-dependent DTN is to represent it using a static graph: nodes in the graph represent individuals, and an edge between any two nodes can be added if they have interacted or met during any one period. Furthermore, certain weight could be assigned to each edge to reflect its interacting frequency (e.g., if two individuals meet each other *m* times among *n* time slots, the influence probability (weight) of the edge between them can be assigned the value *m/n*).

This is a common way to represent the time-evolving DTN by a sequence of static networks. Each static network represents the contact information among users at that time slot. Assuming that the contact information of the dynamic DTN social network within a certain period can be preobtained through certain existing prediction methods or by statistical analysis (so-called predictable DTN), Tang et al. [32] first examined how to select a set of initial source nodes, subject to budget constraints, in order to maximize the total weight of nodes that receive the information at the final stage.

We refer to this as an influence maximization problem with constraints of budget and cost, which can be formally stated as follows. Given a fixed budget b and a random cost function c, find a seed set *S* that fits the budget $\sum_{v_i \in S} c(v_i) < b$ and maximize the number of influenced nodes. This simple greedy algorithm can be directly extended to solve this problem.

Specifically, $\sigma(S)$ denotes the expected weight of the final propagated set under the initial source node *S*; $\sigma(S)$ is submodular monotone and nonnegative for an IC model. Then a simple greedy-based algorithm can be adopted to add the node *v* that can maximize the expected incremental marginal gain; that is, $(\sigma(S \cup \{v\}) - \sigma(S))/c(v)$, until the budget constraint is violated.

More specifically, with the proliferation of influence score services such as PeerIndex (http://www.peerindex.com), one can easily measure an individual's influence in the social sphere and use that to negotiate the price for services they provides. The higher the influence score of a user, the more costly it is to persuade them.

Considering the assumption of equal costs for all seed nodes seldom holds in practice, a generalized version of the influence maximization problem—namely, the budgeted influence maximization (BIM) problem—was seriously investigated in Ref. [33]. First, the authors prove that direct application of the simple greedy algorithm may result in an unbounded performance gap; then a seed selection algorithm is provided that can attain an approximation guarantee of $1 - \sqrt{e}$ (~ 0.394). One critical component of the seed selection process is the determination of the influence spread of a set of seeds. Exact computation of influence spread is known to be NP-complete. By identifying the linkage between the computation of marginal probabilities in Bayesian networks and the influence spread, efficient heuristic algorithms for determining the influence spread of a set of seeds are devised. Empirical study shows that the proposed algorithms can scale up to large-scale graphs with millions of edges with a high degree of accuracy. Moreover, on real-world social network graphs, the methods can achieve an influence spread comparable to that of the greedy algorithm and will incur significantly lower computation costs.

5.4.2 Competitive Information Diffusion

Typically, there always exist multiple innovations within a social network that are competing at same time. This scenario will frequently arise in the real world: multiple companies with comparable products will vie for sales with competing WOM cascades; similarly, many innovations also face the spread of active opposition by WOM.

Suppose there are several firms that would like to advertise competing products via "viral marketing." Each firm initially targets a small subset of users in the hope that the rumor about its product would spread throughout the network. However, a user that adopts one product is reluctant to adopt another product; hence, the campaign of one firm negatively affects the success of another firm's campaign.

Small [34] proposed a model for competitive information diffusion on star graphs, cliques, and trees, which proved the existence of a Nash equilibrium.

In particular, innovation diffusion on strategic decisions made by individuals in a social context can be examined from a game-theoretic perspective. A game-theoretic model of competitive information diffusion on graphs was introduced in Ref. [35]. This model considers the diffusion process as a competitive game taking place on a graph that captures the underlying social structure. The model considers players who wish to spread their idea through the network. This has applications in areas such as viral marketing.

Figure 5.1 shows the diffusion process in which there are two players 1 and 2; the diffusion process starts at time step $t = 0$ and terminates at time step $t = 2$.

The players are initially assigned vertices that they "color" at the first time step in the diffusion process. At each time step that follows, uncolored vertices adjacent to vertices that are already colored are colored according to the following rules. If two or more vertices of different color are neighbors of an uncolored vertex, then

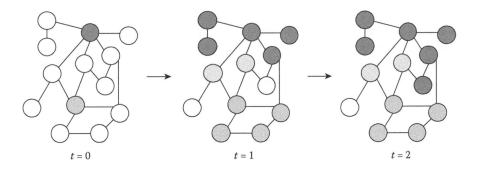

Figure 5.1 The diffusion process (dark grey represents player 1, light grey represents player 2).

in the next time step this vertex is colored gray. This color gray does not propagate through the network. If a vertex is neighbored by vertices colored gray and one other color, then the uncolored vertex takes this color. If an uncolored vertex is adjacent to vertices of only one color, then the uncolored vertex takes this color. All other uncolored vertices remain uncolored.

In other words, if two agents compete for a user at the same time they "cancel out" and the user is removed from the game. And if only one agent competes for a user, the user will adopt the agent.

This represents the spread of an idea through a social network. The diffusion process ends when no further vertices can be colored. The utility of a player is the number of vertices that it has colored at the end of the diffusion process.

Alon et al. [35] give a sufficient condition for a graph of diameter two to have a Nash equilibrium. That is, the following theorem holds: Let G be a graph, $D(G)$ be the diameter of a graph G, and let the set of players be $\{1,...,n\}$ such that $D(G')$ ≤ 2 for every G' is obtained from G by removing $n-1$ vertices along with their neighbors. Then the game admits a Nash equilibrium, which can be found in polynomial time.

Alon et al. also show that a graph of diameter three may not admit a Nash equilibrium and that this result may be extended to graphs of any diameter greater than three or for any number of agents greater than two. However, certain classes of graphs (such as start and clique graphs) are guaranteed to have a Nash equilibrium for any number of players for the diffusion game described here.

5.4.3 Time-Critical Influence Maximization

Recent research reveals that time plays an important role in the influence spread from one user to another and that the time needed for a user to influence another varies. For example, to market a pop vocal concert to be held on a certain date, the marketer would want to maximize the number of users influenced before the concert date. A conventional influence maximization model does not consider that influence among users may depend on the time. For example, some users may only pass the information to others after a rather long period. Consequently, the selected influential users may not spread the influence within a limited time. Indeed, users influenced after the concert would not bring any profit to the marketer. The conventional influence maximization solutions become invalid because the time is not considered in the influence propagation.

However, the family of models considered in literature and follow-ups do not fully incorporate important temporal aspects that often have been observed in the dynamics of influence diffusion:

- The propagation of influence from one person to another may incur a certain amount of time delay, which is evident from recent studies in statistical physics [36].
- The spread of influence may be time-critical in practice. In a certain viral marketing campaign, it might be the case that the company wishes to trigger a large volume of product adoption in a fairly short time frame (e.g., a three-day sale).

Therefore, when we try to maximize the spread of influence for a viral marketing campaign facing these kinds of scenarios, we need to take both the time-delay aspect of influence diffusion and the time-critical constraint of the campaign into consideration.

Specifically, the influence maximization problem was extended having a deadline constraint that reflected the time-critical effect [37]. The authors propose a new propagation model, the independent cascade model with meeting events (IC-M) to capture the delay of propagation in time.

In the IC-M model, each edge $(u,v) \in E$ is also associated with a meeting probability $m(u,v)$ defined by function $m{:}E{\rightarrow}[0,1]$ [if $(u,v){\notin}E$, $m(u,v) = 0$]. As in IC, a seed set S is targeted and activated at step 0. At any time step $t \geq 1$, an active node u meets any of its currently inactive neighbors v independently with probability $m(u,v)$. If a meeting event occurs between u and v for the first time, u is given a single chance to try activating v, with an independent success probability $p(u,v)$. If the attempt succeeds, v becomes active at step t and will start propagating influence at $t + 1$. The diffusion process quiesces when all active nodes have met with all their neighbors and no new nodes can be activated.

Several possibilities can be considered in mapping the meeting events in the IC-M model to real actions in online social networks. For instance, a user u on Facebook posting a message on his or her friend v's wall can be considered as a meeting event. Different pairs of friends may have different frequencies of exchanging messages on each other's walls, which is reflected by the meeting probability. Note that the original IC model is a special case of IC-M with $m(u,v) = 1$ for all edges $(u,v) \in E$. More importantly, for the original influence maximization problem, the meeting probability is not essential because as long as $m(u,v) > 0$, eventually u will meet with v and try to influence v once. Thus, if we only consider the overall influence in the entire run, there would be no need to introduce meeting probabilities. However, if we consider influence within a deadline constraint, then meeting probability is an important factor in determining the optimal seed set.

Formally, for a deadline, $\tau \in \mathbb{Z}^+$, $\delta_\tau : 2^V \to \mathrm{IR}^+$ is defined as the set function such that $\delta_\tau(S)$ with $S \subseteq V$ is the expected number of activated nodes by the end of time step τ under the IC-M model, with S as the seed set. The time-critical influence maximization with a deadline constraint τ is the problem of finding the seed set S with at most k seeds such that the expected number of activated nodes by time step τ is maximization, that is, find $S^* = \arg\max_{S \subseteq V, |S| \leq k} \delta_\tau(S)$.

It is shown that the IC-M model maintains monotonicity and submodularity, which implies a greedy $(1-1/e)$ approximation algorithm to circumvent the NP-hardness of the problem. However, the greedy approximation algorithm is too inefficient to use in practice because it lacks a way to efficiently compute influence spread in general graphs.

To circumvent the inefficiency of the greedy approximation algorithm, two maximum influence arborescence (MIA)–based heuristic algorithms are proposed:

■ Maximum influence arborescence for IC-M (MIA-M), which uses the dynamic programming to compute exact influence of seeds.
■ Maximum influence arborescence with converted propagation probabilities (MIA-C), which first estimates propagation probabilities for pairwise users by combining meeting events, influence events, and the deadline, and then uses MIA for IC to select seeds.

The experiments in Ref. [35] demonstrate that both algorithms produce seed sets with equally good quality as those mined by the approximation algorithm, while being two to three orders of magnitude faster. Moreover, only using standard heuristics such as MIA and disregarding time delays and deadline constraints could result in poor influence spread compared to our heuristics, which are specifically designed for this context.

5.5 Conclusion

The purpose of this chapter is to give a thorough description of information diffusion in MSNs. It is important to understand the characteristics and technologies surrounding these areas before attempting to analyze various dynamic behaviors taking place in MSNs (OSNs).

To begin with, an overview of information diffusion models is presented, including the general threshold model, general cascade model, and the game theory–based model. Then, the problem of influence maximization is explained, which can be roughly defined as how to select the initial seeds such that, based on a specific diffusion model, starting diffusion from those seeds will reach people as much as possible. Influence maximization is extremely important in social networks. Thus, we comprehensively summarize existing schemes approaches (which can be classified as greedy-based and heuristic-based) for solving the problem of influence maximization. Finally, some extensions to diffusion modeling and influence maximization—such as competition, budget, and time criticality—are presented.

References

1. Griliches, Z. Hybrid corn: An exploration of the economics of technological change. *Econometrica* 1957; 25(4): 501–522.
2. Young, H. P. Innovation diffusion in heterogeneous populations: Contagion, social influence, and social learning. *American Economic Review* 2009; 99(5): 1899–1924.
3. Pastor-Satorras, R. and Vespignani, A. Epidemic spreading in scale-free networks. *Physical Review Letters* 2001; 86(14): 3200–3203.
4. López-Pintado, D. Diffusion in complex social networks. *Games and Economic Behavior* 2008; 62(2): 573–590.
5. Jackson, M. O. and L. Yariv. Diffusion on social networks. *Économie Publique* 2005; 16: 3–16.
6. Jackson, M. O. and L. Yariv. Diffusion of behavior and equilibrium properties in network games. *American Economic Review* 2007; 97(2): 92–98.
7. López-Pintado, D. Contagion and coordination in random networks. *International Journal of Game Theory* 2006; 34: 371–381.
8. López-Pintado, D. The spread of free-riding behavior in a social network. *Eastern Economic Journal* 2008; 34(4): 464–479.
9. Kleinberg, J. Cascading behavior in networks: Algorithmic and economic issue. In: *Algorithmic Game Theory*, edited by N. Nisan, T. Roughgarden, É. Tardos, and V. Vazirani. Cambridge University Press, Cambridge, UK, 2007.
10. Kempe, D., J. Kleinberg, and É. Tardos. Maximizing the spread of influence through a social network. In: Proceedings of 9th ACM SIGKDD International Conference on Knowledge Discovery and Data Mining, 2003, pp. 137–146.
11. Domingos, P. and M. Richardson. Mining the network value of customers. In: Proceedings of the 7th International Conference on Knowledge Discovery and Data Mining, San Francisco, CA, August 26–29, 2001.
12. Leskovec, J., A. Krause, C. Guestrin, C. Faloutsos, J. VanBriesen, and N. S. Glance. Cost-effective outbreak detection in networks. In: Proceedings of the 13th ACM SIGKDD Conference on Knowledge Discovery and Data Mining (KDD), San Jose, California, August 12–15, 2007.
13. Goyal, A., W. Lu, and L. V. S. Lakshmanan. CELF++: Optimizing the greedy algorithm for influence maximization in social networks. In: Proceedings of WWW, Hyderabad, India, March 28–April 1, 2011.
14. Narayanam, R. and Y. Narahari. A Shapley value based approach to discover influential nodes in social networks. *IEEE Transactions on Automation Science and Engineering* 2011; 8(1): 130–147.

15. Chen, W., Y. Wang, and S. Yang. Efficient influence maximization in social networks. In: Proceedings of the 15th ACM SIGKDD Conference on Knowledge Discovery and Data Mining (KDD), Paris, France, June 28–July 1, 2009.

16. Chen, W., Y. Wang, and S. Yang. Scalable influence maximization for independent cascade model in large-scale social networks. *Data Mining and Knowledge Discovery* 2012; 25(3): 1–33.

17. Carreras, I., D. Miorandi, G. S. Canright, and K. Engø-Monsen. Eigenvector centrality in highly partitioned mobile networks: Principles and applications. *Advances in Biologically Inspired Information Systems, Studies in Computational Intelligence* 2007; 69: 123–145.

18. Lim, S.-H., S.-W. Kim, S. Park, and J. H. Lee. Determining content power users in a blog network: An approach and its applications. *IEEE Transactions on Systems, Man, and Cybernetics—Part A: Systems and Humans* 2011; 41(5): 853–862.

19. Christakis, N. A. and J. H. Fowler. Social network sensors for early detection of contagious outbreaks. *PLoS One* 2010; 5(9): e12948.

20. Masuda, N. and H. Kori. Dynamics-based centrality for directed networks. *Physical Review E* 2010; 82: 056107.

21. Han, B. and A. Srinivasan. Your friends have more friends than you do: Identifying influential mobile users through random walks. In: Proceedings of the Thirteenth ACM International Symposium on Mobile Ad Hoc Networking and Computing, Hilton Head Island, South Carolina, June 11–14, 2012.

22. Feld, S. L. Why your friends have more friends than you do. *American Journal of Sociology* 1991; 96(6): 1464–1477.

23. Pearson, K. The problem of the random walk. *Nature* 1905; 72(1865): 294.

24. Amori, G., L. Becchetti, G. Persiano, and A. Vitaletti. *Fully-Decentralized Computation of Importance Measures in Dynamic Evolving Networks*. Technical Report no. 6, 2013. Available from: http://www.dis.uniroma1.it/~bihdis/RePEc/aeg/report/2013-06.pdf

25. Newman, M. E. A measure of betweenness centrality based on random walks. *Social Networks* 2005; 27(1): 39–54.

26. Noh, J. D. and H. Rieger. Random walks on complex networks. *Physical Review Letters* 2004; 92(11): 118701.

27. Braginsky, D. and D. Estrin. Rumor routing algorithm for sensor networks. In: Proceedings of MobiCom, Atlanta, Georgia, September 23–28, 2002.

28. Yu, H., M. Kaminsky, P. B. Gibbons, and A. Flaxman. SybilGuard: Defending against sybil attacks via social networks. In: Proceedings of SIGCOMM, Pisa, Italy, September 11–15, 2006.

29. Eubank, S., H. Guclu, V. S. A. Kumar, M. V. Marathe, A. Srinivasan, Z. Toroczkai, and N. Wang. Modelling disease outbreaks in realistic urban social networks. *Nature* 2004; 429(6988): 180–184.

30. Kotz, D., T. Henderson. Crawdad: A community resource for archiving wireless data at dartmouth. *IEEE Pervasive Computing* 2005; 4(4): 12–14.

31. Zyba, G., G. M. Voelker, S. Ioannidis, and C. Diot. Dissemination in opportunistic mobile ad-hoc networks: The power of the crowd. In: Proceedings of INFOCOM Shanghai, China, April 10–15, 2011; pp. 1179–1187.

32. Tang, S., J. Yuan, X.-Y. Li, Y. Wang, C. Wang, and X. Liu. MINT: Maximizing information propagation in predictable delay-tolerant network. In: Proceedings of the Fourteenth ACM International Symposium on Mobile Ad Hoc Networking and Computing (MobiHoc), Bangalore, India, July 29–August 01, 2013.

33. Nguyen, H. and R. Zheng. On budgeted influence maximization in social networks. *IEEE Journal on Selected Areas in Communications* 2013; 31(6): 1084–1094.
34. Small, L. Information diffusion on social networks. Master thesis, National University of Ireland, Maynooth, 2012.
35. Alon, N., M. Feldman, A. D. Procaccia, and M. Tennenholtz. A note on competitive diffusion through social networks. *Information Processing Letters* 2010; 110(6): 221–225.
36. Karsai, M., M. Kivelä, R. K. Pan, K. Kaski, J. Kertész, A. L. Barabási, and J. Saramäki. Small but slow world: How network topology and burstiness slow down spreading. *Physics Review E* 2011; 83: 025102.
37. Chen, W., W. Lu, and N. Zhang. Time-critical influence maximization in social networks with time-delayed diffusion process. In: Proceedings of the Twenty-Sixth AAAI Conference on Artificial Intelligence, Toronto, Canada, July 22–26, 2012.

Chapter 6

Mobile Search and Ranking

6.1 Introduction

Social media has greatly changed ways of communicating and interacting by capturing our social interactions and utterances in machine-readable format. Searching and analyzing massive and frequently updated social media data bring significant and diverse rewards across many different application domains, from politics and business to social science and epidemiology.

The ever-increasing availability and affordability of mobile broadband connections and smartphones have further propelled the market for digital content that, though still fragmented, is rapidly growing. This growth of mobile contents, in turn, has triggered the emergence of a mobile search market, which represents a major business opportunity in the mobile sector. Search applications on mobile devices span over a continuum that ranges from a mere adaptation of existing Web search services to the mobile environment on one extreme to novel services leveraging on the unique features of mobile devices and the way they are used on the other. In the first case, users submit a query in a "traditional" way by typing some keywords to get a list of results matching the search criteria. In the second case, contextual information is used in the search to retrieve results that are highly relevant for the user, overcoming the approach of a conventional Web search and paving the way to mobile augmented reality. The more a mobile search evolves to become a gateway to the mobile social network (MSN), the more it is expected to respond to information seeking patterns that demand a strong link between informational and physical worlds.

Although it has promising prospects, the mobile search market is still in its infancy and has not met the expectations forecasted some years ago. Its evolution remains difficult to predict, and its impact is difficult to quantify. This is due to an incomplete understanding of the factors necessary for a successful deployment, generally presented as an extended list of drivers and threats. Because there is hardly quantifiable data on the importance of each of these factors, it is not possible to properly discriminate the barriers lying ahead.

The remainder of this chapter is organized as follows. Section 6.2 presents some challenges of search and ranking in MSNs, including technological and socioeconomic factors. Section 6.3 describes existing schemes of search and ranking in MSNs consisting of preference-enabled querying mechanisms, social search browsers, social network document ranking, and mobile decentralized search. Finally, Section 6.4 briefly concludes this chapter.

6.2 Some Challenges of Search and Ranking in MSNs

Generally, an MSN offers anytime, anywhere access to a wealth of information to billions of users around the globe. However, an MSN represents a challenging information access platform due to the inherent limitations of mobile environments—limitations that go beyond simple screen size and network issues. Following are some special challenges [1].

6.2.1 Technological Factors

Previous investigations by the Institute for Prospective Technological Studies (IPTS) show that technology is only a minor barrier to the deployment of mobile search applications [2]. An April 2010 IPTS mobile search survey confirmed this. Specifically, 78% of respondents in this survey believe that the main technological bricks (devices, networks, applications) for mobile search are already available. However, development work is still greatly needed. About half of the respondents state that technical aspects form a major barrier to the full deployment of mobile search services. In brief, in spite of its general availability, the technology is not yet mature and not reliable for high performance in a mass user environment.

In particular, experts consider that some enabling technology necessary for further boosting mobile search deployment (such as wireless networks, sensor networks, smart devices, and the cloud computing framework) is slowly entering the market but not yet in a pervasive manner. In addition, they think that current user tracking for geopositioning is not sufficiently fine grained for optimal service provision/fruition. Technologies to enhance interaction need further improvement. Gesture-based interfaces are not robust enough to sustain interaction in real-world environments. Furthermore, they are not as "natural" as they have been promoted to be. Similarly, although many agree that voice may become the main way to interface in future mobile search applications and that it may be the most appropriate interface for illiterate users, voice recognition accuracy is advancing very slowly. Voice-based interfaces are often disappointing from a user experience viewpoint. Moreover, a technology driven approach could not compete in a complex ecosystem. No single technological advance or improvement in human computer interaction is likely to have a disruptive effect on mobile search uptake, even though important changes are observed in the landscape due, for instance, to the penetration of touch-based interfaces, which are shaping a new wave in the mobile content trajectory [3].

The issue of standards also deserves to be debated. Whereas it is important to monitor the emergence of enabling technologies, the importance of standardization cannot be underestimated when projecting the potential penetration of cutting edge advancements. Many efforts are being made by groups such as the World Wide Web Consortium (W3C) to reach a consensus about ontologies (both media ontology and context of delivery ontology), metadata formats that enhance discovery, time stamps for the Web, interactions with social networks, speech synthesis, or gesture-based interaction. However, experts were consensual in acknowledging that complex topics such as context awareness or privacy in relation to user personal identity data need further research before a standardization road map can be created. However, regarding the availability of personal information for service provision and the related privacy concerns, literature does not converge to a consensual formulation of the problem and its implications.

6.2.2 Socioeconomic Factors

The availability of mobile broadband connections coupled with a large deployment of smartphones and their operation in 3G networks is rapidly increasing the consumption of digital content from mobiles [4]. However, for mobile search, analysts observe that there is still a gap between the diffusion (potential usage) and the adoption (actual usage) of mobile search services. It is reported that URLs and bookmarks are the best two ways to find contents, and mobile search comes in a close third.

There is still potential. Whereas 80%–100% of PC Internet browsers use *search*, depending on the EU country, only around 40% of mobile browsers use search. In this case, their portal is more and more often made up by Google. However, though this may lead our thinking that the future of search lays in the adaptation of existing Web search services to the specificity of mobile user interfaces, the case of Apple demonstrates that alternative mobile business models can also be sustainable. Apple has built a set of elements that together make an ecosystem and thus are boosting (the potential of) the mobile Internet. This set includes powerful mobile devices (iPhone and iTouch), an easy-to-use payment/distribution system (iTunes), and a developer-friendly environment for new applications (App Store). This creates fertile soil for the growth of mobile-specific Internet usage. Recently, we have witnessed shifts in user behaviors. More and more people access content online on their mobile phones, especially with the significant growth in smartphone adoption and fast data networks.

The emerging mobile search market is very dynamic and new developments happen almost every day. A number of innovative search applications lever on mobile specificity; different business models are currently being explored with the objective of providing customers with a satisfactory search experience. Here, user satisfaction is going to be benchmarked against the speed, the relevance, and the usefulness of retrieved information.

It has been argued that the major challenge for mobile search is economic in nature. Naturally, this statement has launched a discussion on business and revenue models in the domain of mobile search. However, disruptive business models have not yet emerged. An evolution of marketing promotion strategies probably is to be expected; advertising is a concept that does not fit today's media. Advertising is the model that makes the most out of broadcasting communication channels. Mobile networks enable narrowcasting, potentially reaching the level of one-to-one marketing and allowing for a customized mix.

Today's technology enables a paradigm shift in product/service promotion leading to a model where customer engagement is the main key to success. However, solutions have to be tailored to the type of service as well as to the target audience. It is unlikely that a single winning revenue model will emerge that fits all business models.

6.3 Existing Schemes of Search and Ranking in MSNs

There exist many schemes related to search and ranking in MSNs. In this section, we briefly introduce several typical schemes.

6.3.1 A Preference-Enabled Querying Mechanism

Advances in spoken dialog systems (SDSs) have allowed the provision of context-dependent information and services for mobile users in ubiquitous environments. For example, a tourist (Maria) can ask his or her smartphone to recommend the points of interest (POIs) that may be of relevance for him or her, such as suggesting which museums to visit or which restaurants to go to. One of the critical issues is how to provide desirable answers to the user by exploiting preferential information given his or her contextual situation. Two crucial subtasks are: (1) modeling the user preferences by exploiting background knowledge and (2) reasoning and query answering by the continuous processing of social semantic data [5]. Most of the current search engines and SDSs focus on delivering information according to a single search criterion specified by the user, but few of them provide efficient solutions for proactively providing relevant answers given a user's preferences and the social context. Let us consider a motivating scenario—a tourist (Maria) asks her dialog system to recommend the best restaurant near the Eiffel Tower in Paris. First of all, the system infers her preferences given multiple attributes from the dialog history (e.g., she prefers French restaurants at medium prices). Then, according to her current location, the system can select a list of nearby restaurants that might be of interest to Maria. In order to determine the most relevant answers, the system also uses the opinions (e.g., positive) of Maria's friends collected from social media streams to influence the restaurants' ranking. For this scenario, what the system needs do is to support advanced inferences and reasoning over background knowledge and streams of changing information.

The European PARLANCE project,* which aims to design and develop mobile, interactive, "hyper-local" search through speech, has been involved in tackling this challenge. It focuses on presenting a preference-enabled querying mechanism, which is able to infer the weighted interests by exploiting the dialog history and further offer relevant answers by considering the ratings of the user's followers as well as following POIs. The main results in Ref. [5] are twofold. First, it presents an architecture for preference-enabled querying that can infer the weighted interests from the dialog history and encode those interests into the queries; second, it proposes a mechanism that can determine the most relevant answers by exploring both persistent knowledge base and streaming data from social networking Web sites such as Twitter.

6.3.1.1 Related Work

6.3.1.1.1 Preference Queries

From a quantitative perspective, preferences are ratings and are defined as a function μ that captures the satisfaction or "appealingness" of an item $i \in I$ to user $u \in U$ within a scale of numerical values, usually the real interval $[-1, 1]$; that is, $\mu: U \times I \rightarrow [-1,1]$ [6]. On the contrary, preferences can also be expressed as qualitative descriptions of a set of properties that specify user interests, which can be added to the queries as constraints. To query the Semantic Web with preferences, the PREFERRING clause has been provided in the SPARQL, which states preferences among values of variables [7]. However, this clause cannot encode the social context that may impact the ranking. In contrast to the extension of the standard query language, a scheme proposed by Hu et al. [5] focused on modeling preferences as a type of weighted interest according to a user's dialog history.

6.3.1.1.2 Continuous Querying Social Semantic Data

Social data as feeds and microblogs are adapted to publish information in a real-time stream through a social networking Web site [8]. This trend toward the interlinking of the social Web with semantics uses vocabularies such as Semantically Interlinked Online Communities (SIOC),† Friend-of-a-Friend (FOAF),‡ and the Simple

---

* http://sites.google.com/site/parlanceprojectofficial/
† http://www.sioc-project.org/
‡ xmlns.com/foaf/0.1/

Knowledge Organization System (SKOS).* Much research focuses on the linked stream data [9] and stream reasoning,† with the aim of realizing transient streams and continuous reasoning tasks. C-SPARQL, as an extension of SPARQL, allows existing reasoning mechanisms to be further extended in order to support continuous reasoning over data streams and rich background knowledge [10]. A mobile search method proposed in Ref. [11] aims to compute novel and effective features for ranking fresh Uniform Resource Identifiers (URIs) by using Twitter data. Also, a stream reasoner named BOTTARI [8] has been developed to continuously analyze Twitter to understand how the social users collectively perceive the POIs in a given area. This scheme aims to investigate how to use the ratings from the social users to influence the relevance of retrieved items.

6.3.1.2 Architecture and Components

This scheme in Ref. [5] designed a preference-enabled querying mechanism for personalized mobile search (PQMPMS), which supports providing personalized answers tailored to user preferences by taking into account the social context. The main components and their interactions are elaborated here (and shown in Figure 6.1).

6.3.1.2.1 Context-Aware Recommender

This component fulfills the core functionality of recommending the desirable items in two steps. First, it generates the preference-enabled queries and further refines them according to a user's feedback, relying on the query refinement mechanism. Referring to the earlier motivating scenario, Maria's request is translated to a query, which encodes her preferred cuisine (recorded in the user profile) and her current location (represented by the context model). Second, the relevance of the retrieved items is assessed based on the ranking schema. To represent the relevance of a given item for a specific user, the concept item has the rating object with its properties *isForUser* and *hasRating*. The ranking of item i for user j is represented as:

$$R_{ij} = \alpha \frac{1}{n} \sum_{z=1}^{n} \omega_z + \beta\gamma F + 1 - \gamma P(0 < \gamma < 1, \alpha + \beta = 1) \tag{6.1}$$

where ω_z represents the inferred interest scores, F is the normalized number of followers/following, and P is the percentage of positive opinions. Once a new relevance value is computed as Equation 6.1 by considering the weighted interests and the ratings from user's friends, the SPARQL CONSTRUCT queries are used as rules to infer new triples for updating the relevance of a given item.

---

* www.w3.org/2004/02/skos/
† http://streamreasoning.org/

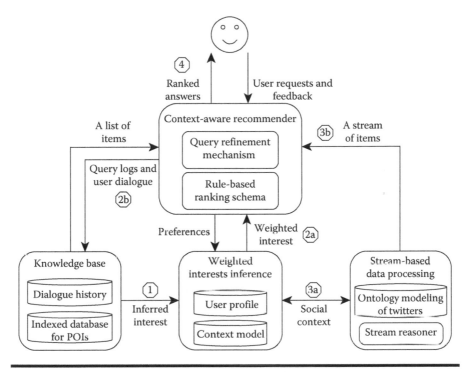

Figure 6.1 Preference-enabled querying mechanism for personalized mobile search (PQMPMS) architecture. (From Hu, B. et al., PQMPMS: A preference-enabled querying mechanism for personalized mobile search, In: Proceedings of the 7th International Conference Web Reasoning and Rule Systems (LNCS 7994), 2013. With permission.)

6.3.1.2.2 Knowledge Base

The knowledge base is composed of the dialogue history and an indexed database containing POIs. The dialogue history consists of an abstract representation (known as dialog act units) of all past dialogue between the user and the system.

The second part of the knowledge base is the indexed database with POIs. The preference-enabled query, being the query extended with the interests of the user obtained from the user profile component, will be executed to search in the database for the set of POIs that are in correspondence with the query.

6.3.1.2.3 Weighted Interests Inference

Weighted interests interference consists of a user profile and a context model. The user profile contains modular ontologies that are annotated with scores representing the interests of a user on different levels. The history will be used to infer the interests of the user on several levels. First, the interest of the user on the attribute type level can be inferred. For example, if a user in his or her past interactions with the system has asked a lot about the food type of a restaurant but not about the price, it infers that the food type attribute is of more interest to the user than the price. Second, it infers the interests of the user on the attribute value level, meaning that for one specific attribute type such as food type, the user prefers Italian or Chinese food over Mexican. Third, the interests on these two levels are combined to obtain the interest of a user at the instance level (i.e., his or her interest in specific restaurants). It should be noted that these interests evolve over time. When the user has interacted with the system, the dialog is analyzed and the weights in the model are updated accordingly. Each ontology module corresponds to one domain of interest, such as restaurants or transportation. The recommender component calls the appropriate modular ontology, depending on the subject of the dialog (e.g., searching for restaurants) and analyzes the associated scores to build the preference-enabled queries. The context model consists of characteristics of the user that can be relevant for recommendation and of the current location of the user. Also the social context of the user is included in FOAF format, representing the followers of the user on Twitter.

6.3.1.2.4 Stream-Based Data Processing

To model the Twitter data, stream-based data processing adapts the SIOC vocabulary, which defines Twitter User as a special case of User Account. The model defines the notion of POI as named place, which is enriched with a categorization and count of positive/negative/neutral ratings. It defines the object property talks about and its subproperties for positive, negative, and neutral opinions so that the opinions of a tweet about POIs can be represented. It uses C-SPARQL queries to continuously analyze the Twitter data for collecting the items rated by the user's followers/following.

The architecture works as follows. First, the weighted interests are inferred from the user's dialog history and are modeled in the user profile (noted as 1 in Figure 6.1). Then the user's requests are translated to the corresponding SPARQL queries by considering user preferences and contextual information (noted as 2a). Accordingly, a set of items defined as I_q is retrieved by sending preference-enabled queries to our knowledge base to get POIs (2b). Next, the component of stream-based data processing exploits the social context (3a) to provide a stream of the items (defined as I_s) that have been positively rated by the user's friends by executing the C-SPARQL queries (3b). Then the intersection of I_q and I_s is computed, and the relevance rating of those items needs to be obtained. The final relevance rating of a certain item is determined by the weighted interests and opinions of the user's friends (4). Finally, the top item with the highest relevance rating is recommended to the user. By exploiting the user's dialog history, the weighted user preferences and interests can be inferred. To further compute personalized answers, stream-based data processing aims to continuously collect the ratings given by the user's friends regarding relevant topics from stream-based data sources such as Twitter. The experiment shows that this approach allows the computation of the most relevant answers, providing an increased quality of search experience for the user.

6.3.2 Social Search Browser

There are more mobile phones in the world than personal computers. They allow communication across remote locations and ubiquitous access to a wealth of information sources. Yet most of the information retrieval systems that are available within the mobile space today are simple adaptations of their desktop counterparts and as such are not well suited for handling the complexities of mobile contexts. In particular, geographical information is intimately related to mobile devices because they are typically carried by their owners in their daily lives. Contextual information of this nature is too fine grained and too dynamic to be captured by Web technologies. Often, this scheme seeks and shares local information through other communication channels, and word-of-mouth appears to be the most reliable and efficient communication medium in certain information seeking tasks. For instance, standard search tools have difficulty providing answers to questions such as finding information about an upcoming friend's birthday or what the most populated club in a city is after a major sporting event.

Furthermore, humans are social beings who often seek new and improved ways of sharing information with their peers. In the past few years, several research projects have attempted to exploit the social dimension of search by designing interfaces that allow users to collaboratively help each other in finding the information they need. However, these prototypes were designed with the desktop experience in mind. Conversely, prototypes that were built to improve mobile search did not exploit the social dimension of information seeking.

In this regard, there has been interest in understanding whether people's information needs while on-the-go could be addressed by providing a readily available connection to a user's social network. Friends and family, who are trusted information sources, are likely to be able to draw on their experiences to provide interesting, valuable, and relevant answers to the user's on-the-go queries.

Social search browser (SSB) [12] is a proof-of-concept, map-based mobile search prototype that was designed to enhance the search and information discovery experience of mobile users by allowing them (1) to see the queries and interactions of peers and (2) to pose queries of their own (see a detailed description in Section 6.3.2.1). SSB gives users the ability to connect with friends or family members while on-the-go and to ask them questions. Interactions are handled on a rich map-based interface, which enables the use of deictic gestures among remote peers [13]. Furthermore, SSB provides novel methods for filtering the queries displayed based on the level of friendship among users.

This prototype is similar in nature to the Questions not Answers (QnA) [14] prototype. The QnA system tags queries with a location. These queries are displayed on a map-based interface enabling users to visualize the search space. The QnA prototype does not, however, provide any means for a user to filter queries, other than by location. Given that the volume of queries at specific locations is likely to be high and that there is no means for filtering queries, the QnA prototype raises a new interface/presentation challenge. The SSB prototype addresses this issue by offering the user three types of filters. In addition, although QnA pioneered the proactive display of queries on a map, the application did not allow users to issue their own queries or to communicate with the author(s) who generated the queries being displayed. Thus, the realism of their user study is limited in some regards. The proposed SSB prototype allows its users to add and respond to queries and to interact with other users.

In addition, this work in Ref. [12] is related to the CityFlocks approach, although the SSB prototype focuses the interaction on the group of peers that can provide trusted information to the user [15]. In this regard, a key contribution of this work is to explore the social side to mobile search, not only through a social query filter but also by studying mobile search–mediated interactions with peers (e.g., members of the user's social network).

In summary, relevant research questions are: (1) How do mobile users interact with proactive mobile information access applications? (2) What are the implications of these types of applications on the design of mobile search services?

6.3.2.1 SSB Prototype

SSB is designed to enhance information search and discovery by displaying what other users have been searching for on an interactive map-based interface. The software architecture of SSB consists of three components: (1) an iPhone application that allows users to browse, answer, and add queries; (2) a Facebook application that allows a given user's social network to browse queries and add new answers to those queries; and (3) a server that synchronizes and stores all queries in the SSB database. The server feeds applications (1) and (2) with an up-to-date list of all queries and answers. When a new query is issued by a user, the server submits the query to the Google Local Search Application Programming Interface (API) for a set of possible search results. The server also has a short message service (SMS) notification facility that informs members of the appropriate social network about new queries and new human-generated answers. In addition, the server logs all the interactions between the user and the graphical user interface (GUI) of the iPhone application for off-line analysis of user behavior.

6.3.2.2 Mobile Application

The main interface of the SSB mobile application consists of a Google Maps visualization of the user's current location with overlaid queries. This map-based interface provides users with a sense of place at a glance by allowing them to visualize the kind of queries that other users have issued while they were at the same location. The map indicates the user's current location by a blue circle. A tiny red circle positioned at the center of the map marks the location of new queries. As users pan or zoom within the map, the set of visible queries is updated.

An icon is assigned to each query to indicate what kind of information is available about the query. This icon is displayed to the left of the query text (long query texts are truncated on the map interface). A small magnifying glass icon is assigned to a query that does not result in the selection of any search result. A query that resulted in the selection of at least one search result is identified by the globe icon. If a query has been answered by a user of the SSB application, the associated icon is augmented with a small image that depicts a user.

A semitransparent background color conveys the origin of the query; queries are either issued by the user (green background), his or her friends (red background), or other people (blue background). Furthermore, the size of the query icon reflects the popularity of a query based on the number of answers that the query has received and the number of times the details of a query have been accessed by users.

6.3.2.3 Interactive Filters

Three filters positioned above the map allow the user to control the queries that are displayed. The filters may be used to reduce the number of queries on a crowded map or to focus on queries that fulfill specific criteria. The time filter enables a form of temporal visualization of queries. For example, users can view queries that have been submitted in the last two hours (setting: now), in the last 24 hours (setting: today), the last week, and so on.

A similar principle applies to the friendship filter. This filter allows users to show queries submitted by everyone or only by friends. This social filter has several degrees of "levels of friendship." The premise behind this filter is that, in some situations, users might prefer to see queries that have been generated by friends. The level of friendship is determined by SSB with number of wall posts, tags, and comments that have been exchanged between the two peers within social networks (e.g., Facebook). In the current implementation, the authors use Facebook as a source of social network information; however, other online social network information could also be exploited. Upon registration, users grant SSB access to their social network. This information is stored in SSB's server and the user's original social network is never accessed again.

Finally, the similarity filter allows users to limit the displayed queries to those that are similar to queries that have been previously entered by the user himself or herself. For a given user and a given query, SSB calculates the similarity as the number of words that match between the query in question and all other queries issued by the user in question. Queries are displayed only if the number of matching words is equal or greater than the threshold value that is set through the similarity filter.

6.3.2.4 Query Details and Answers

Double-tapping on a query brings the user to the query details. The query details screen consists of four components:

- Full query details: Showing the complete query string, the time stamp when the query was issued, and the name of the user who issued the query (only if this user is a friend of the current user).
- Answers: A list of all answers that have been submitted by other users. Answers can be submitted by users via the mobile application or via the Facebook application (see following).
- Local search results: A list of localized search results extracted from Google's local search and place service (https://developers.google.com/places/).
- Event search results: A list of related events extracted from the Eventful API (http://eventful.com/).

Tapping on the "plus" icon beside any of the individual query details expands additional information about the associated answer, local search result, or the event search result. These details include a map with a location, phone numbers, and a "more info" link to an external Web page.

SSB has proven to be a useful showcase of proactive map-based mobile social search and has allowed evaluation of new search paradigms with real mobile users. The results have showed that mobile users enjoyed the proactive, social nature of the SSB prototype.

6.3.3 Social Network Document Ranking

In search engines, ranking algorithms measure the importance and relevance of documents mainly based on the contents and relationships between documents. User attributes are usually not considered in ranking. This user-neutral approach, however, may not meet the diverse interests of users, who may demand different documents even with the same queries. To satisfy this need for more personalized ranking, a ranking framework called social network document rank (SNDocRank) is proposed in Ref. [16], which considers document contents and the relationship between a searcher and document owners in a social network. This method combines the traditional Term Frequency–Inverse Document Frequency (TF-IDF) [17] ranking for document contents with the Multilevel Actor Similarity (MAS) algorithm to measure to what extent document owners and the searcher are structurally similar in a social network. As the proposed ranking method, SNDocRank was implemented in a simulated video social network based on data extracted from YouTube and its effectiveness was evaluated on video search. The results show that compared with a traditional ranking method like tf-idf, the SNDocRank algorithm returns more relevant documents.

More specifically, a searcher can get significantly better results by being in a larger social network, having more friends, and by being associated with larger local communities in a social network.

A core component in search engines is the ranking algorithm that ranks the relevance of documents. Interest in improving the performance of ranking algorithms will probably always continue. Some of these algorithms are user-neutral and measure the relevance of documents primarily based on the contents and relationships of documents. Although this approach is effective in determining the relevance of documents to queries, it ignores one important factor involved in search—users who initiate queries. Users often have diverse interests and may want different documents, even when they use the same queries. A user-neutral ranking approach will not address this need because of the lack of user data in the ranking.

Personalized ranking algorithms have been proposed that include various types of user information in ranking. To enhance ranking performance and improve search results, algorithms use various information as a user's search context, such as geographical location and searching histories, click-through logs, topics of interest, and personal bookmarks. Some algorithms consider the information needs of a user's friends. However, these algorithms largely focus on local activities of the user, and they fail to embrace the large social contexts of the user.

In reality, users are involved in different social communities. Users are increasingly engaged in social networks through online social networks (OSN) such as Facebook, Flickr, and YouTube in order to communicate with their friends, family, and colleagues and to share documents, images, and videos. Their social networks may provide richer and more reliable clues about the purposes and interests of their information search. For example, when a devoted animal lover searches for information about snow leopards and uses a query "snow leopard," search engines often cannot tell whether required information is about an endangered species or an operating system and, as such, search results may not match the user's need. If the information about the user's social networks is available, search engines can disambiguate the query based on such information as which networks the user belongs to and who the user's friends are and then possibly deliver more relevant information.

Based on the aforementioned consideration, the framework SNDocRank considers a searcher's social network when ranking the relevancy of documents. The premise of our methodology is that "birds of a feather flock together" [18]. That is, users tend to be friends with those who share common interests, and users are more interested in information from friends than from others. Therefore, an MAS method was proposed in Ref. [16] to calculate actor similarity in social networks. By dividing a large social network into network modules at multiple scale levels, the MAS approach can dramatically accelerate the computation process.

Figure 6.2 illustrates major components of the SNDocRank framework, which is composed of three core components: an actor similarity module to compute actor similarity scores, a document-matching module to match user queries with indexed

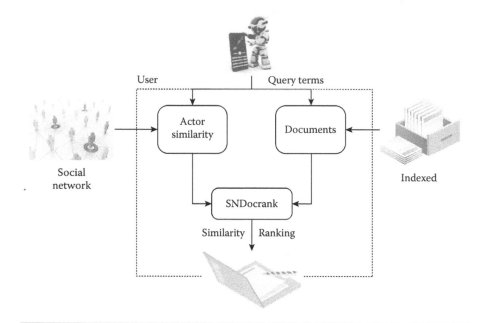

Figure 6.2 Social network document rank (SNDocRank) framework. (From Gou, L. et al., Social network document ranking, In: Proceedings of the 10th Annual Joint Conference on Digital Libraries (JCDL), 2010. With permission.)

documents, and a SNDocRank module to produce the final ranking by combining document relevance scores with actor similarity scores. Particularly, the document-matching module is a typical term-based search engine, and is not introduced here. The actor similarity module and the implementation of the SNDocRank module are discussed in the following sections.

6.3.3.1 Multilevel Actor Similarity

One way to expand the scope of social relationships in a social network for a personalized searcher is to consider actor similarity in ranking algorithms. The actor similarity of social networks measures the similarity of two actors in a social network based on the structural information of that network. Several approaches are introduced in Ref. [19]. In this subsection, the concept of actor similarity and the MAS algorithm are introduced.

6.3.3.1.1 Actor Similarity Algorithms

One of the simplest ways to measure actor similarity in social works is cosine similarity based on structural equivalence [20]. In this approach, two actors are regarded as similar if they share many neighbors in a social network. A basic method is cosine similarity, which only considers directly connected actors as neighbors but fails to integrate larger social contexts of a social network.

Leicht, Holme, Newman (LHN) vertex similarity [21] improves the similarity measure by considering the similarity scores of the whole neighborhood rather than direct connection. For two nodes in a network, if their immediate neighbors are similar, then these two nodes are similar, even though they are not directly connected. The calculation of the similarity of two nodes with the LHN vertex similarity is a recursive process, which involves not only their direct neighbors but also all nodes connected to the neighbors. Consequently, the similarity score of the LHN method integrates both the local connectivity of these two nodes (e.g., the direct connections) and their global connectivity (e.g., the nodes indirectly linked to them).

However, the LHN vertex similarity approach faces one challenge: scalability. Because of the recursive process that involves expensive matrix multiplication, its computational complexity is extremely high, which becomes impractical in processing large social networks with thousands or millions of nodes and edges.

6.3.3.1.2 MAS Approach for Actor Similarity

To address the issue of computation complexity, an approach was developed that first clusters a large social network into smaller ones at multiple scale levels and then computes the similarity within and between network clusters. This hierarchical approach, called multilevel actor similarity, preserves the primary benefit of the LHN vertex similarity, which is the availability of the global structural information in similarity scores and, at the same time, significantly reduces computation complexity, making the SNDocRank approach a feasible algorithm for various applications.

Specifically, the MAS approach includes three steps. The first step is an algorithm to cluster and aggregate a social network at multiple levels and to create a node cluster hierarchy. Each network node can belong to one and only one node cluster in the hierarchy. The network of node clusters at each hierarchical level captures the main structural characteristics of the network and serves as a backbone of the network at that level. Then, a weighted LHN vertex similarity method is applied to compute the similarity among these node clusters in the hierarchy. The similarity between two node clusters in the backbone network offers contextual information for the similarity among network nodes within them. Finally, the similarity of any two network nodes is computed by considering the similarity between each node and its parent node cluster, and the similarity between their parent node clusters. Figure 6.3 shows an example of computing the similarity of two nodes in a social network with the proposed MAS method. Assume the goal is to calculate the similarity between nodes 5 and 9 (shown as Figure 6.3a). The first step is to generate a hierarchy of node clusters based on the structural characteristics of the network. Assume three node clusters can be identified (Figure 6.3b), and networks and node clusters form a hierarchy (Figure 6.3c). Then, the whole network can be

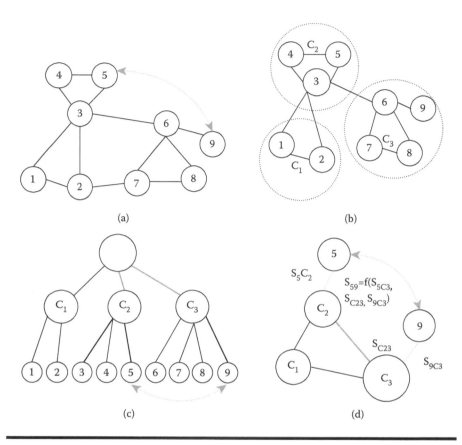

Figure 6.3 An example of multilevel actor similarity. (From Gou, L. et al., Social network document ranking, In: Proceedings of the 10th Annual Joint Conference on Digital Libraries (JCDL), 2010. With permission.)

simplified and represented as a backbone network with these three node clusters, C_1, C_2, and C_3. Each cluster contains aggregated information of its member nodes, and the relationship among them is also the aggregation of the relationships among their members (Figure 6.3d). The similarity values among these node clusters can be calculated by applying the LHN method on the backbone network. Then, the similarity value between nodes 5 and 9 is calculated by considering the similarity values between node 5 and its parent cluster C_2 (S_{5C_2}), between node 9 and its parent cluster C_3 (S_{9C_3}), and between two clusters C_2 and C_3 (S_{C23}).

Social network clustering, or community detection, is a continuing topic of research in social networks. Various algorithms have been proposed, such as the fast community detection algorithm by Clauset et al. [22]. This algorithm is based on a quality measurement for clustering a network—modularity [23]. A good clustering of a network, which has a high value of modality, maximizes the number of edges within clusters and minimizes the number of edges between clusters. Clauset et al. [22] provide more details about the measure of modularity. The clustering method is based on this modularity-oriented approach.

The result of the clustering process is a hierarchical tree in which each leaf node is an actor, and each nonleaf node is a group of actors. One of the concerns of this clustering process is that the output is an unbalanced tree, in which some clusters are very large while some are very small. The unbalanced tree could have serious computational effects because large clusters may become a scaling bottleneck for the whole process. The way to address this issue is to set a size limit for node clusters so that large clusters are broken into smaller ones. Aggregation is an approach to describe the main structural features of a node cluster. After aggregation, a node cluster can be treated as a network node and the edges between the node cluster and another node (network node or node cluster) can also be grouped as one single meaningful connection.

Various methods can be applied to node aggregation and edge aggregation. The implementation of SNDocRank uses the node with highest degree in a cluster to represent the whole cluster and adds the edges between a node cluster and another node as the aggregated edge that now indicates the connection strength of two entities.

The result of network clustering and aggregation is a weighted hierarchy of network nodes and clusters. In the hierarchy, leaf nodes are the nodes from the original social network. Nonleaf nodes are the clusters that contain nodes that share some common structural characteristics. As a result of edge aggregation, the clustered networks are weighted. The clustered networks preserve the key structural features of the social network, and the structure at different levels of hierarchy serves as a backbone for further network analysis.

6.3.3.2 Implementation of Social Network Document Rank

SNDocRank considers the relevance of documents to a query and the similarity between document owners and a searcher. Thus, a composite ranking can be written as

$$SND(u,q,d) = f(R(q,d), S(u,o)) \qquad (6.2)$$

where $SND(u,q,d)$ is the ranking score of a document d when a user u is conducting a search with query q; $R(q,d)$ is the value that indicates the relevance of d to q based on a term-document similarity function, such as tf-idf and BM25 [24]; and $S(u,o)$ is the similarity value between the user u and the owner of the document o. Both u and o are in the same social network.

The implementation of SNDocRank is flexible. Different document similarity functions (e.g., tf-idf or BM25) can be used. The similarity component can also be selected based on the size and nature of social networks.

Evaluation results indicate that, overall, the SNDocRank framework can return better search results than the traditional tf-idf ranking algorithm in terms of relevance, the matching of interests with searchers, and the ranking effectiveness of returned results. Compared with the cosine similarity method, the proposed MAS method outperforms the cosine similarity algorithm consistently across different sizes of social networks, different degrees of searchers, and different sizes of interest groups in a social network. This indicates that the structure of a searcher's social network can provide clues about the user's information needs and then can be used to help improve the performance of ranking algorithms.

These results also indicate the sensitivity of the SNDocRank approach to certain characteristics of a searcher's social network. The search results from the SNDocRank method (both MAS and cosine) vary with the size of a searcher's social network, a searcher's degree, and the size of a searcher community in a social network. Although the SNDocRank method considers the global information of a social network, it becomes effective only as the size of a network reaches a certain magnitude. This problem is believed to be similar to that of the cold-start problem in collaborative filtering [25], which for good performance needs sufficient information about new users.

The degree of a searcher in a social network can also affect the performance of the SNDocRank framework. Generally speaking, both MAS and cosine methods benefit high-degree searchers more than they do low-degree searchers. This may be due to the fact that higher-degree searchers leave more clues and traces about themselves in the social network, which can then be used to improve document rankings. This MAS method is more effective than the cosine approach because MAS considers the global information of social networks, while the cosine approach only focuses on that from direct neighbors. The size of local communities in a social network also affects the SNDocRank results. Both MAS and cosine algorithms favor larger interest groups. This may again be related to the availability of information about searchers. Larger communities tend to spread information about themselves and their members more broadly than smaller communities.

These results suggest actions that users can pursue to improve searching results. First, a user should join large social networks because SNDocRank methods benefit more from large social networks than from small networks. Second, a user should try to be connected to as many people as possible to increase their degree, which for the proposed method leads to better search results. Finally, in a social network, a user should be connected to large communities or interest groups.

6.3.4 Mobile Decentralized Search

As mobile phones have become pervasive in daily life, mobile applications have transcended from basic communication and entertainment services into enablers of societal and political transformation. Social networks such as Twitter and Facebook, as well as search services such as Google and Bing, have been used to help coordinate mass uprisings and revolutions throughout the world. However, centralized systems—whether controlled by a government or a business—rely on a few nodes that can be easily subverted or censored. If a service provider does not cooperate with such censoring entities, access to the service might be denied entirely. For example, in some countries, the Facebook group meeting service was used to help organize the places and times of protest meetings. In several cases, governments disabled local access to the Internet to hinder the organization of such meetings.

A decentralized search and retrieval system where multiple nodes or peers in the network share documents, metadata, and queries can better withstand temporary or sustained network blocking and shutdowns. Peers can reroute network traffic away from nonoperational or nonresponsive nodes and can retrieve documents from one of several alternative sources.

A distributed search and retrieval system called iTrust was proposed that does not rely on a centralized search engine such as Google, Yahoo! or Bing; thus, it is resistant to censorship by central administrators. The first implementation of iTrust, named iTrust over HTTP [26], is based on the Hypertext Transfer Protocol (HTTP) and is most appropriate for desktop or laptop computers on the Internet. However, modern day users expect mobile phones to have many of the same capabilities that more traditional computers have. The modern user wants a computer that fits in his (her) pocket (purse) and that is network enabled. In many countries, mobile phones are the only computing platform generally available; thus, it is appropriate to provide the iTrust system on mobile phones.

Lombera et al. [27] have extended the traditional HTTP-based iTrust search and retrieval system so that it relies not only on the Internet but can use the cellular telephony network. First, the iTrust over HTTP system is extended to allow users of mobile phones to connect to iTrust over HTTP via SMS so that they can benefit from the decentralized search and retrieval service provided by iTrust. This system is named iTrust with SMS by the authors. Its objective is not to supplant HTTP but instead to have SMS work alongside it, to increase accessibility during dynamic situations where mobile phones are used. Second, the authors completely reimplemented the iTrust over HTTP system to work only over SMS, thus creating the iTrust over SMS system. Whereas iTrust with SMS allows mobile phones to send text messages to the iTrust over HTTP network, iTrust over SMS allows mobile phones to form self-contained peer-to-peer networks that are not necessarily connected to the Internet. iTrust over SMS allows mobile phones to search for and retrieve documents entirely within the cellular telephony network; an Internet connection is not required.

6.3.4.1 Mobile Search and SMS

Due to the form factor, the limited bandwidth, and the battery life of a mobile device, mobile search is fundamentally different from desktop search, as Sohn et al. [28] have observed. In desktop search, users can use a simple search interface to enter keyword queries. The accuracy is generally satisfactory if the desired results are within the first 10 URLs returned; if not, the user can interactively refine his or her queries in subsequent search rounds. In mobile search, it is expensive and tedious for a user to explore even the two most relevant pages returned by a traditional centralized search engine. Moreover, the information sought tends to focus on narrower topics, and the queries often are shorter (e.g., requests for phone numbers, addresses, times, directions). Kamvar and Baluja [29] have found that most mobile search users have a specific topic in mind, use the search service for a short period of time, and do not engage in exploration. In a subsequent study [30], they found that the diversity of search topics for low-end mobile phone searches is much less than that for desktop searches. Several other researchers have also focused on the needs of the users rather than on the mechanisms involved. The SMS works on low-end mobile phones and is available worldwide. Global SMS traffic is expected to reach 8.7 trillion messages by 2015. In developing countries, SMS is the most ubiquitous protocol for information exchange after human voice.

In SMS-based search, the query and the response are limited to 140 bytes each. Moreover, the user has to specify a query and obtain a response in one round of search. Significant work has been undertaken to improve mobile search using SMS text messages [31]. In iTrust, an SMS request (query) consists of a list of keywords, which are typically less than 140 bytes. An SMS response simply returns the requested information if it is small (less than 140 bytes). If the requested information or document is larger than 140 bytes, it is fragmented into multipart SMS messages. Alternatively, the SMS response can return a URL, which is typically less than 140 bytes.

Finally, the short message size and transmission frequency of SMS messages accustoms users to utilizing the service for almost real-time momentary or fleeting communication. After an hour, or even several minutes, most SMS messages are no longer important to the user; in many cases, even important messages are meaningless without surrounding context such as time, circumstances, or information not recorded directly by the mobile device. For this reason, the temporal integrity of an SMS query result is relevant only if a search hit is returned relatively quickly (within minutes); otherwise, the information is meaningless without context provided by the user.

6.3.4.2 Traditional Design of the iTrust Search and Retrieval System

The iTrust search and retrieval system involves no centralized mechanisms and no centralized control. The nodes that participate in the iTrust network are called the participating nodes or the membership. Multiple iTrust networks may exist at any point in time, and a node may participate in several different iTrust networks at the same time.

In an iTrust network shown in Figure 6.4a, some nodes—the source nodes—produce information and make that information available to other participating nodes. The source nodes produce metadata that describes their information and

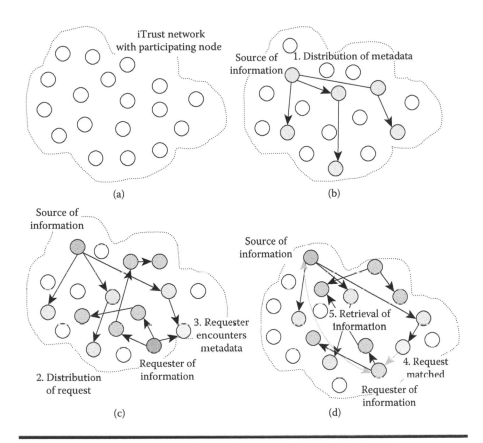

Figure 6.4 **(a) An iTrust network with participating nodes. (b) A source node distributes metadata to randomly selected nodes. (c) A requesting node distributes its request to randomly selected nodes. (d) A node matches the metadata and the request and reports the match to the requesting node, which then retrieves the information from the source node. (With kind permission from Springer Science+Business Media: *Mobile Network and Applications*, Mobile decentralized search and retrieval using SMS and HTTP to support social change, 18, 2013, 22–41, Lombera, I. M. et al.)**

distribute that metadata to a subset of participating nodes that are chosen at random, as shown in Figure 6.4b. The metadata are distinct from the information that they describe and include a list of keywords and the URL of the source of the information.

Other nodes (the requesting nodes) request and retrieve information. Such nodes generate requests (queries) that refer to the metadata and distribute their requests to a subset of the participating nodes that are chosen at random, as shown in Figure 6.4c.

The participating nodes compare the metadata in the requests that they receive with the metadata that they hold. If such a node finds a match, which is called an encounter, the matching node returns the URL of the associated information to the requesting node. The requesting node then uses the URL to retrieve the information from the source node, as shown in Figure 6.4d.

Distribution of the metadata and the requests to relatively few nodes suffices to achieve a high probability of a match. Moreover, the strategy is robust. Even if some of the randomly chosen nodes are subverted or nonoperational, the probability of a match is high. Moreover, it is not easy for a small group of nodes to subvert the iTrust mechanisms to censor, filter, or subvert information.

6.3.4.3 Search and Retrieval for iTrust over SMS

Search and retrieval of resources involves four types of iTrust over SMS messages: two query messages and two response/retrieval messages. These four types of messages are described in the following sections. Examples are given of message passing between nodes, and then there is an example of actual SMS text messages sent for search and retrieval of resources.

6.3.4.3.1 SEND QUERY

The SEND QUERY message is used to query a node for resources in the iTrust network. The message contains three parameters: call number, query id, and query text. The call number is the mobile phone number of the node that is issuing the query. The query id is any text string that nodes in the iTrust membership use to track the query. The same query id is used for all four iTrust over SMS messages pertaining to the search request; in effect, it is a global identifier. The query text should be checked by the user application to ensure that it is within the Global System for Mobile Communications (GSM) default alphabet table.

If a node is originating the query, it creates these three parameters and then sends the message. If a node is relaying the query, it relays the message without modification; it can use the query id to prevent relaying the same message more than once (and prevents network flooding). On receiving a SEND QUERY message, a node immediately checks whether an encounter has occurred by comparing the query text against its available resources. If an encounter has indeed occurred, it sends a NOTIFY MATCH message; otherwise, it takes no further action.

6.3.4.3.2 NOTIFY MATCH

When a node has an encounter, it responds to the original querying node with a NOTIFY MATCH message. The message is sent directly to the call number in the first parameter of the SEND QUERY message; it is not sent to the node that relayed the query. The NOTIFY MATCH message contains three parameters: the source phone number, the query id, and the resource id. The query id is the same as that in the SEND QUERY message; again, it is a global identifier for the query and may be used by the application for various purposes. For example, an application might ignore a query that it did not originate to protect against rogue nodes that send spurious NOTIFY MATCH messages.

If the resource is stored locally, the source phone number is the mobile phone number of the node at which the encounter occurred (i.e., the node that received the SEND QUERY message and is about to send the NOTIFY MATCH message), and the resource id is from the local resource table. If the resource is not stored locally, the source phone number is the mobile phone number of the node where the resource is stored, and the resource id is from the resource table of that node.

Receiving the NOTIFY MATCH message requires relatively little processing. Because the node that sent the NOTIFY MATCH message did the processing required, finding the node on which the resource is located and set the message parameters accordingly, the only required action is to save the values for further processing before discarding the message. The user or application can decide when to retrieve the message at the source phone number; retrieval of the document is not mandatory and is done at the convenience of the user or the application, using the REQUEST RESOURCE message.

6.3.4.3.3 REQUEST RESOURCE

When a node wants to retrieve a particular resource, it directly contacts the node that stores the resource, using the REQUEST RESOURCE message. The message contains three parameters: the string literal now, the query id, and the resource id of the stored resource on the receiving node. Although the query id is not strictly needed in this case (it is possible that the associated SEND QUERY message was never relayed to the receiving node), it is still sent for possible use by the application. On receiving a REQUEST RESOURCE message, a node immediately looks up the resource using the resource id and sends it using the SEND RESOURCE message. If the resource id does not exist in its table, the node ignores the message and stops processing.

6.3.4.3.4 SEND RESOURCE

When a node receives a REQUEST RESOURCE message, it immediately gets the resource data and packages it for transmission in the SEND RESOURCE message. The SEND RESOURCE message has three parameters: the string literal data, the query id, and the data itself. Again, the query id is sent only for optional tracking by the user application that interfaces with iTrust over SMS. Transmitted data in the third parameter of the SEND RESOURCE message can be in any format suitable for the application as long as it fits within the GSM default alphabet table. The iTrust over SMS API provides several convenience functions for inserting (extracting) plain text data into (from) the message, which make sending (receiving) plain text trivially simple. To send (receive) custom data apart from plain text, the user application simply needs to escape (unescape) that custom data.

Note that only the original querying node is involved with each of the four message types. Intermediate nodes may relay queries or send notifications of a match, but their involvement ends immediately thereafter.

6.3.4.4 Performance Evaluation of iTrust

The diagrams in Figure 6.5 depict the flow of messages between nodes during the search and retrieval of resources in the iTrust over SMS network. Parts A, B, and C represent independent interactions; messages in the three parts do not follow or precede one another chronologically. Node S is the source node that has the resources locally stored; node Q is the querying node that sends the original search query; and nodes Z and Y are other nodes in the iTrust network. Again, the lines and arrows show the directions and destinations of the messages; the numbers preceding the messages signify the order in which the messages are sent.

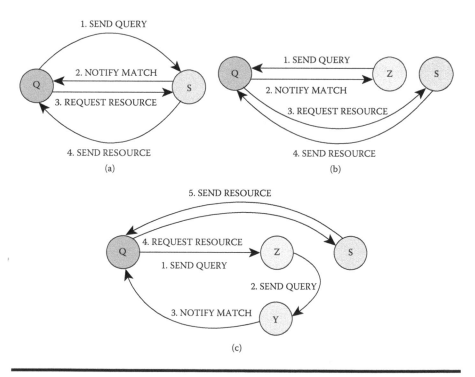

Figure 6.5 Search and retrieval message flow for iTrust over SMS. (With kind permission from Springer Science+Business Media: *Mobile Network and Applications*, Mobile decentralized search and retrieval using SMS and HTTP to support social change, 18, 2013, 22–41, Lombera, I. M. et al.)

6.3.4.4.1 Part A: Search and Retrieval between Two Nodes

Node Q sends a SEND QUERY message to node S. S has an encounter and responds to Q with a NOTIFY MATCH message. When it is convenient, Q sends a REQUEST RESOURCE message to S. On receiving the REQUEST RESOURCE message, S sends the resource to Q in the SEND RESOURCE message.

6.3.4.4.2 Part B: An Intermediate Node Has an Encounter

At some prior time, node S distributed metadata to node Z. Node Q sends a SEND QUERY message to Z; Z immediately has an encounter as a result of Q's query and the metadata distributed by S. Z sends a NOTIFY MATCH message to Q. When convenient, Q then sends a REQUEST RESOURCE message to S. S sends the resource to Q in the SEND RESOURCE message.

6.3.4.4.3 Part C: A Search Query Is Relayed

At some prior time, node S distributed metadata to node Y. Node Q sends a SEND QUERY message to node Z as shown by message 1. Z does not have a match but relays the SEND QUERY message to Y as shown by message 2. Y immediately has an encounter between Q's query and the metadata distributed by S and sends a NOTIFY MATCH message to Q. At its convenience, Q sends a REQUEST RESOURCE message to S. S sends the resource to Q in the SEND RESOURCE message.

6.4 Conclusion

This chapter describes some challenges and several typical schemes about mobile search and ranking on MSNs. The challenges mainly consist of technological and socioeconomic factors. Several typical schemes are respectively summarized. Specifically, PQMPMS can provide personalized answers by inferring user preferences from dialog history and can explore the ratings from the social users given POIs to determine the ranking of answers. SSB incorporates social networking capabilities with key mobile contexts to improve the search and information discovery experience of mobile subscribers, providing users with proactive access to interesting content. SNDocRank incorporates both the term-document similarity and the actor similarity in associated social networks to rank search results to fit the interests of searchers. To deal with the complexity of similarity computation in large social networks, it has developed a MAS method for the SNDocRank framework. The iTrust over SMS system enables SMS-capable mobile phones to communicate directly over the cellular telephony network to distribute metadata and to search for and retrieve information. Information stored locally on any iTrust over SMS mobile device can be sent directly to another such mobile device by instant messages; Internet access is not required. In the iTrust over SMS network, mobile devices can send information to each other using different platforms, such as Android, iOS, and so on, as long as each platform implements the iTrust over SMS protocol.

In mobile search and ranking, personalization will play an important role in how search results are tailored to a user's profile and in evolving media behavior. Furthermore, socially driven search will segment users according to their interests and affiliations. Arguably, no algorithms can replace the objectivity of a human being: let the user define the sites they feel are relevant, leverage their social networks, and—over time—see their results become highly personalized.

References

1. Bacigalupo, M., S. G. Nikolov, and R. Compañó. Is mobile search finally going mainstream? An attempt to reconcile expert views. In: 2010 NEM Summit Proceedings, Barcelona, Spain, October 13–15, pp. 1–6.
2. Gómez-Barroso, J. L. et al. *Prospects of Mobile Search*. Technical report EUR 24148 EN. Institute for Prospective Technological Studies (IPTS), European Commission, Seville, Spain, 2010.
3. Taptu. Exploring the touch-friendly web. White paper. Cambridge, UK, 2009.
4. Westlund, O., J.-L. Gomez-Barroso, R. Compano, and C. Feijoo. Exploring the logic of mobile search. *Behaviour & Information Technology* 2011; 30(5): 691–703.
5. Hu, B., Y. Vanrompay, and M.-A. Aufaure. PQMPMS: A preference-enabled querying mechanism for personalized mobile search. In: Proceedings of the 7th International Conference Web Reasoning and Rule Systems (LNCS 7994), Mannheim, Germany, July 27–29, 2013.
6. Le-Phuoc, D., J. X. Parreira, and M. Hauswirth. Linked stream data processing. In: Eiter, T. and Krennwallner, T. (eds.), *Reasoning Web*. 2012. Vol. 7487. LNCS, Springer, Heidelberg, pp. 245–289.
7. Siberski, W., J. Z. Pan, and U. Thaden. Querying the semantic web with preferences. In: Proceedings of ISWC, Athens, GA, November 5–9, 2006; pp. 612–624.
8. Barbieri, D. F., D. Braga, S. Ceri, V. E. Della, and M. Grossniklaus. Incremental reasoning on streams and rich background knowledge. In: Aroyo, L., Antoniou, G., Hyvönen, E., ten Teije, A., Stuckenschmidt, H., Cabral, L., and Tudorache, T. (eds.), *ESWC*. 2010, *Part I*. Vol. 6088. LNCS, Springer, Heidelberg, pp. 1–15.
9. Polo, L., I. Mínguez, D. Berrueta, C. Ruiz, and J. Gómez. User preferences in the web of data. *Semantic Web Journal* 2013; 5(1): 67–75.
10. Barbieri, D. F., D. Braga, S. Ceri, V. E. Della, and M. Grossniklaus. Querying RDF streams with C-SPARQL. *SIGMOD Record* 2010; 39(1): 20–26.
11. Dong, A. et al. Time is of the essence: Improving recency ranking using twitter data. In: Proceedings of the 19th International Conference on World Wide Web, Raleigh, North Carolina, April 26–30, 2010, pp. 331–340.
12. Church, K., J. Neumann, M. Cherubini, and N. Oliver. Social search browser: A novel mobile search and information discover tool. In: Proceedings of the 15th International Conference on Intelligent User Interfaces, Hong Kong, China, February 07–10, 2010.
13. Cherubini, M. Annotations of maps in collaborative work at a distance. PhD thesis, Swiss Federal Institute of Technology (EPFL), Lausanne, Switzerland, 2008.

14. Jones, M., G. Buchanan, R. Harper, and P. L. Xech. Questions not answers: A novel mobile search technique. In: Proceedings of CHI, San Jose, California, April 28–May 3, 2007.

15. Heath, T. Information-seeking on the web with trusted social networks: From theory to systems. PhD thesis, The Open University, Milton Keynes, UK, 2008.

16. Gou, L., X. Zhang, H.-H. Chen, J.-H. Kim, and C. L. Giles. Social network document ranking. In: Proceedings of the 10th Annual Joint Conference on Digital Libraries (JCDL), Gold Coast, Australia, June 21–25, 2010.

17. Salton, G. and C. Buckley. Term-weighting approaches in automatic text retrieval. *Information Processing and Management* 1988; 24(5): 513–523.

18. McPherson, M., L. Smith-Lovin, and J. Cook. Birds of a feather: Homophily in social networks. *Annual Review of Sociology* 2001; 27(415): 444–465.

19. Wasserman, S. and K. Faust. *Social Network Analysis: Methods and Applications.* Cambridge University Press, MA, 1994.

20. Lorrain, F. and H. C. White. Structural equivalence of individuals in social networks. *Journal of Mathematical Sociology* 1971; 1: 49–80.

21. Leicht, E. A., P. Holme, and M. E. J. Newman. Vertex similarity in networks. *Physical Review E* 2006; 73(2): 026120.

22. Clauset, A., M. E. J. Newman, and C. Moore. Finding community structure in very large networks. *Physical Review E* 2004; 70(6): 066111.

23. Newman, M. E. J. and M. Girvan. Finding and evaluating community structure in networks. *Physical Review E* 2004; 69(2): 026113.

24. Jones, K. S., S. Walker, and S. E. Robertson. A probabilistic model of information retrieval: Development and comparative experiments. *Information Processing and Management* 2000; 36(6): 779–808.

25. Schein, A. I., A. Popescul, L. H. Ungar, and D. M. Pennock. Methods and metrics for cold-start recommendations. In: Proceedings of the ACM Conference on Information Retrieval (SIGIR), Tampere, Finland, August 11–15, 2002.

26. Lombera, I. M., Y.-T. Chuang, P. M. Melliar-Smith, and L. E. Moser. Trustworthy distribution and retrieval of information over HTTP and the Internet. In: Proceedings of the 3rd International Conference on the Evolving Internet, Luxembourg City, Luxembourg, June 19–24, 2011.

27. Lombera, I. M., L. E. Moser, P. M. Melliar-Smith, and Y.-T. Chuang. Mobile decentralized search and retrieval using SMS and HTTP to support social change. *Mobile Network and Applications* 2013; 18(1): 22–41.

28. Sohn, T., K. A. Li, W. G. Griswold, and J. D. Hollan. A diary study of mobile information needs. In: Proceedings of the 26th ACM SIGCHI Conference on Human Factors in Computing Systems, Florence, Italy, April 5–10, 2008.

29. Kamvar, M. and S. Baluja. A large scale study of wireless search behavior: Google mobile search. In: Proceedings of the 24th ACM SIGCHI Conference on Human Factors in Computing Systems, Montréal, Canada, April 22–27, 2006.

30. Kamvar, M., M. Kellar, R. Patel, and Y. Xu. Computers and iPhones and mobile phones, oh my!: A log-based comparison of search users on different devices. In: Proceedings of the 18th International Conference on the World Wide Web, Madrid, Spain, April 20–24, 2009.

31. Schusteritsch, R., S. Rao, and K. Rodden. Mobile search with text messages: Designing the user experience for Google SMS. In: Proceedings of the CHI '05 Extended Abstracts on Human Factors in Computing (CHIEA), Portland, OR, April 2–7, 2005.

Chapter 7

Energy-Efficient Mechanisms in Mobile Social Networks

7.1 Introduction

With the great improvements in information and communication technology (ICT), smartphones have become more and more available for people. Meanwhile, an increasing number of applications and services also have become available on mobile devices—such as watching videos, Global Positioning System (GPS) location, map guides, playing games, social network services, and so forth. These applications and services enrich our lives and bring us convenience and fun. In recent years, mobile social networks (MSNs) have become the focus of much attention. For example, LoKast* is an iPhone application that provides mobile social networking services by discovering and sharing media content among users in proximity. Apple's iGroups† allows groups of friends or colleagues attending such events as a concert, a trade show, a business meeting, a wedding, or a rally to stay in communication with each other as a group to share information or reactions to live events as they are occurring.

Unfortunately, the restricted battery power of mobile devices has a serious impact on users' experiences. The amount of energy that can be stored in a typical battery is increasing by only 5% annually [1]. Although there are other factors

---

* http://www.lokast.com/
† http://www.igroupnet.com/

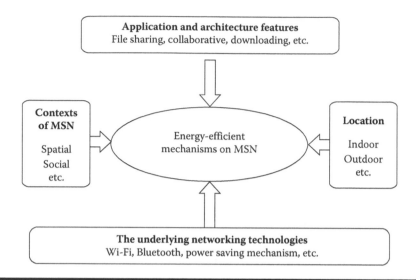

Figure 7.1   Framework of energy-efficient mechanism on mobile social network (MSN).

of importance, energy consumption is one of the most fundamental issues that should be carefully considered by all MSN applications and services. The focus of this chapter is research and applications related to energy-efficient MSN schemes, which are shown in Figure 7.1.

This chapter consists of the following parts. Section 7.2 presents the power saving mechanism (PSM) for mobile location and novel positioning technology. Energy efficiency on the underlying network technologies will be introduced in Section 7.3, which includes information on Wi-Fi, Bluetooth, and power saving models. Context-aware energy-efficient schemes are discussed in Section 7.4. Section 7.5 provides several frameworks that save energy through incorporating special features of applications and architecture in MSNs. Tools for developers to design energy-efficient application are presented in Section 7.6. Finally, conclusions are drawn in Section 7.7.

7.2 Energy Efficiency on Location Technology

7.2.1 Energy Efficiency on Outdoor Location Technologies

Location-based applications (LBAs) are among the most typical applications in MSNs. LBA obtains a user's current position and provides various user position–related services (e.g., social network, health care, mobile commerce, transportation, entertainment). The locating technologies used today mainly include GPS, Wi-Fi, and Global System for Mobile Communications (GSM). These technologies can vary widely in energy consumption and localization accuracy. Experiments have shown that GPS is able to run continuously for only 9 hours, although Wi-Fi and GSM can be sustained for 40 and 60 hours, respectively [2]. At the same time, the corresponding localization accuracies are about 10 m, 40 m, and 400 m, respectively. Today most LBAs prefer GPS for its accuracy, although it is also perceived as extremely power-hungry. What is worse is that phones currently only offer a black box interface to the GPS for the request of location estimates, and the lack of sensor control makes energy consumption more inefficient. In addition, many LBAs require continuous localization over reasonably long time scales. Therefore, energy-efficient location-sensing methods must be adopted to obtain accurate position information while expending minimal energy.

Several methods of energy-efficient locating sensing have been proposed in recent years, which have been proved to be useful (Table 7.1).

EnTracked [3] uses an accelerometer alone to detect movement. It proposes an energy model to dynamically estimate parameters such as delays and consumption, which can describe the power consumption of a real phone with much higher precision. EnTracked estimates the user's speed using the speed and accuracy provided by the GPS module. Then, using these parameters and the device model, EnTracked is able to calculate the point at which to power features (mainly GPS and radio) on and off.

However, this method has several limitations. First, the accelerometer would not be able to power off when EnTracked is running. In some scenarios, the power used by the accelerometer may be higher than occasionally making it up for a simple position update and for calculating a new sleeping period. Second, the movement detection algorithm is not clever and accurate enough; in some cases, this algorithm might be misled.

EnTrackedT [7] extends the EnTracked system in several aspects. It proposes the idea of trajectory tracking corresponding to position tracking in EnTracked. The former refers to a sequence of continuous positions, while the latter focuses on a current position. First, EnTrackedT adopts a heading-aware strategy, which employs the compass as a turn point sensor and significantly reduces power consumption of trajectory tracking. EnTrackedT calculates the accumulated distance traveled orthogonal to the initial heading given by the compass and compares this to the prescribed trajectory error threshold. Intervals between GPS usage can be much larger than in EnTracked. Second, EnTrackedT uses adaptive duty-cycling strategies for the accelerometer and compass sensors, which make the system more efficient. Third, EnTrackedT uses a speed threshold–based strategy together with an accelerometer-based strategy for movement detection. This strategy enables the system to handle different transportation modes (e.g., walking, running, biking, or commuting by a car). Fourth, it explores algorithms of a simplified motion trajectory to reduce data size and communication costs caused by sending motion information.

Table 7.1   Energy-Efficient Location Sensing

Paper	Target	Sensors	Scheme
EnTracked [3]	Position tracking	GPS accelerometer	Dynamic prediction with less power-intensive sensors
EnLoc [4]	Position tracking	GPS, Wi-Fi, GSM, compass, accelerometer	Dynamic selection among alternative location-sensing mechanisms, dynamic prediction with less power-intensive sensors and historical data
LBAs [5]	Position tracking	GPS, GSM, accelerometer	Dynamic selection among alternative location-sensing mechanisms, battery level considering
CAPS [6]	Trajectory tracking	GSM, GPS	Cell-ID sequence matching with historical sequences
EnTrackedT [7]	Trajectory tracking	GPS, compass, accelerometer	Dynamic prediction with less power-intensive sensors, trajectory simplification
Ctrack [8]	Trajectory mapping	GSM, compass, accelerometer, GPS	Building a database with GPS, GSM when training and mapping in database using GSM when working
a-Loc [9]	Position tracking	GPS, Wi-Fi, Bluetooth, cell tower	Dynamically select among alternative location-sensing mechanisms with prediction
RAPS [10]	Position tracking	GPS, accelerometer, Bluetooth, cell tower	Dynamic prediction with less power-intensive sensors

However, the error percentage of the EnTrackedT system is relatively high when the requested error threshold is small, although the power consumption is much lower at the same time. Although EnTrackedT claims to have joint trajectory and position tracking, it seems to work better for trajectory-based applications.

RAPS [10] is based on the observation that GPS is generally less accurate in urban areas. It introduces the concept of activity ratio, which is the fraction of time that the user is in motion between two position updates. It uses an accelerometer to detect movement while measuring the activity ratio at the same time. It then uses this activity ratio along with the history of velocity information to estimate the current velocity of the user. RAPS duty-cycles the accelerometer carefully, using a duty-cycling parameter deduced empirically. A significant portion of the energy savings of RAPS comes from avoiding GPS activation when it is likely to be unavailable in order to use cell tower–received signal strength (RSS) blacklisting. It records the current cell tower ID and RSS information and associates with the success or failure of GPS. In addition, RAPS uses Bluetooth to share the newly updated position information to save more energy. RAPS uses a combination of spatiotemporal location history, user activity, and cell tower RSS blacklisting, and it also proposes sharing position readings among nearby devices (which is a different approach from the former two options). However, it has limitations as well. First, RAPS is mainly designed for pedestrians in urban areas. Second, the user space-time history and the cell tower RSS blacklist must be populated for RAPS to work efficiently. Third, its velocity estimation based on activity ratio can be misled by handset activity not related to human motion. Fourth, accelerometers on smartphones may need a one-time, per-device calibration of the offset and scaling before running RAPS. Moreover, context sharing using Bluetooth raises privacy and security concerns.

CAPS [6] presents a cell-ID-aided positioning system based on the consistency of traveled routes and on consistent cell-ID transition points. It stores the history of cell-ID and GPS position sequences, and then it senses the cell-ID sequences to estimate the current position using a cell-ID sequence-matching technique. According to the observation, for mobile users with consistent routes, the cell-ID transition point for each user can often uniquely represent the current user position. CAPS consists of three core components—sequence learning, sequence matching and selection, and position estimation. CAPS opportunistically learns and builds the history of a user's route for future usage. Using a small memory footprint, CAPS maintains the user's past routes and triggers GPS, if necessary. For each cell-ID in a sequence, CAPS maintains a list of 2-tuple as < position, time stamp >, where position represents a GPS reading, and time stamp is the time at which that reading was taken. It uses a modified Smith–Waterman algorithm for cell-ID sequence matching. CAPS is designed for highly mobile users who travel long distances in a predictable fashion. It will not work in some cases where GPS is not available (such as indoors), and the size of the historical database may be very large if the user travels much. Also, it is evaluated only in urban areas where cell tower density is high. At the time of this writing, CAPS does not make use of the underlying geography.

EnLoc [4] also explored how to make use of the spatiotemporal consistency in user mobility. When exploiting habitual mobility, EnLoc uses the logical mobility tree (LMT) to record the person's actual mobility paths. The vertices of the LMT are also referred to as uncertainty points. The basic idea is to sample the activity at a few uncertainty points, and EnLoc predicts the rest. The scheme mentioned here highly relies on, as well as limits, the spatial–temporal consistency in user mobility. It cannot handle users' deviations from habits. So EnLoc further exploits mobility of large populations as a potential indicator of the individual's mobility.

EnLoc hypothesizes that a "probability map" can be generated for a given area from the statistical behavior of large populations. Then an individual's mobility in that area can be predicted. For example, considering a person approaching a traffic intersection of Street A, because the person has never visited this street, it is difficult to predict how he or she will behave at the imminent intersection. However, if most people take a left turn to Street B, the person's movement can be inferred accordingly. EnLoc is evaluated using traces collected from a University of Illinois at Urbana-Champaign (UIUC) campus, which is not representative of EnLoc's actual service territory. In addition, it does not describe the detailed implementation. These two issues suggest room for improvement. However, there is also the potential for the heuristic prediction in energy saving.

Appealing to the requirement of energy savings, many approaches of energy-efficient locating sensing have been explored. Methods beyond the action of locating are somehow auxiliary, and most of the attentions are focused on locating sensing–based methods. A class of lightweight positioning systems has been developed to explore a large part of the energy-accuracy trade-off space. These systems either reduce accuracy requirements or aggressively use other cues to determine when and where to turn on GPS. Implicitly or explicitly, these systems generally make several assumptions about the environment or about user activity.

7.2.2 Energy Efficiency on Indoor Location Technologies

Due to GPS limitations in indoor environments, many indoor localization systems using Wi-Fi received signal strength indicator (RSSI) updates, GSM signal readings, short-range scans via Bluetooth, and infrared, Ultra-wideBand (UWB), or Radio Frequency IDentification (RFID) readings were introduced [11]. Wi-Fi-based indoor localization continues to be one of the most attractive techniques due to its reliance on ubiquitously deployed infrastructure.

GreenLoc [12] proposed an energy-efficient indoor localization architecture that leverages mobile sensors, short-range wireless technologies, and group mobility patterns. GreenLoc aims to minimize the overall energy consumed on localization services by filtering out unnecessary wireless measurements, with an acceptable trade-off in accuracy. In addition, it is designed to be easily integrated with any Wi-Fi-based indoor localization system. GreenLoc allows for the adoption of various energy reduction strategies based on sensed data at the client's side including RSSIs, accelerometers, and neighboring devices detected via short-range wireless technologies. Ultimately, based on the strategy invoked, the best localization sensing rate of the client could be computed.

And furthermore, a proximity-based clustering algorithm (CLoc) is designed and implemented as a representative strategy integrated with GreenLoc. CLoc uses low energy wireless technologies, such as Bluetooth, to detect and cluster individuals moving together. It then assigns a group representative to act as a designated cluster head (CH) that would be constantly tracked. The location of other group members is

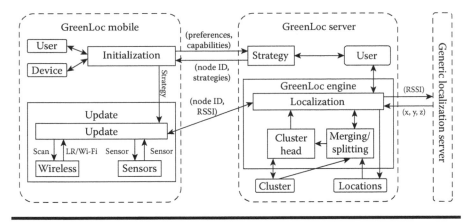

Figure 7.2   GreenLoc system architecture.

then inferred so long as they remain within proximity of the corresponding CH. CLoc dynamically handles the merger or splitting of clusters as a result of mobility.

GreenLoc is an energy-efficient architecture that communicates with an existing generic localization server as shown in Figure 7.2. GreenLoc is a client-server system connecting a server to multiple mobile clients running an indoor localization application. It then acts as a proxy that connects a mobile indoor localization application to its generic server. Its goal is to mainly filter out unnecessary wireless measurements and consequently reduce the overall energy consumption on the mobile devices. GreenLoc also allows the generic localization server to act more efficiently because the server will only be tracking a subset of the nodes that need to be localized. In general, mobile clients connect to the GreenLoc server and send different sensed data (RSSIs, accelerometers, neighboring devices, etc.). The server receives such data and decides, according to one or many strategies, whether the client should keep sensing, reduce the sensing frequency, or stop sensing. The server dynamically delegates a node representative to act as a designated CH. The location of other group members can be inferred so long as they remain within proximity of the CHs.

The details of the client and server components, the interactions between them, and the system operation for GreenLoc are detailed here.

1. GreenLoc Server: It is responsible for communicating with its registered clients. It stores and maintains information about clients' locations, localization strategies, and the set of current selected CHs. The GreenLoc server consists of two main components: the strategy determinator and the GreenLoc engine.
   a. Strategy Determinator: New mobile clients contact the strategy determinator in order to request the strategy that best performs within a given area. It collects data about the network capabilities and client preferences to notify the new clients about the running strategy.

b. GreenLoc Engine: The GreenLoc engine dynamically detects a group of devices moving together and assigns a group representative to act as a designated CH that would be constantly tracked (CH selector). Moreover, it handles the merging or splitting of clusters as a result of mobility. The GreenLoc engine consists of three main components: localization updater, CH selector, and merging/splitting engine.

The localization updater controls the two other modules. It receives RSSI updates from the clients and, based on the strategy used, triggers the merging/splitting engine and forwards the RSSIs to the generic localization server. Upon receiving the client location, it updates the corresponding database. The merging/splitting engine detects group modification or formation and asks the CH selector to cluster a set of nodes and elect a CH. It will then notify the CH clients that they should keep sending RSSI updates while other group members remain idle.

2. GreenLoc Client: Mobile clients receive explicit requests from the server to send RSSI data and other sensing information—such as neighboring devices detected via short-range technology (e.g., Bluetooth). Mobile clients have two main components: initialization module and update dispatcher.
   a. Initialization Module: The initialization module is responsible for registering new nodes with the GreenLoc server. Whenever a new node needs to join GreenLoc, it sends its user preferences and device capabilities to the GreenLoc server. The server responds with a registration ID with which the node would be identified at the server, and informs the client of the strategy to be used to conduct the localization process. This strategy choice will dictate which sensory data the client needs to continue to send and which ones it can halt.
   b. Update Dispatcher: The update dispatcher receives the strategy used from the initialization module and triggers the corresponding sensor readings and algorithms to perform RSSI measurements. It then sends the RSSI measurements to the server. For instance, it can trigger the motion detection readings to compute the mobility state of the mobile client or Bluetooth scans to identify the set of neighboring devices in the vicinity. These readings are collected by the update dispatcher, processed, and sent to the GreenLoc server, which then uses them to estimate the client's location.

7.3 The Underlying Networking Technology

7.3.1 Power Saving Mechanisms for Mobile Stations

A mobile station can provide users with a variety of broadband multimedia and data services such as voice/video communications, Web browsing, video services, augmented services, location-based services, and social network services. While a user enjoys such services actively, the mobile station (MS) may be put in idle state without receiving or transmitting any traffic from or to a Base Station (BS) most of time. For example, when an individual uses a voice communication service actively, there is a silent time that the user does not talk and so no traffic is transmitted or received during the time. Another example is when a user looks through an article or contents from a Web page—there is traffic downloading time and idle time when the user read contents from the page without receiving any data. PSMs attempt to save the battery power of an MS during idle time (i.e., when the MS does not actively operate).

Basic operation of a PSM consists of two modes: wake mode and sleep mode. In wake mode, an MS actively transmits/receives traffic to/from its serving BS. In sleep mode, an MS does not transmit or receive traffic, keeps its battery usage as low as possible, and waits for service requests. To optimize the battery usage, it is essential to control the transition between wake mode and sleep mode with careful consideration of the states of an MS. An MS may go into sleep mode from wake mode according to the following two cases. When an MS itself does not use any services and wants to be in sleep mode, it sends a BS a sleep request message (MOB_SLP_REQ) with operating parameters that control sleep mode operation. After the BS receives the MOB_SLP_REQ message, it sends the MS a sleep response message (MOB_SLP_REQ) with required information for sleep mode. After the MS receives the MOB_SLP_RSP message, the MS begins sleep mode. This MS initiation of sleep mode is illustrated in Figure 7.3. On the contrary, when a BS wants an MS to be in sleep mode, it sends the MS an MOB_SLP_RSP message. This is in the case of BS initiation of sleep mode. In either case, after the MS receives the MOB_SLP_RSP message, the MS starts sleep mode. The MOB_SLR_RSP message is sent from the BS to an MS on its basic connection identifier (CID) in response to the MOB_SLP_REQ message or broadcast CID, or it may be sent unsolicited.

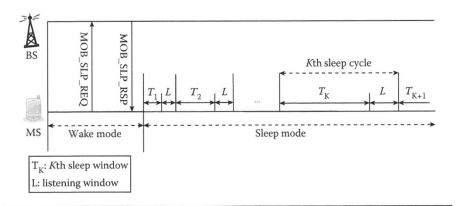

Figure 7.3 Sleep-mode operation initiated by a mobile station.

Sleep mode contains a group of sleep cycles. Each sleep cycle consists of a listening window and a sleep window. In sleep mode, an MS temporarily wakes up in a very short time (listening window) to listen to any service requests (MOB_TRF-IND message) from its serving BS. This message indicates whether there was any traffic addressed to the MS during the previous sleep cycle. If there was no request, the next sleep window follows. Likewise, an MS periodically transits between listening window and sleep window until the sleep mode ends. During sleep mode operation, there are two key operating parameters—the minimum (or initial) sleep window and the maximum (or final) sleep window. The former is the length of a sleep window in the first sleep cycle, and the latter limits the maximum length of a sleep window in sleep mode. The manipulation of the two parameters can be variable for the application services and the purpose of PSMs, which will be discussed in detail in the following section. An MS powers down its major components in order to save battery energy as much as possible during sleep cycles.

7.3.2 Adaptive Power Saving Mechanism for 802.16e

The Institute of Electrical and Electronics Engineers (IEEE) 802.16e standard, which specifies mobile worldwide interoperability for microwave access (WiMAX), is an extension of WiMAX technology enabling fixed and mobile convergence through broadband wireless access (BWA) technology for providing mobile services. With legacy voice-centric communication services, the traffic pattern is rather simple so that the PSM sleep mode is well operated. However, in 4G technology for various data-centric services, PSM essentially needs to be adaptive to changing traffic condition in order to achieve optimal energy efficiency.

As described earlier, during sleep mode operation, there are two key operating parameters—the minimum sleep window and the maximum sleep window. The manipulation of the two parameters can be variable for the application services and for the purpose of PSMs. The standard document IEEE 802.16e defines the power saving class (PSC) for distinguishing service requirement levels in PSMs [13]. There are three types of PSCs according to target services and different QoS requirements, as is illustrated in Table 7.2.

Table 7.2 Power Saving Classes

Types	Services	Operating Parameters
PSC I	BE (best-effort), NRT-VR (non–real-time variable rate)	Listening window, T_{min}, T_{max}
PSC II	UGS (unsolicited grant service), RT-VR (real-time variable rate)	Listening window, T_{min}
PSC III	Multicast management operation	T_{max}

PSC I is defined for delay-insensitive services such as non–real-time and best-effort (BE) services. In PSC I, when there is no request for service in the listening window, the sleep window doubles from the preceding sleeping window. The process is repeated until the sleep window reaches the maximum sleep window (T_{max}). The duration of kth sleep window (T_k) is given by:

$$T_k = \begin{cases} T_{\min} & k = 1 \\ \min\left(2^{k-1}T_{\min}, T_{\max}\right), & k > 1 \end{cases} \qquad (7.1)$$

where $T_{\min}$ is the minimum sleep window. PSC II is defined for UGS-supporting real-time applications generating fixed-size packets on a periodic basis and real-time variable-rate (RT-VR) service. Because these services require guaranteed service quality, the sleep window is constantly set to the minimum sleep window in PSC II. It sacrifices energy saving to reduce the response delay for delay-sensitive services. The final PSC, PSC III, is reserved for multicast connections and management purposes. An MS automatically switches over from sleep mode to wake mode without any requests to initiate wake mode because the period of the sleep state without the listening state is predefined. That is, after the expiration of the sleep window, PSC automatically becomes inactive. The sleep window is set to the $T_{\max}$ in PSC III.

7.3.3 Power Saving Mechanism for Wi-Fi Direct

Wi-Fi Direct, initially called Wi-Fi P2P, is a Wi-Fi standard that enables devices to connect easily with each other without requiring a wireless access point and allows them to communicate at typical Wi-Fi speeds for everything from file transfer to Internet connectivity. One advantage of Wi-Fi Direct is the ability to connect devices even if they are from different manufacturers. Only one of the Wi-Fi devices needs to be compliant with Wi-Fi Direct to establish a peer-to-peer connection that transfers data directly from one to the other with greatly reduced setup. Wi-Fi Direct negotiates the link with a Wi-Fi Protected Setup (WPS) system that assigns each device a limited wireless access point. The "pairing" of Wi-Fi Direct devices can be set up to require the proximity of a near-field communication, a Bluetooth signal, or a button press on one or all the devices. Wi-Fi Direct may not only replace the need for routers but may also replace Bluetooth's need for applications that do not rely on low energy.

Using Wi-Fi Direct, battery-constrained devices may typically act as peer-to-peer group owner (P2P GO) (a software-enabled access point, or soft AP); therefore, energy efficiency is of capital importance. However, PSMs in current Wi-Fi networks are not defined for APs but only for clients. Notice that with Wi-Fi Direct, a P2P client can benefit from the existing Wi-Fi power saving protocols, such as legacy power save mode [14] or Unscheduled Automatic Power Save Delivery (UAPSD) [15]. In order to support energy savings for the soft AP, Wi-Fi Direct defines two new PSMs: the opportunistic power save (OPS) protocol and the notice of absence (NoA) protocol.

7.3.3.1 Opportunistic Power Save

The basic idea of OPS is to leverage the sleeping periods of P2P clients. The mechanism assumes the existence of a legacy power-saving protocol and works as follows. The P2P GO advertises a time window, denoted as the CTWindow, within each beacon and probe response frames. This window specifies the minimum amount of time after the reception of a beacon during which the P2P GO will stay awake; therefore, P2P clients in power saving can send their frames. If after the CTWindow the P2P GO determines that all connected clients are in doze state, either because they announced a switch to that state by sending a frame with the power management (PM) bit set to 1 or because they were already in the doze state during the previous beacon interval, the P2P GO can enter sleep mode until the next beacon is scheduled; otherwise, if a P2P client leaves the power saving mode (which is announced by sending a frame with the PM bit set to 0), the P2P GO is forced to stay awake until all P2P clients return to power saving mode. Figure 7.4 provides an example of the operation of the OPS protocol for a scenario consisting of one P2P GO and one P2P client. Notice that, using this mechanism, a P2P GO does not have the final decision on whether to switch to sleep mode or not because this depends on the activity of the associated P2P clients. To give a P2P GO higher control of its own energy consumption, Wi-Fi Direct specifies the NoA protocol, which is described next.

7.3.3.2 Notice of Absence Protocol

Protocol allows a P2P GO to announce time intervals, referred to as absence periods, where P2P clients are not allowed to access the channel, regardless of whether they are in power save or in active mode. In this way, a P2P GO can autonomously decide to power down its radio to save energy. Such as in the OPS protocol, in the case of a NoA, the P2P GO defines absence periods with a signaling element included in beacon frames and probe responses. In particular, a P2P GO defines a NoA schedule using four parameters: (i) duration that specifies the length of each absence period, (ii) interval that specifies the time between consecutive absence periods,

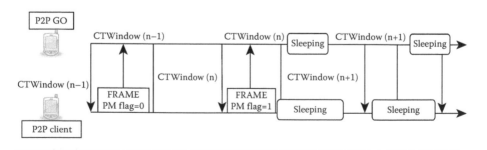

Figure 7.4   Example of opportunistic power save.

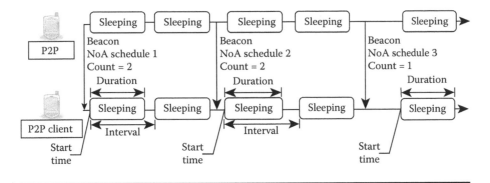

Figure 7.5 Example of notice of absence.

(iii) start time that specifies the start time of the first absence period after the current beacon frame, and (iv) count that specifies how many absence periods will be scheduled during the current NoA scheduled. A P2P GO can either cancel or update the current NoA schedule at any time by respectively omitting or modifying the signaling element. P2P clients always adhere to the most recently received NoA schedule. Figure 7.5 depicts an example operation of the NoA protocol. In order to foster vendor differentiation, the Wi-Fi Direct specification does not define any mechanism to compute the CTWindow in the OPS protocol or the schedule of absence periods in the NoA protocol.

7.3.4 Energy Efficiency on Bluetooth

Opportunistic communications have been widely explored in delay-tolerant networks, mobile social applications, and mobile advertising, to facilitate message forwarding, media sharing, and location-based services. Device discovery is essentially the first step of opportunistic communications. Under the condition of no equipment around, the average power of Bluetooth inquiry is ~162 mw, and during inquiry intervals, the power is ~16 mw. On the contrary, as shown in Table 7.3, the power of Wi-Fi inquiry is ~836 mw, and during inquiry intervals, the power is ~791 mw [16]. Obviously, it is more energy efficient to choose Bluetooth as the wireless interface to discover mobility devices instead of Wi-Fi. Bluetooth operates in the 2.4 GHz industrial, scientific, and medical (ISM) frequency band that is shared with other devices such as IEEE802.11 stations, baby monitors, and microwave ovens. Bluetooth has 79 frequency bands (1 MHz width) in the range 2402–2480 MHz.

Suppose the power is P_{idle} for the idle state and P_{probe} for the inquiry state of Bluetooth devices, the duration of Bluetooth inquiry is T_{probe} and the inquiry interval is P_{idle}. Then the estimated energy consumption is: $E = T_{\text{idle}} \cdot P_{\text{idle}} + T_{\text{probe}} \cdot P_{\text{probe}}$.

Table 7.3 Average Power (in mW) of Bluetooth and Wi-Fi Device Discovery

	P_{idle}	P_{probe}
Bluetooth	16.54	253.05
Wi-Fi	791.02	836.65

In the traditional Bluetooth protocol, T_{idle} and T_{probe} is constant. As a result, the efficiency of the traditional Bluetooth protocol is low. As seen from the formula, by adjusting the T_{idle} and T_{probe} we can reduce the power consumption of Bluetooth. eDiscovery, an energy-efficient device discovery protocol dynamically adjusts these two parameters based on the number of surrounding devices. There are four important parameters in eDiscovery: inquiry window and inquiry interval, which control the duration and interval of Bluetooth inquiry, N the initial number of devices and num_responses the number of devices that have searched. Energy is saved by comparing N and num_responses and dynamically adjusting T_{idle} and T_{probe}.

7.4 Energy Efficiency on Mobile User Context

7.4.1 Sensors for Detecting User Context

The tremendous growth of sensor technology in smartphones increases daily, and even greater growth will be seen over the next few years. The success of smartphones is leading to an increasing amount of Micro-Electro-Mechanical System (MEMS) and sensors in mobile phones, which provide new features/services to end users that reduce cost through more integration or that improve hardware performance. We provide a brief and basic discussion about these sensor technologies here, which are listed in Figure 7.6.

7.4.1.1 Ambient Light Sensor

An ambient light sensor (ALS) in portable devices such as tablets, smartphones, and laptops extends battery life and enables easy-to-view displays that are optimized to the environment. According to the spectral sensitivity of light measurement in human eyes, modern ALSs attempts to live with an incomplete match to the photonic international commission on illumination (CIE) curve [17]. They instead use the principle of superposition to calculate the ambient light brightness. Most light sensors on the market today use two or more different types of photodiodes, each sensitive to a different portion of the light spectrum. By combining these photodiode outputs mathematically, each with a suitably adjusted gain, the sensor can be made to output a fairly accurate measurement of ambient brightness for the light sources

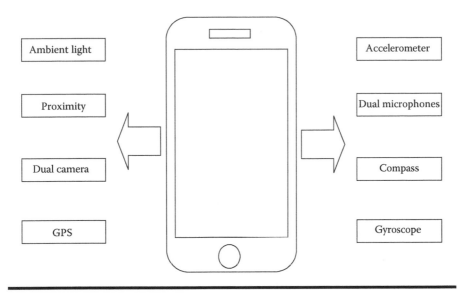

Figure 7.6 Smartphone with different sensors.

commonly available. Basically, an ALS adjusts the display brightness, which in turn saves battery power in a smartphone; it saves power by adjusting the brightness of the display based on how much ambient light is present. As an example, providing ambient light sensing and proximity detection operation in environments over 60,000 lux (sunlight), TSL2x72 series creates a dynamic range of operation necessary for mobile products to adapt to all lighting conditions. Dynamic range is enabled by a device's programmable gain modes, including reduced gain mode. Proximity detection features include signal-to-noise ratio (SNR) performance and a programmable register, which allow compensation for optical signal cross talk.

7.4.1.2 Proximity Sensor

A proximity sensor is very useful in a smartphone. It detects how close the screen of the phone is to your body. This allows the phone to sense when you have brought the phone up to your ear. At that point, the display turns off in order to save the battery. It also stops detecting touches, so as to avoid unwanted input, until you take the phone away from your ear. The proximity sensor can turn off the screen to avoid accidentally activating the screen if it is touched by the ear. It is also useful for detecting towers and sources of interference so that one can amplify them or filter them out using beam-forming techniques. In the iPhone, the proximity sensor shuts off the screen and touch-sensitive circuitry when the iPhone is brought close to the face, both to save battery and to prevent unintentional touches. The proximity sensor in a smartphone senses how close the phone is to the user's face so that it can pause whatever activity it is in the middle of (playing music or browsing the Web, for example) so the user can take a phone call. When the phone is removed from the ear after the call, the phone resumes its previous activity.

7.4.1.3 Global Positioning System

GPS was originally intended for military applications, but in the 1980s, the government made the system available for civilian use. It is a navigation tracking system, often with a map "picture" in the background. It shows where you have been and allows "routes" to be preprogramed, giving a line you can follow on the screen of a smartphone. GPS satellites circle the Earth twice a day in a very precise orbit and transmit signal information. GPS receivers take this information and use triangulation to calculate the user's exact location. As an example, iPhone models use A-GPS (or "Assisted GPS"), which accesses an intermediary server when it is not possible to connect directly via satellite (indoors, for example), and this server provides the nearest satellite with additional information to make it possible to more accurately determine a user's position. The iPhone 3G, 3GS, and 4 employ A-GPS, and the iPhone 3GS and 4 also have a digital compass. iPhone 4S supports GLONASS global positioning system in addition to GPS.

7.4.1.4 Accelerometer

The accelerometer allows the smartphone to detect the orientation of the device and adapts the content to suit the new orientation. For example, when you rotate your device sideways, the Web browser automatically switches the screen to landscape mode so that you now have a wider viewing space.

7.4.1.5 Compass

Basically compasses are attracted to the Earth's poles using magnets. But the modern smartphone is not using magnets. Magnetic interference would render the smartphone's cellular capabilities useless. Once introduced, magnetic interferences drop signal strength parabolically. Frequency and ranges are consistent with GSM bands. As an example, the AK8973 chip on the iPhone is a small sensor that "listens" for an ultra-low frequency signal. If that signal comes from a specific spot such as from the north, paired with the accelerometer the device can calculate the orientation and direction.

7.4.1.6 Gyros

A gyroscope is a device for measuring or maintaining orientation, based on the principles of angular momentum. Gyroscopic sensors are used in navigation systems and gesture recognition systems in smartphones and tablet PCs. Gyroscopes are used in smartphones and tablet PCs for finding the position and orientation of devices.

7.4.1.7 A New Sensor—Back-Illuminated Sensor

A back-illuminated sensor, also known as a backside illumination (BSI or BI) sensor, is a type of digital image sensor that uses a novel arrangement of the imaging elements to increase the amount of light captured and thereby improves low-light performance. The technique was used for some time in specialized roles such as in low-light security cameras and astronomy sensors but was complex to build and required further refinement to become widely used.

As we see the application and programs of sensor technology increase geometrically in most smartphones, this rapid growth may give us a new direction with a smart planet in our hand.

7.4.2 User Context

Context is any information that can be used to characterize the situation of an entity. An entity is a person, place, or object that is considered relevant to the interaction between the user and the application, including the user and applications themselves. The term "context-aware" appeared in Ref. [18] for the first time, when the authors described context as "location, identities of nearby people, objects, and changes to those objects."

On one hand, context consists of spatial contexts and social contexts. Spatial contexts are mainly related to position and proximity. They allow context-aware applications to provide input for LBS. For example, an iPhone can obtain location information and provide Foursquare software so that the user can play online games. Social contexts are contexts that have been explored in social network analysis threads. These contexts can facilitate users' interactions in the context of mobile computing.

On the other hand, contexts can also be broadly categorized as network contexts and mobile terminal (MT) contexts (see Figure 7.7) [19].

Five categories of MT context information have been identified:

- Context information about device capabilities is composed of data describing the device itself, comprising hardware and software (central processing unit [CPU], operating system, memory, display capabilities), with a particular focus on the network interfaces. These data are usually constant over time. Most of the other context data changes over time.
- Mobility context provides information about the current location of the MT and its velocity (comprising speed and direction of mobility). This information is used to predict future location of MTs and can be used to determine if a given vertical handover (VHO) or short-range cooperation can be useful for energy saving. Moreover, a network can also use this context information about MT mobility to predict its own load, based on the MT going in and out of its coverage.

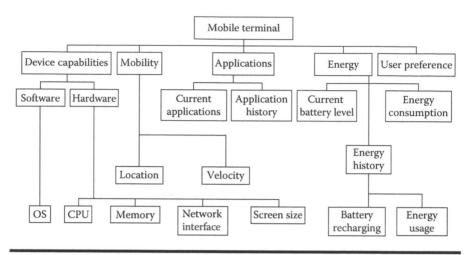

Figure 7.7 Context framework related to mobile terminal.

- The application category contains information about applications currently in use by the MT (the component of "current applications" shown in the Figure 7.7), which determine the QoS requirements. Moreover, a log of application history is maintained to profile the common use of the MT and to predict the behavior of the user on a short-term basis.
- The system we describe focuses on energy saving and, in fact, an important set of context information regarding the energy of the MTs. The energy category comprises the current battery level, the rate of energy consumption, and the energy history of the terminal. Energy history provides quantitative data on energy usage at different times of the day and on the frequency of battery recharging. The energy information is used to predict the user needs for energy in the short and medium term.
- The last category of context information is user preferences, which are a set of parameters that are defined by the user to specify what the user wants or expects. The user can set his or her priorities to be energy saving, price minimizing, or performance maximizing. These user preferences can be a complex structure that bears data for different scenarios. For example, even for a user mainly concerned with energy saving, his or her priorities can be considered to be temporarily shifted to "performance maximization" if the user is executing an emergency call.

7.4.3 Context-Aware Services

There is a growing desire among telecommunication operators to increase traffic volume by offering value-added services to customers in addition to traditional voice and data communication. Some of these services must be enabled or disabled depending on the actual user context. For example, if the user is driving, then a service could automatically provide an estimate for the rush-hour delay to reach the desired destination, or given certain sound and lighting conditions, a service could be configured not to disturb or interact with the user.

Stuedi et al. [20] presented WhereStore, a location-based data store for smartphones interacting with the cloud. The key property of WhereStore is that it uses the phone's location history to determine what data to replicate locally. The main goal of caching cloud data on the phone is to decrease the overall data access latency and also to reduce the probability of data becoming unavailable in periods of no connectivity. Furthermore, WhereStore is a shared resource for different applications and exchanges data with the cloud in batches, thus potentially reducing the overall energy consumption on the phone. Angin et al. [21] proposed a mobile-cloud collaborative approach for context-aware navigation by exploiting the computational power of resources as well as location-specific resources available on the Internet. The authors design an extensible system architecture that minimizes reliance on infrastructure, thus allowing for wide usability.

Kovachev et al. [22] proposed Mobile Community Cloud Platform (MCCP) as a cloud computing system that can leverage the full potential of mobile community growth. Also, the authors analyze the requirements of mobile communities, propose a cloud computing model for mobile communities, and discuss the technical settings of this cloud infrastructure. Lan Zhang et al. [39] designed and constructed a multihop networking system named MoNet based on Wi-Fi and a privacy-aware geosocial networking service. Also the authors designed a distributed content-sharing protocol that can significantly shorten the relay path, reduce conflicts, and improve data persistence and availability. A role strategy is designed to encourage users to collaborate in the network. Furthermore, a key management and an authorization mechanism were developed to prevent some attacks and to protect privacy. Eric Jung et al. [40] proposed cooperatively exploiting the potential of smartphones in proximity, using their resources to reduce the demand on the cellular infrastructure. The author introduces Resource Aware Collaborative Execution (RACE), a Markov decision process (MDP) optimization framework that takes user profiles and user preferences to determine the degree of collaboration. Then RACE can enable the use of other mobile devices in the proximity as mobile data relays.

C2POWER [23], on the contrary, targets the problem of energy saving using cognitive networks by improving their energy usage while providing the QoS required by the applications in use. Both vertical handoffs (VHOs) and cooperation between MTs can help the system achieve longer lifetime and system efficiency. In this sense, C2POWER has to use context information to make informed decisions about the actions that can be taken to improve the system efficiency. Given a set of active MTs connected to radio access technologies (RATs), C2POWER determines which ones perform VHO toward different RATs and which ones get their traffic relayed by other MTs, with the goal of energy saving.

7.4.4 Sensor PM

Urban sensing, participatory sensing, and user activity recognition can provide rich contextual information for mobile applications such as social networking and location-based services. However, continuously capturing this contextual information on mobile devices consumes huge amount of energy. A novel design framework for an energy-efficient mobile sensing system (EEMSS) is proposed by Wang et al. [24], which uses hierarchical sensor management strategy to recognize user states as well as to detect state transitions.

In EEMSS implementation, the state description subsystem defines the following states: "Walking," "Vehicle," "Resting," "Home talking," "Home entertaining," "Working," "Meeting," "Office loud," "Place quiet," "Place speech," and "Place loud." The sensors used to recognize these states are an accelerometer, Wi-Fi detector, GPS, and microphone.

The core component of an EEMSS is a sensor management scheme that uniquely describes the features of each user state by a particular sensing criteria, and state transition will only take place once the criteria is satisfied. An example would be that "meeting in office" requires the sensors to detect the existence of speech and the fact that the user is currently located in the office area.

Sensor assignment is achieved by specifying an XML-format state descriptor as system input that contains all the states to be automatically classified, as well as sensor management rules for each state. The system will parse the XML file as input and automatically generate a sensor management module that serves as the core component of EEMSS and controls sensors based on real-time system feedback. In essence, the state descriptor consists of a set of state names, sensors to be monitored, and conditions for state transitions. It is important to note that the system designer must be very familiar with the operation of each sensor and how a user state can be detected by a set of sensors. State description must therefore be done with care so as to not include all the available sensors to detect each state because such a gross simplification in state description will essentially nullify any energy-saving potential of an EEMSS.

Figure 7.8 illustrates the general format of a state descriptor and the corresponding state transition process. It can be seen that a user state is defined between the "<State>" and "</State>" tags. For each state, the sensor(s) to be monitored is specified by "<Sensor>" tags. The hierarchical sensor management is achieved by assigning new sensors based on previous sensor readings in order to detect state transition. If the state transition criteria has been satisfied, the user will be considered as entering a new state (denoted by "<NextState>"in the descriptor), and the sensor management algorithm will restart from the new state.

The system can be viewed as a layered architecture that consists of a sensor management module, a classification module, and a sensor control interface that is responsible for turning sensors on and off and obtaining sensed data, real-time user state updates, logging, and user interfaces. Figure 7.9 illustrates the design of

```
<stateDescriptor>
        <state>
        <stateName>state1</stateName>
        <sensor>
                <sensorName>sensor1</sensorName>
                <case>
                        <condition>sensor reading 1</condition>
                        <nextState>state2</nextState>
                </case>
                <case>
                        .
                        .
                </case>
        </sensor>
        </state>
        <state>
        <stateName>state2</stateName>
            ..
            ..
        </state>
</stateDescriptor>
```

Figure 7.8   The format of an XML-based state descriptor and its implication of state transition.

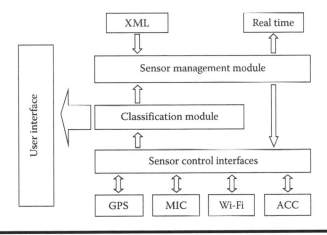

Figure 7.9   System architecture of energy-efficient mobile sensing system (EEMSS) implementation.

the system architecture and the interactions among the components. The sensor management module is the major control unit of the system. First, it parses a state description file that describes the sensor management scheme, and then it controls the sensors based on the sensing criteria of each user state and state transition conditions by specifying the minimum set of sensors to be monitored under different scenarios (states). The sensor management module configures the sensors in real time according to the intermediate classification result acquired from the classification module and informs the sensor control interface what sensors are to be turned on and off in the following step.

The classification module first processes the raw sensing data into the desired format. For example, the magnitude of three-axis accelerometer sensing data is computed, and fast Fourier transform (FFT) is performed on sound clips to conduct frequency domain signal analysis. The classification module returns user activity and position features such as "moving fast," "walking," "home wireless access point detected," and "loud environment" by running classification algorithms on processed sensing data. The resulting user activity and position information are considered as intermediate states that will be forwarded to the sensor management module.

The sensor management module then determines whether the sensing results satisfy the sensing criteria and makes sensor assignments according to the sensor management algorithm. The sensor interface contains APIs that provide direct access to the sensors. Through these APIs, application can obtain the sensor readings and instruct sensors to switch on/off for a given duty cycle, as well as change the sample rate. Due to J2ME limitations, GPS and embedded microphones are operated through J2ME APIs, while accelerometers and Wi-Fi detectors are operated through python APIs.

Mobile device–based sensing is able to provide rich contextual information about users and their environment for higher layer applications and services. However, the energy consumption by these sensors, coupled with limited battery capacities, makes it infeasible to be continuously running such sensors. The sensor management scheme for mobile devices operates sensors hierarchically by selectively turning on the minimum set of sensors to monitor user state, and it triggers new sets of sensors (if necessary) to achieve state transition detection. Energy consumption can be reduced by shutting down unnecessary sensors at any particular time.

7.5 Application and Architecture Features

7.5.1 Energy-Efficient File Sharing

In the MSNs, file sharing is a very important application—such as sharing photos, documents, and videos between mobile devices. Studies presented in Zhang et al. [25] show that transmitting one bit over the wireless card requires 1000 times more energy than a single 32-bit computation. Therefore, compressing data before transmission may be able to reduce energy consumption. However, data compression requires additional computation, which may consume more energy. Studies presented in Ref. [26] propose a lightweight adaptive compression scheme for energy-efficient mobile-to-mobile file-sharing applications. The proposed scheme monitors the signal strength level during the file transfer process and compresses data blocks on the fly only whenever energy reduction gain is expected. The proposed scheme exploits the trade-off between spending energy to compress a file and transmitting less data and spending energy sending the file uncompressed but without additional computations before transmission. By applying data compression, the intended information is sent with a lower number of bits and, thus, less transmission energy.

The authors compared compression ratio and time for the following compression methods: Bzip2, Gzip, Deflate, Zip (level 1), and Zip (level 9). It is shown that Zip (level 1) is the optimal compression method for file sharing between mobile devices, through taking transmission experiments of compressed and uncompressed files under different signal strength and comparing their energy consumption. And furthermore, it is better to send data uncompressed beyond a signal strength level of around –79 dB. A flowchart that shows the different steps of the proposed adaptive scheme is presented in Figure 7.10.

Data are divided into several blocks before transmission, and the size of block has impact on energy consumption. In order to study the impact of the data block size on energy consumption, energy measurements were conducted with four block sizes: 4, 8, 12, and 16 KB. Experimental results show that the energy consumed decreases as the block size increases whenever the signal strength is stable (either weak or strong). In our opinion, the authors may measure a data block of a bigger size, such as 32KB or 64KB. Perhaps more optimal results can be obtained.

Compression ratio should vary with different redundancy of files. To achieve this objective, an experiment is designed and conducted using four generated arbitrary binary data files with different entropies. Experimental results show that total compression before transmission will increase energy efficiency for high-redundancy files. That is because the compressed file is far smaller than the uncompressed file;

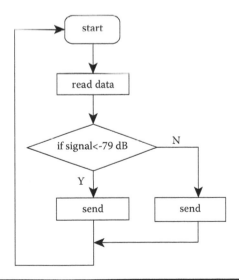

Figure 7.10 Adaptive compression scheme flowchart.

as a result, a compressed file can reduce transmission time and energy consumption significantly. On the contrary, the compressed files do not increase energy efficiency for low-redundancy files. That is because the size of a compressed file and an uncompressed file is similar. The proposed adaptive scheme has the following characteristics: It divides the data file into multiple blocks, decides dynamically whether to transfer a given block as compressed based on the estimated signal strength level between the devices, and performs compression for a given block on the fly while transmitting the previous data block. These characteristics result in notable energy consumption reduction that can be translated to a longer battery lifetime.

7.5.2 Collaborative Downloading

For examples, students in the same classroom each use their phones to download the same teaching PPT—this is a traditional method. Maybe they can work together to get the job done because they want to download the same content. Each student can just download some of the information and then share with other students via Bluetooth or the Wi-Fi network—this is a collaborative approach method. There are many benefits, for example. This approach can save the phone battery because time and energy consumption via General Packet Radio Service (GPRS) is much greater than the energy consumed via Bluetooth. Besides, it can save the cost of the Internet and can reduce network congestion.

Traditionally, each mobile user communicates only with a server and downloads the desired content via mobile network technology. We refer to this standard approach of the mobile telecom service provisioning process as the individual approach. A novel approach presented by Vrdoljak et al. [27] is that mobile users, who are interested in the same content, collaborate and download the desired content together.

As shown in Figure 7.11, the main idea of the collaborative approach is to identify a set of *n* mobile users who are physically at a smaller distance to each other (e.g., 10 meters—the range of Bluetooth) and who are interested in the same content. Each of *n* mobile users downloads only a part of the requested content via a mobile network (e.g., GPRS or Universal Mobile Telecommunications System [UMTS]) and shares it with other mobile users within the newly formed mobile ad hoc network via short-range communication technology (e.g., Bluetooth or Wi-Fi). The collaborative approach is enabled by filtering mobile users by their physical location and clustering those mobile users into groups of similar interest (i.e., swarms). The most significant benefit of the proposed service refers to the fact that the Bluetooth (Wi-Fi) network is more energy efficient than the GPRS (UMTS) network. Furthermore, communication in the Bluetooth (Wi-Fi) network is virtually free of charge for mobile users, as opposed to communication in the GPRS (UMTS) network. Also, switching communication from the GPRS (UMTS) network to the Bluetooth (Wi-Fi) network releases GPRS channels (UMTS bandwidth). Those resources can then be put to further use for other mobile users, thus decreasing the mobile telecom service provisioning cost for both telecom operators and mobile users.

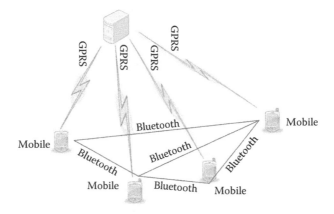

Figure 7.11   The collaborative downloading scenario.

Collaborative downloading service can be used as a real-world application that is based on Bluetooth or Wi-Fi communication. For instance, at a cinema viewing before the actual film starts, the audience is shown several movie trailers on the cinema screen. After they see a certain movie trailer, some of them might decide to download that trailer to their mobile devices and afterwards share it with their friends or family. By using the collaborative approach, they could save some energy in their mobile device batteries as well as lower their expenses.

7.6 Developers and Users

Battery life is a critical performance and user experience metric in mobile devices. However, it is difficult for app developers to measure the energy used by their apps and to explore how energy usage might change with conditions that vary outside of the developer's control such as network congestion, choice of mobile operator, types of MTs, and user settings for screen brightness.

WattsOn, presented by Mittal et al. [28], is a measurement system that allows a developer to estimate the energy consumed by their app in the development environment itself. WattsOn can (i) identify energy hungry segments during the app run and (ii) determine which component (display, network, or CPU) consumes the most energy.

WattsOn scales down the emulation environment (network, CPU, display) to mimic the phone and applies empirically derived power models to estimate app energy consumption. This approach gives developers the flexibility to test an app's energy consumption under various scenarios and operating conditions.

A block diagram of WattsOn is shown in Figure 7.12. The leftmost blocks represent the measurement of real device power characteristics required for power model generation, which may be developed for all mobile devices of interest

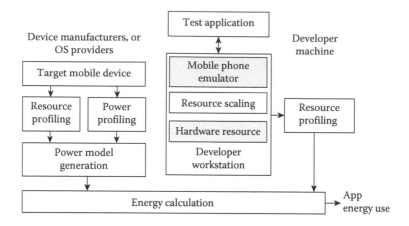

Figure 7.12 Block diagram of WattsOn components (blocks shaded gray are used in existing systems; WattsOn adds the other blocks for energy emulation).

(the number of models required may be reduced by considering representative devices in various device classes with different screen sizes and cellular network types). These measurements may be performed by the smartphone manufacturers, mobile OS platform developers, or even by volunteers using automated modeling methods. The mobile app developer simply downloads the appropriate models.

On the developer machine, the app code for the mobile device runs in a mobile device emulator. Resource-scaling techniques are inserted between the emulator and the actual hardware. As the app is executed on the emulator, its resource consumption can be monitored using resource-profiling methods available on the developer machine, which is then used in the energy calculation block to estimate the app energy using power models.

Energy consumption of mobile applications, devices, and networks has been measured from a variety of perspectives, including different access networks [29], context-aware battery management [30], voice-over IP [31], location-aware applications [32], a video-streaming application [33], data-sorting algorithms [34], and ad hoc and peer-to-peer systems.

However, user attitudes and behavior were not directly assessed by these studies. McCalley and Midden [35] found that products giving feedback about their energy consumption and having a means to set an energy conservation goal could motivate their users to save energy. Vallerio et al. [36] demonstrated how graphical user interface (GUI) design can improve the energy efficiency of a mobile device. Abrahamse et al. [37] note that giving consumers more information about their energy consumption may result in energy savings.

On the contrary, better understanding of user behavior helps in multiple ways in developing new technical solutions that better match the user expectations. First, developing more energy-efficient devices and applications is an obvious engineering problem. Knowledge on the influence of energy consumption on application usage allows allocation of development resources to the most important issues. Second, making energy consumption more explicit to users can increase their satisfaction. With additional feedback, users can adjust their behavior and take proactive measures to manage the energy consumption of their devices and the applications running in them.

Common user behavior and attitudes can be examined with two approaches: questionnaires and usage monitoring. Surveys conducted by Heikkinen [38] indicate that users are interested in optimizing the energy consumption of advanced mobile handsets. Users demand more detailed and accurate battery level indicators and power saving settings. They want to understand and control the energy consumption of their advanced mobile handsets.

7.7 Conclusion

Smartphones have become commonplace in recent years. Applications and services using Internet technologies previously available only for devices with limited mobility, such as laptops, are now readily accessible on devices fitting in our pockets. Unfortunately, the battery performance and energy efficiency of small-scale mobile devices have not progressed at a desired pace, resulting in users having to frequently recharge their devices and to pay considerable attention to battery consumption. The purpose of this chapter is to understand various methods of improving energy efficiency. Appealing to the requirement of energy savings, many approaches of energy-efficient locating sensing have been explored. Better understanding of user behavior helps in multiple ways for developing new technical solutions that better match the user expectations. A lightweight adaptive compression scheme has been proposed for file-sharing applications among mobile devices. Energy efficiency of the underlying network technologies has attracted broad attention including Wi-Fi, Bluetooth, and power saving models.

Although energy can be a challenge for an MSN, it is also an opportunity. Energy efficiency is therefore a fruitful area for further research. Moving forward, with the development of energy-efficient methods and increased battery energy amounts, we believe that these efforts will significantly increase the battery lifetime of MTs.

References

1. Robinson, S. Cellphone energy gap: Desperately seeking solutions. Strategy Analytics technical report 2009; p. 28.
2. Maa, X., Y. Cuia, and I. Stojmenovic. Energy efficiency on location based applications in mobile cloud computing: A survey. *Procedia Computer Science* 2012; 10: 577–584.

3. Kjærgaard, M., J. Langdal, T. Godsk, and T. Toftkjær. Entracked: Energy-efficient robust position tracking for mobile devices. In: Proceedings of the 7th International Conference on Mobile Systems, Applications, and Services, ACM, Kraków, Poland, June 22–25, 2009, pp. 221–234.

4. Constandache, I., S. Gaonkar, M. Sayler, R. Choudhury, and L. Cox. Enloc: Energy-efficient localization for mobile phones. In: INFOCOM, Rio de Janeiro, Brazil, April 19–25, 2009; pp. 2716–2720.

5. Zhuang, Z., K. Kim, and J. Singh. Improving energy efficiency of location sensing on smartphones. In: Proceedings of the 8th International Conference on Mobile Systems, Applications, and Services, San Francisco, CA, June 15–18, 2010.

6. Paek, J., K. Kim, J. Singh, and R. Govindan. Energy-efficient positioning for smartphones using cell-id sequence matching. In: Proceedings of the 9th International Conference on Mobile Systems, Applications, and Services, Bethesda, MD, June 28–July 1, 2011.

7. Kjærgaard, M., S. Bhattacharya, H. Blunck, and P. Nurmi. Energy-efficient trajectory tracking for mobile devices. In: Proceedings of the 9th International Conference on Mobile Systems, Applications, and Services, Bethesda, MD, June 28–July 01, 2011.

8. Thiagarajan, A., L. Ravindranath, H. Balakrishnan, S. Madden, and L. Girod. Accurate, low-energy trajectory mapping for mobile devices. In: Proceedings of 8th USENIX Symposium on Networked Systems Design and Implementation (NSDI), Boston, MA, March 30–April 1, 2011.

9. Lin, K., A. Kansal, D. Lymberopoulos, and F. Zhao. Energy-accuracy trade-off for continuous mobile device location. In: Proceedings of the 8th International Conference on Mobile Systems, Applications, and Services, San Francisco, CA, June 15–18, 2010.

10. Paek, J., J. Kim, and R. Govindan. Energy-efficient rate-adaptive GPS-based positioning for smartphones. In: Proceedings of the 8th International Conference on Mobile Systems, Applications, and Services, San Francisco, CA, June 15–18, 2010.

11. Liu, H., H. Darabi, P. Banerjee, and J. Liu. Survey of wireless indoor positioning techniques and systems. *IEEE Transactions on Systems, Man, and Cybernetics* 2007; 37(6): 1067–1080.

12. Abdellatif, M., A. Mtibaa, K. A Harras, and M. Youssef. GreenLoc: An energy efficient architecture for WiFi-based indoor localization on mobile phones. In: Proceedings of IEEE ICC, Budapest, Hungary, June 9–13, 2013.

13. Choi, J., M.-G. Kim, H. Jeong, and H.-S. Park. Power-saving mechanisms for energy efficient IEEE 802.16e/m. *Journal of Network and Computer Applications* 2012; 35(6): 1728–1739.

14. IEEE 802.11-2007 Standard, Wireless LAN medium access control (MAC) and physical layer (PHY) specifications, 2007.

15. Wi-Fi Alliance, Quality of Service (QoS) Task Group, Wi-Fi Multi-media (including WMM PowerSave) Specification v1.1, 2005.

16. Han, B. and A. Srinivasan. eDiscovery: Energy efficient device discovery for mobile opportunistic communications. In: Proceedings of the 20th IEEE International Conference on Network Protocols (ICNP), Austin, TX, October 30–November 2, 2012.

17. Holenarsipur, P., and Mehta, A. Ambient-light sensing optimizes visibility and battery life of portable displays. *Issue of Digi-Key's TechZone magazine.* 2011.

18. Schilit, B. and M. Theimer. Disseminating active map information to mobile hosts. *IEEE Network* 1994; 8(5): 22–32.

19. Alam, M., M. Albano, A. R. Radwan, and J. Rodriguez. Context parameter prediction to prolong mobile terminal battery life. In: Proceedings of International Mobile Multimedia Communications Conference, Lisbon, Portugal, September 6–8, 2010.
20. Stuedi, P., I. Mohomed, and D. Terry. WhereStore: Location based data storage for mobile devices interacting with the cloud. In: Proceedings of the 1st ACM Workshop on Mobile Cloud Computing & Services: Social Networks and Beyond, San Francisco, CA, June 15, 2010.
21. Angin, P., B. Bhargava, and S. Helal. A mobile-cloud collaborative traffic lights detector for blind navigation. In: Proceedings of the Eleventh International Conference on Mobile Data Management (MDM), Kansas City, MO, May 23–26, 2010.
22. Kovachev, D., D. Renzel, R. Klamma, and Y. Cao. Mobile community cloud computing: Emerges and evolves. In: Proceedings of the Eleventh International Conference on Mobile Data Management (MDM), Kansas City, MO, May 23–26, 2010.
23. Radwan, A., M. Albano, J. Rodriguez, and C. Verikoukis. Analysis of energy saving using cooperation use-case: WiFi and WiMedia. In: Future Networks & Mobile Summit (FUNEMS), Berlin, Germany, July 4–6, 2012.
24. Wang, Y., J. Lin, M. Annavaram, Q. A. Jacobson, J. Hong, B. Krishnamachari, and N. Sadeh. A framework of energy efficient mobile sensing for automatic user state recognition. In: Proceedings of the 7th International Conference on Mobile Systems, Applications, and Services (MobiSys), Kraków, Poland, June 22–25, 2009.
25. Zhang, Y., W. Liu, W. Lou, and Y. Fang. Location-based compromise tolerant security mechanisms in wireless sensor networks. *IEEE Journal on Selected Areas in Communications* 2006; 24: 247–260.
26. Sharafeddine, S. and R. Maddah. A lightweight adaptive compression scheme for energy-efficient mobile-to-mobile file sharing applications. *Journal of Network and Computer Applications* 2011; 34: 52–61.
27. Vrdoljak, L., I. Bojic, V. Podobnik, G. Jezic, and M. Kusek. Group-oriented services: A shift towards consumer–managed relationships in the telecom industry. In Nguyen, N. T., and Kowalczyk, R. (Eds.), *Transactions on CCI II*, Lecture notes in Computer Science. Vol. 6450. Springer; 2010, pp. 70–89.
28. Mittal, R., A. Kansal, and R. Chandra. Empowering developers to estimate app energy consumption. In: Proceedings of the 18th Annual International Conference on Mobile Computing and Networking, Istanbul, Turkey, August 22–26, 2012.
29. Balasubramanian, N., A. Balasubramanian, and A. Venkataramani. Energy consumption in mobile phones: A measurement study and implications for network applications. In: Proceedings ACM Internet Measurement Conference, Chicago, IL, November 4–6, 2009; pp. 280–293.
30. Ravi, N., J. Scott, L. Han, and L. Iftode. Context-aware battery management for mobile phones. In: Proceedings of the Sixth Annual IEEE International Conference on Pervasive Computing and Communications, Hong Kong, China, March 17–21, 2008.
31. Gupta, A. and P. Mohapatra. Energy consumption and conservation in WiFi based phones: A measurement-based study. In: Proceedings of the Fourth Annual IEEE Communications Society Conference on Sensor, Mesh and Ad Hoc Communications and Networks, San Diego, CA, June 18–21, 2007.
32. Anand, A., C. Manikopoulos, Q. Jones, and C. Borcea. A quantitative analysis of power consumption for location-aware applications on smart phones. In: Proceedings of the IEEE International Symposium on Industrial, Electronics, Vigo, Spain, June 4–7, 2007.

33. Xiao, Y., R. S. Kalyanaraman, and A. Yla-Jaaski. Energy consumption of mobile YouTube: Quantitative measurement and analysis. In: Proceedings of the Second International Conference on Next Generation Mobile Applications, Services and Technologies, Cardiff, September 16–19, 2008.

34. Bunse, C., H. Hopfner, E. Mansour, and S. Roychoudhury. Exploring the energy consumption of data sorting algorithms in embedded and mobile environments. In: Proceedings of the 10th International Conference on Mobile Data Management (MDM), Kansas City, MO, May 23–26, 2010.

35. McCalley, L. T. and C. J. H. Midden. Energy conservation through product-integrated feedback: The roles of goal-setting and social orientation. *Journal of Economic Psychology* 2002; 23: 589–603.

36. Vallerio, K. S., L. Zhong, and N. K. Jha. Energy-efficient graphical user interface design. *IEEE Transactions on Mobile Computing* 2006; 5: 846–859.

37. Abrahamse, W., L. Steg, C. Vlek, and T. Rothengatter. The effect of tailored information, goal setting, and tailored feedback on household energy use, energy-related behaviors, and behavioral antecedents. *Journal of Environmental Psychology* 2007; 27: 265–276.

38. Heikkinen, M. V. J. Energy efficiency of mobile handsets: Measuring user attitudes and behavior. *Telematics and Informatics* 2012; 29(4): 387–399.

39. Zhang, L., X. Ding, Z G. Wang, M. Gu, and X. Y. Li. WiFace: A secure geosocial networking system using WiFi-based multi-hop MANET. In: Proceedings of the 1st ACM Workshop on Mobile Cloud Computing & Services: Social Networks and Beyond (MCS), San Francisco, USA, June 15, 2010.

40. Jung, E., Y. C. Wang, I. Prilepov, F. Maker, X. Liu, and V. Akella. User-profile-driven collaborative bandwidth sharing on mobile phones. In: Proceedings of the 1st ACM Workshop on Mobile Cloud Computing & Services: Social Networks and Beyond (MCS), San Francisco, USA, June 15, 2010.

Chapter 8

Privacy, Trust, and Reputation in Mobile Networking and Computing

8.1 Privacy

8.1.1 Introduction

Social networks that extend the social circles of users have become an integral part of our daily lives. With social networking tools, we are able to easily share information, images, and videos with our friends and to search for desirable service information with their recommendations.

Online social networks (OSNs) such as Facebook, Twitter, and Foursquare are central to the lives of millions of users and are still growing. As reported by ComScore, social networking sites such as Facebook and Twitter have reached 82% of the world's online population, representing 1.2 billion users around the world.

In the meantime, fueled by the dramatic advancements of smartphones and the ubiquitous connections of the Internet, social networking is becoming further available to mobile users and is keeping those users posted on up-to-date worldwide news and messages from their friends and families anytime, anywhere. The eMarketer estimates that up to 46% of mobile users will access their social networks with smartphones in 2014, although this number was merely 16% in 2010.

With the growing number of smartphone users, it is envisioned that a pervasive and omnipotent communication platform, the mobile social network (MSN), will become mainstream and will provide smartphone users with extensive methods for obtaining desired information, from browsing over the Internet to querying nearby peers.

This boom in MSNs is fostering a large volume of promising and smart mobile applications. Apple, Inc., has greatly increased the number of its mobile applications—from 800 in July 2008 to more than 825,000 in April 2013. Nowadays, as many smartphone users indulge themselves in enjoying various mobile social applications, they can no longer live or work effectively without using the applications. Despite the tremendous benefits brought about by the MSN and its applications, the MSN still faces many security and privacy challenges. Even though applications normally require access to users' personal information in order to serve them better, not much attention has been paid to security and privacy preservation in many application designs. For example, in most social applications, users need to register with personal profiles such as name, birthday, home address, and phone number, which are very likely to be disclosed due to the lack of protection. In the United Kingdom, the number of complaints and alleged crimes associated with Facebook and Twitter has increased by 780% in the last four years, resulting in about 650 people being charged in 2012. InternetSafety states that, in 2012, 29% of Internet-related sex crimes originated from social networking sites. As mobile applications enable smartphone users to interact with social networks more pervasively, there will be more severe security and privacy concerns for users [1].

In a survey of 2253 adult OSN users, 65% had changed their privacy settings to limit what information they share with others, 36% had deleted comments from their profile, and 33% expressed concern over the amount of information about them online [2]. In a survey of 1000 young adults, 55% of respondents reported being more concerned about privacy issues on the Internet than they were five years ago [3].

The remainder of this section is organized as follows. Section 8.1.2 categorizes MSN architecture and presents it analytically. Section 8.1.3 enumerates the taxonomy of MSN privacy issues. Section 8.1.4 concentrates on privacy as the main category of safety challenges and solutions for MSNs. Future research directions are outlined in Section 8.1.5, and Section 8.1.6 concludes the chapter.

8.1.2 Architecture and Classification of MSNs

According to the way users are able to inject and access information, MSN architecture can be separated into three categories: centralized networks, distributed networks, and hybrid networks.

8.1.2.1 Centralized MSNs

This is the most common architecture used for the deployment of MSNs. All the information concerning the members of the social network is preserved in remote servers belonging to a service provider. End nodes use the deployed wireless infrastructure of a cellular network, Wi-Fi, or other technologies, to gain access to the remote service providers in order to communicate with each other or to access and change personal information. This client–server structure can be further enhanced by the use of third-party application servers (Simple Mail Transfer Protocol [SMTP], Voice over Internet Protocol [VoIP], etc.). The benefits of the centralized architecture are obvious and indisputable—it can provide high-quality services through the currently available infrastructure. Moreover, it bridges the gap between the physical location of the user and the global social community. Kemp and Reynolds [4] argue though that this "hub and spoke" model of clients and servers may cause bottlenecks because all traffic should pass through the hub. In addition, each transaction should pass up one spoke of the hub and then down another, even when the transaction's two endpoints are physically close. A major disadvantage has to do with the huge amount of personal data stored in a few physical locations under the authority of an alien entity.

Centralized MSNs are mainly an extension of their Web-based predecessors. By using specific mobile applications or by simply using a mobile browser, users are able to connect to a centralized server for sharing and exchanging information. In that sense, when a mobile user accesses his or her Facebook account from his or her mobile, he or she is a part of a greater MSN. However, there are plenty more new mobile applications that exclusively target mobile users, such as JuiceCaster. Research in this area mainly focuses on creating a unified middleware that is able to provide the basis of future mobile social networking applications. Such an attempt was made by Borcea and colleagues [5], who presented MobiSoC, a mobile social computing middleware providing the necessary functionality for the deployment of mobile social computing applications (MSCAs). The main goal of this attempt was to provide a unified solution for people-centric and place-centric applications divided into several modules able to be distributed in multiple servers. Another interesting proposal was made by Ma et al. [6], who introduced an application called MoViShare. This centralized application is able to provide a location-aware video-sharing platform. A survey on the most important middleware projects so far is presented by Karam and Mohamed [7]. Among the interesting features of this survey are the middleware challenges in the MSN environment as well as a framework for comparing the various projects.

One more method of deploying centralized MSNs is through the use of wireless sensor networks (WSNs). Sensors have already infiltrated daily life, and there is a plethora of advantages from the integration of mobile social and sensor networks. By using contextual information, WSNs can provide much more detailed and personalized social services. One prominent example is presented by Miluzzo et al. [8]. The authors designed and developed a smartphone application able to gather information through the sensors of a mobile device. The information is then transferred to back-end software. This is able to support push- and

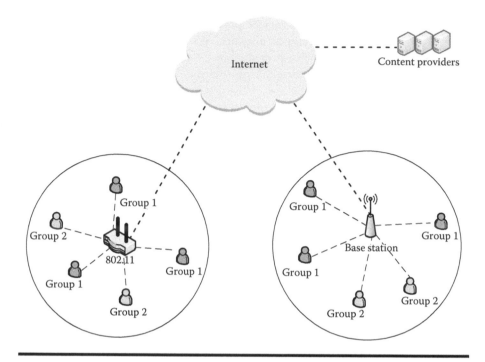

Figure 8.1 Centralized system architecture for mobile social network services.

pull-based data publishing via a hypertext transfer protocol (HTTP) markup language and a standard Web service–based application program interface (API), respectively.

A general centralized architecture for MSN services in the context of social services is given by Chang et al. [9]. The main advantage of this proposal is its ability to provide location-based interactive services without the installation of an additional location module. An overview of this architecture is presented in Figure 8.1. In that figure, users are split into two different social groups and are able to access the social information through the deployed wireless infrastructure. A hand-over mechanism can be employed to provide connection continuity.

8.1.2.2 Distributed MSNs

The key feature of the distributed MSNs is the total absence of centralized servers. Mobile users are able to communicate and access social information only by connecting to other users. Therefore, the network devices themselves have to store and route the social data until the correct destination is found. However, this does not necessarily mean that there is no infrastructure at all. Cellular networks and access points can be used to provide interconnection between two nodes. Nevertheless, the mobile terminals nowadays are further equipped with multiple wireless interfaces such as Bluetooth and Wi-Fi, providing them with the capability of connecting in an ad hoc manner with no need for any kind of infrastructure.

The various ways in which distributed MSNs can be deployed are depicted in Figure 8.2. In that figure, every node is characterized by an integer ID while intercommunication is provided by three different techniques, namely a Wi-Fi access point, a 3G network base station, and an ad hoc network. The fact that nodes are mobile and can roam through the three offered communication technologies is also depicted. Three nonoverlapping social groups are being considered, characterized by the color of each user.

Keeping in mind their capabilities, mobile devices all over the world can be seen as a vast mobile ad hoc network with enormous potentials. Pietiläinen et al. introduced a middleware called MobiClique [10] presenting the utilization of such a network. MobiClique is able to create and preserve a mobile ad hoc social network, providing content exchange by using the user mobility. However, it lacks the ability to predict user contacts and therefore performs a form of flooding as a means of content dissemination, leading to low efficiency and very high resource usage. As a general idea, though, it may provide a valuable insight on how distributed MSNs may function in the near future. Although the distributed MSNs provide many advantages, such as low cost deployment and maintenance, they also pose many challenges. These include high latency, low delivery rate compared to the centralized architecture, and privacy concerns.

Distributed MSNs can be further divided into two different categories. In the first, users share content directly or by using some deployed (but noncentral) infrastructure. In the second, which is the most challenging, intermediate nodes might

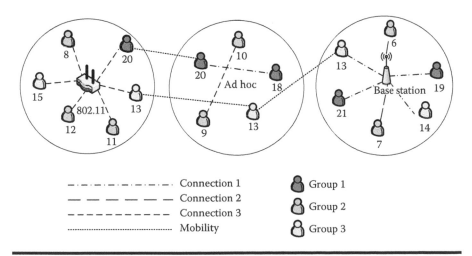

Figure 8.2 Generic layout of distributed mobile social networks.

be required to provide intercommunication between two end users. Because no infrastructure is considered, the mobile nodes themselves are designated to route or carry the messages until the final destination. This is why most of the research in the field of MSNs targets this area and the inherited problems that appear due to the dynamic nature of the underlying ad hoc networks. A characteristic example of a middleware belonging to this category, AdSocial, is discussed by Sarigöl et al. [11]. This is a software platform that is able to support social networking applications such as video calls and gaming over ad hoc networks. Nevertheless, it is widely accepted that delay-tolerant networking (DTN) is preferable for overcoming the dynamic and sparse nature of the underlying network; DTN provides intercommunication because it is not mandatory for it to have an a priori knowledge of the network topology [12].

8.1.2.3 Hybrid MSNs

Hybrid MSNs are an extension of the two major types of architecture mentioned previously. They are created by a combination of both in the sense that, despite the existence of centralized servers containing all the social information, direct data exchange between nodes is also possible, as presented in Figure 8.3. This assumes the existence of the necessary data locally as well as remotely. A typical hybrid scenario can be described as follows. Nodes asking for specific information can access it via the centralized server using the cellular infrastructure and therefore might as well share information with each other in an ad hoc manner.

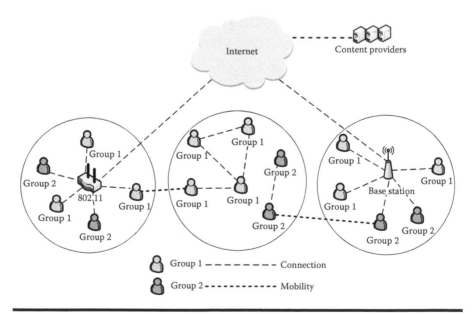

Figure 8.3   Generic layout of hybrid mobile social networks.

One big advantage of the hybrid architecture is that it can be established very easily by extending almost every existing centralized or distributed MSN. Han et al. [13] propose the exploitation of opportunistic forwarding in MSNs as a method for offloading mobile data from 3G networks. By using complementary technologies already available in mobile devices, mobile data traffic is reduced up to 73.66%. Three algorithmic approaches are examined by the authors—a greedy approach, a random approach, and a heuristic approach. However, the heuristic approach raises privacy concerns because the everyday contact information of users is required to be uploaded to the service providers [14].

8.1.3 Privacy Leakage Information in MSNs

MSNs have recently grown in popularity. With the ubiquitous use of mobile devices and a rapid shift in technology and access to MSNs, it is important to examine the impact of MSNs from a privacy standpoint. Here, we present a taxonomy of ways to study privacy leakage, and we report on the current status of known leakages. All MSNs exhibit some leakage of private information to third parties. With this leakage in MSNs, more concern should be focused on the combination of new features unique to mobile access.

There are two classes of MSNs: (1) traditional OSNs (such as Facebook and MySpace) that have expanded to embrace access via mobile devices; and (2) applications and MSNs that were created largely to deal with the new mobile context. The latter class forms a majority. The taxonomy in this section may differ between the two classes. Privacy issues that were a concern in traditional OSNs, such as permissive sharing of personal information to all MSN users and leakage of private information to third parties, remain relevant to the former class, although they need to be examined anew for the latter class.

Specifically, in addition to privacy issues observed for traditional OSNs, which may be exacerbated as a result of the new features in mobile online social network, new privacy issues exist in the mobile context. The concepts that are either novel or that play a predominant role in MSNs include presence and location, which will be explained in more depth here. These concepts have played less of a role in traditional OSNs; however, determining a user's presence has become more and more important in MSNs such as Facebook that seek to provide an instant messaging service to its users. Twitter has recently allowed users to add their location information even when users access their traditional site.

Figure 8.4 lists a few of the MSN user's pieces of private information and some of the entities (inside and outside the mobile online social network) to which information might leak. Privacy leakage is considered from two perspectives: the personal information that may be leaked and the destination to which it could be leaked. The latter is important in the context of MSNs because there is a larger set of possible destinations due to MSNs' expanded features. There are at least three possible destinations: (1) internally—to entities within the mobile

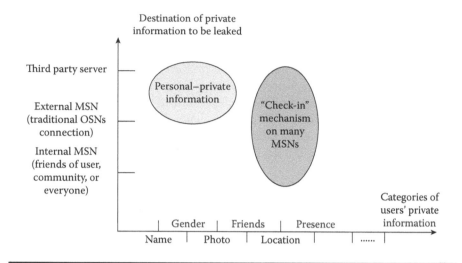

Figure 8.4   Potential privacy leakage in mobile social networks.

online social network (e.g., to a user's friends, networks/communities, or everyone), (2) externally—to other traditional OSNs through the connection feature (and thus to the user's contacts in those traditional OSNs, which can be limited to the user's friends or can be accessible to everyone), or (3) to third-party aggregators and advertisers.

Many MSNs do provide a range of privacy settings. However, the multidimensional nature of the issue makes the problem of protecting information significantly difficult. Consider the amount of information a user has to keep track of while interacting with an MSN. They have to be aware of the duration of any privacy setting they have made. When they allow some information, such as location, to be used by the MSN for a legitimate purpose (locating them on a map, etc.), they have to be aware that this information might be handed over to third parties. They have to keep track of what subset of users has access to which subset of their private information: their friends, their friends who are currently online on this MSN, their friends in other MSNs, and so on. In addition, popular atomic actions on MSNs such as "checking in" at a location reveal much about the user—their presence, their location, and the current time stamp. The richer the features of a mobile online social network, the more complex the results of a single action could be.

Presence on an MSN is not a new concept, but in most traditional OSNs users were not automatically made aware of the presence of their friends (or any other users). Such a feature has been long available in instant messaging systems. Many MSNs, on the contrary, allow users to indicate their presence via a "check-in" mechanism, where users establish their location at a particular time. Presence is an important notion in MSNs because one of their key features is the notion of the checking physical colocation of users. Users who are not present on an MSN are not likely to participate in any dynamic interactions. The indication of presence allows a user's friends to expect a quick response. Sharing one's presence more broadly than just with friends allows one to meet new people who are members of the same MSN.

A user's availability to communicate is indicated by presence, and presence exists independent of a closely related notion: location. Location is a widely used feature in MSNs that until recently was not an available feature in traditional OSNs. The ubiquity of Global Positioning System and the ability to automatically locate oneself have led to location being considered a basic feature of many MSNs. However, a number of MSNs would have limited functionality if users did not disclose their location. In a sense, location might be viewed as essential for the proper functioning of an MSN. However, users may not want to disclose their location beyond their set of friends to avoid the potential pitfalls from preying users. Many MSNs allow such disclosure to be limited to friends or to friends that are within a given distance from the user. It is important to be aware that users can indicate their presence on an MSN without disclosing their exact location.

There is additional private data at risk of being leaked in MSNs, including information related to the mobile device. For example, mobile devices typically have a unique identifier that is used for various purposes, such as installing approved applications. There is a potential privacy issue if this unique identifier is leaked to a third party via an application that has access to the identifier through the device's API. If leaked, this identifier could be associated with a user's identity and could be used to track an unknowing user's actions across different applications.

Perhaps the most interesting issue that raises significant new privacy concerns is the interaction potential between MSNs and traditional OSNs. Such interaction has already been made available in many MSNs to increase their popularity. Mobile OSNs encourage users to link their activities on MSNs with traditional OSNs such as Facebook and Twitter. Such connections are useful to users who, while interacting with an MSN, can expect some of their actions to show up on traditional OSNs and be visible to their friends there. The information supplied by users and the degree of interconnection based on API connections vary across MSNs. For example, if a user discloses their location to an MSN and is automatically connected to Facebook or Twitter, then friends on those OSNs may also be able to see this information. And moreover, the location information posted on the user's Facebook wall or Twitter timeline would be available by default to all users [15].

8.1.4 Privacy Protection Classification

Privacy protection implies the ability of an individual or group to reveal its information selectively and to remain unnoticed or unidentified over an MSN. It encompasses anonymity, information blurring, and furtiveness of exchanged messages between intermediate nodes. The level of privacy in MSNs depends on the application, the point of view (sender, receiver, intermediate node, and outside observer), and the level of the trust among entities. In the case of context-based forwarding, the context of the message is directly linked to the profile of the destination and is considered private. The context should therefore be protected (encrypted) even though the information about the shared context should be revealed. This raises the problem of computation on encrypted data. In the case of content-based forwarding, preserving the privacy of users mainly consists of protecting their interests. In this case, users want to receive content corresponding to their interests without revealing themselves. However, user privacy and forwarding present conflicting requirements. The first requires the encryption of the interests, while the second requires access to the filters. This raises another problem of computation on encrypted data. In the following subsections, privacy challenges in MSNs will be partitioned into the following different classes according to their behavior: (1) obfuscation, (2) private matching, (3) location privacy, and (4) communication privacy.

8.1.4.1 Obfuscation

Privacy preferences are generally specified to govern context exchange among nodes in ubiquitous environments. Aside from who has rights to see what information, a user's privacy preference could also designate who has rights to have what obfuscated information. By obfuscation, people are able to present their private information in a coarser granularity or simply in a falsified manner, depending on the specific situations. In other words, obfuscation is a form of data masking where data are purposely scrambled to prevent unauthorized access to sensitive materials. This form of encryption results in unintelligible or confusing data.

A popular and traditional mechanism for privacy enhancement and anonymous communication over a network is onion routing. Using this mechanism, messages are repeatedly encrypted and routed through a group of collaborating nodes to prevent the intermediary nodes from knowing the origin, destination, and content of the message. Like someone peeling an onion, each onion router removes a layer of encryption to uncover routing instructions and then sends the message to the next router where this is repeated.

Aside from onion-based routing schemes, one of the important privacy methods is called none of your business (NOYB) [16], which uses obfuscation to enhance privacy in social network sites. It combines users' information from different sites to prevent an attacker from profiling an individual user. In order to provide privacy, this method is equipped with several mechanisms such as the marginal distribution of the cipher text, atom compartmentalization, steganography, public dictionary, random nonce, standard ciphers, and communication across different channels. NOYB was extended by a defense strategy called hide-and-lie [17], which obfuscates users' interests in an opportunistic publish–subscribe application. Using this strategy, the success probability of an attacker can be equalized to the success probability of simple guessing. Furthermore, in some scenarios, the hide-and-lie strategy increases the message delivery ratio.

Another obfuscation-based privacy method for social networks was presented in Ref. [18]. This work studies the feasibility of perceivable social networks through the comparison of an anonymous data set to another available social network data set. The first problem with this method is that, to randomize data, it is necessary to keep the statistics close to the origin, which will reveal the hidden data themselves. The second problem is that this method focuses on one-time releases. In other words, the republication of dynamic social network data has not been considered in this method.

In order to resolve the first problem and keep the data out of attackers' access, Wondracek et al. [19] suggested another solution. They proved that it is also possible to reveal social network data if group membership information is public. Later, an approach consisting of two complementary methods dealing with this subject was presented by Parris et al. [20]. These methods enhance privacy in social network routing by obfuscating the friends' lists in order to inform routing decisions. To do this, a one-way hashing technique is used that is independent from any kind of key management schemes. Using three real-world data sets, this work evaluates the proposed methods and shows that it is feasible to use such methods without any reduction in routing performance.

In order to resolve the second problem, Xuan et al. [21] show that, by using correlations between sequential releases, the adversary can achieve high precision in the deanonymization of the released data. It lets enemies suppress the uncertainty of reidentifying each release separately and synthesize the results afterward. In addition, this work suggests a combination of structural knowledge with node attributes to compromise graph modification–based defenses.

8.1.4.2 Private Matching

Matchmaking is a key component of mobile social networking. It notifies users of nearby people who fulfill some criteria, such as having shared interests, and who are therefore good candidates for being added to a user's social network.

Generally, there are two straightforward ways to implement matchmaking in mobile social networking. One way is for a device to broadcast its owner's profile information to the public, for example, using Bluetooth. MobiClique is an example of the applications that take this approach. MobiClique users download their profile information from Facebook to their device and send this information to any Bluetooth device nearby. After receiving a piece of profile information, a device matches the received profile information and its owner's profile information and decides whether the other party is of interest to the owner according to the matching result. This approach is risky because it leaks users' private information to anyone in the users' proximity. The other way to implement matchmaking in mobile social networking is to introduce a trusted server for the matchmaking operation; looptmix and Gatsby are examples that use this approach [22]. Here, the server stores all users' personal information and tracks their location. The server informs two users if they are nearby and could become friends based on their requirements. This approach has the limitation that it requires the server to be always available in order to find a friend. From a privacy point of view, the server learns all users' personal information, all pairs of users who meet each other, and the locations of all users.

Based on their mechanisms, private-matching schemes can be divided into three subgroups: secret-sharing, coarse-grained schemes, and fine-grained schemes. These subgroups are described in the following three subsections.

8.1.4.2.1 Secret Sharing

The first set of secret-sharing approaches adopted homomorphic encryption. The first attempt to implement such encryption dates back to Freedman et al. [23]. Using this scheme, a client and a server compute the intersection of their sets while the client gets the result and the server learns nothing. In other words, FNP takes advantage of homomorphic encryption to represent the client's input obliviously. A complimentary scheme not only enables set intersection, union cardinality, and over-threshold operations but also extends FNP to multiple players.

The problem with all these methods is that, due to their strong dependence on homomorphic encryption and their inability to implement linear computational complexity, almost none of them are practical enough to be applied to MSNs. To make such schemes compatible, some other studies have pursued information about theoretical security and try to follow secret-sharing techniques. Narayanan et al. [24] use a secret-sharing scheme to provide a distributed solution for the FNP. In this scheme, the authors use a polynomial to represent one party's data set as in FNP and then distribute the polynomial to multiple servers; they then extend their solution to the distributed set intersection and the cardinality of the intersection. Another example of secret sharing–based private matching suggests an unconditional, secure, multiparty set intersection scheme in which the inputs are shared among all parties using threshold secret sharing. Computations are done on those shares to obtain the shares of the outputs [25].

8.1.4.2.2 Coarse-Grained Schemes

The basic aim of coarse-grained private-matching schemes is to match two users according to the privacy-preserving computation of the intersection (cardinality) of their attribute sets. Almost all these schemes implicitly assume that each user's personal profile consists of multiple public sets of characteristics derived from a public set of attributes. FindU [26] was the first coarse-grained, privacy-preserving, personal profile–matching scheme for MSNs. It fulfills the primary privacy demand for a personal profile. An initiating user can find best matches according to their desired attributes even though the actual set of matching attributes between the initiating user and any other user is hidden from all participants. FindU also contains different levels of user privacy. While leveraging secure multiparty computation techniques, it defines protocols to realize increasing levels of user privacy protection.

Some coarse-grained approaches focus on friendship discovery using friend-of-friend detection algorithms. These algorithms mainly define an intermediate node called matchmaker that is responsible for establishing interconnections between nodes that have similar interests and with which they are in contact (see Figure 8.5).

An example of mobile social networking platforms that implement the friend-of-friend detection algorithm is a privacy-preserving, personal profile–matching scheme called VENETA [27]. Rather than only exploiting information about the users of the system, the method relies on real friends and adequately addresses the arising privacy issues. It makes use of some notations, including commutative and homomorphic encryption, and passive and active adversaries to exhibit features such as contact matching, decentralized messaging, server-bound messaging, and user location tracking.

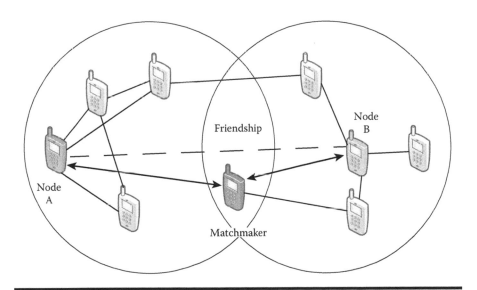

Figure 8.5 Example of a friend-of-friend detection scheme in mobile social networks, showing a matchmaking scenario where an intermediate node, which is in interconnection with different nodes and shares their profile information, tries to maintain friendship between them.

Wei et al. [28] focus on developing some techniques and protocols to compute social proximity between two users in order to discover potential friends. The authors identify potential attacks against friend discovery to develop a solution for secure proximity estimation with privacy and variability considerations. In this way, their proposed approach does not match two users using the cardinality of their attribute sets. Instead, they proposed using the social proximity between two users as the matching metric, which measures the distance between their social coordinates with each being a vector precomputed by a trusted central server to represent the location of a user in an MSN.

To expand matchmaking protocols for MSNs, a privacy-preserving matchmaking protocol [22] has been proposed. It lets a potentially malicious user learn only the shared common interests with a nearby user. This protocol is distributed and does not require a trusted server to track users or to be involved in any matchmaking operation. The mentioned mechanism offers the capability to defy passive and active attacks such as the user's interest exploration, impersonation, and eavesdropping.

8.1.4.2.3 Fine-Grained Schemes

Generally, the schemes based on coarse-grained private matching are unable to further differentiate users with the same attributes. To deal with this problem, fine-grained private-matching protocols were proposed [29] that enable two users to perform-profile matching without any need for information disclosures. In contrast to coarse-grained private-matching schemes, these protocols allow finer differentiation among users and support a wide range of matching metrics at different privacy levels.

8.1.4.3 Location Privacy

Location sharing is a fundamental component of MSNs, which also raises significant privacy concerns. Location privacy gives individuals or group users the ability to selectively seclude or reveal their location. The following four subsections discuss different types of schemes designed to maintain this kind of privacy, including obfuscation-based schemes, social-based schemes, dynamic pseudonymity, and key anonymity.

8.1.4.3.1 Obfuscation-Based Schemes

Obfuscation was probably the first scheme to be employed to achieve location privacy. One of the first attempts in this area was proposed by Duckham and Kulik [30]. They use obfuscation in a formal framework to protect location privacy within a pervasive computing environment. The proposed framework provides a computationally efficient mechanism for balancing a user's need for high-quality information services against the user's need for location privacy. Negotiation is used to ensure that a location-based service provider receives only the information that it needs to know in order to provide satisfactory quality of service.

Ardagna et al. [31] propose a different obfuscation technique to protect the location privacy of users. Specifically, various obfuscation operators are presented by changing their location information. In addition, they introduce an adversary model and provide an analysis of the proposed obfuscation operators to evaluate their robustness against adversaries.

Another example of obfuscation-based location privacy approaches is social-based location privacy protocol [32]. It offers location privacy through a request/reply location obfuscation technique. This protocol uses the nodes' own social network to drive the forwarding heuristic and uses social ties among nodes to ensure k-anonymity. This can result in a noticeable improvement in location privacy for users accessing location-based services.

8.1.4.3.2 Social-Based Schemes

Location privacy is considered a big concern in social networks. This impacts some of the key features of these types of networks such as information sharing and content distributions. Fundamental research in this area has been done in Ref. [33], where privacy aspects in geosocial networks (GeoSNs) are studied. These types of social networks provide context-aware services that help associate location with users and content. Privacy aspects were classified in four categories (including location, absence, colocation, and identity privacy), and possible means of protecting privacy in these circumstances were described.

Another study was proposed by Li et al. [34], which addresses the location privacy issue for the nearby friend alert service, a common and fundamental service in mobile GeoSNs. In this paper, the grid-and-hashing paradigm was adopted, and an optimal grid overlay and multilevel grids were developed to increase the detection accuracy while saving the wireless bandwidth. Based on these techniques, the client-side location update scheme and the server-side update handling procedure for continuous proximity detection were devised.

Wei et al. [35] demonstrate a new type of user privacy attack and propose a solution for it. They used a fake location reporting technique to prevent an enemy from combining the location and friendship information found in MSNs. This solution does not require any additional trusted third-party deployment and can enhance location privacy.

A common problem with most MSN location privacy-preserving approaches is that the data-forwarding process can be interrupted or even disabled when the privacy preservations of users are applied. This is because users become unrecognizable to each other and the social ties and interactions are no longer traceable to facilitate cooperative data forwarding. Another problem with these approaches is that, to apply user cooperation, an intrusion must be exerted on user privacy. In order to solve such problems, social morality [36] was proposed. It is a protocol suite that achieves both privacy preservation and cooperative data forwarding in three steps. The first step is to provide a user's anonymized mobility information to the public using a privacy-preserving, route-based authentication scheme. The second step measures the proximity of the user's mobility information to a specific packet's destination and evaluates the user's forwarding capacity for the packet. The third step determines the optimal data-forwarding strategy according to morality level and payoff, using a game-theoretic approach.

8.1.4.3.3 Dynamic Pseudonymity

Anonymity can be provided via the frequent changing of pseudonyms, in order to make it difficult for adversaries to detect a user's movement, information, location, and so on. This technique has been widely adopted to provide location privacy. One of the finest approaches in this area was proposed by Magkos et al. [37]. In this paper, a distributed scheme is suggested that deals with both security and historical privacy in MSNs. In other words, it establishes access control while protecting the privacy of a user in sporadic and continuous queries. To maintain security, this scheme employs a hybrid network architecture that gives users an ability to communicate with a location-based service provider through a network operator. Using this architecture, users are able to create wireless ad hoc networks with other users in order to obtain privacy against an adversary that performs traffic analysis. To maintain historical privacy, this scheme adopts the generic approach of using multiple pseudonyms that are changed frequently. Messages are not sent directly to the cellular operator but are distributed among mobile neighbors, so they can re-encrypt the messages before sending them to the location-based service provider via the cellular operator. This makes messages untraceable against traffic analysis attacks.

Another important work regarding dynamic pseudonymity is a privacy-preserving location proof updating system for location-based services called APPLAUS [38]. It uses colocated Bluetooth devices, which enables mobile devices to mutually generate location proofs and updates to a location proof server. This work contains a user-centric location privacy model to evaluate user location privacy levels in real time and gives users the ability to accept a location proof exchange request based on their location privacy levels. In this method, mobile devices use periodically changed pseudonyms to protect source location privacy.

8.1.4.3.4 Key Anonymity

An important challenge in the wide deployment of location-based services is to provide safeguards for the location privacy of mobile clients against vulnerabilities for abuse. In order to achieve such a goal, key anonymity schemes have been widely deployed. For example, Gedik and Ling [39] develop a personalized location anonymization model and a suite of location perturbation algorithms to protect location privacy in the deployment of location-based services. This architecture takes advantage of a flexible privacy personalization framework to support location k-anonymity for a wide range of users with context-sensitive privacy requirements. The proposed framework enables each node to specify the minimum level of anonymity that it desires and the maximum temporal and spatial tolerances that it is willing to accept when requesting k-anonymity-preserving location-based services. The model is able to be run by the anonymity server on a trusted platform and can perform location anonymization on identity removal and spatiotemporal cloaking of the location information.

Traditional approaches to k-anonymity guarantee privacy over publicly released data sets with specified quasi-identifiers. However, due to the fact that common public releases of personal data are done through social networks and their API, these approaches are barely responsible for today's needs. In other words, k-anonymity in social networks does not allow clear assumptions about quasi-identifiers, which makes it impossible for traditional approaches to be responsible for their privacy needs. In order to address this problem, Social-K [40] suggests a new definition of k-anonymity. In this definition, social networks guarantee privacy in real time to users of their API. In order to achieve privacy while improving the key update efficiency of location-based services in vehicular ad hoc networks, a dynamic privacy-preserving key management scheme called DIKE [41] was proposed. It uses a particular type of privacy-preserving authentication technique that not only provides the vehicle user's anonymous authentication but also enables double-registration detection as well. DIKE updates keys by dividing the session of a location-based service into several time slots so that each time slot holds a different session key. When no vehicle user departs from the service session, each joined user can use a one-way hash function to autonomously update the new session key for achieving forward secrecy. In addition, this scheme integrates a dynamic threshold technique to achieve the session key's backward secrecy.

8.1.4.4 Communication Privacy

Preserving an individual's privacy when communicating is another issue in every type of network. In order to give proper solutions, different privacy-enhancing technologies have been developed. That includes technologies such as cryptography, authentication, and digital signatures. These technologies have various algorithms and protocols, which are used to a large extent in computer networks [42]. One example of the approaches in this area is Privacy-Enhanced yet Accountable seCurity framEwork (PEACE). It is a privacy-enhanced security framework consisting of authentication and key agreement protocols for wireless mesh networks (WMNs). These types of networks contain nodes with the ability to disseminate their own data and to propagate the data in the network. PEACE enforces strict user access control to cope with free riders and malicious users. In addition, it offers user privacy protection against adversaries and various other network entities.

Another example is the work proposed by Ardagna et al. [43], which presents a hybrid communication protocol to ensure mobile users' anonymity against various adversaries. It exploits the capability of handheld devices to connect to Wi-Fi and cellular networks. The authors consider all parties that can intercept communications between a mobile user and a server as potential privacy threats. In addition, they describe how a micropayment scheme that suits their mobile scenario can provide incentives for peers to collaborate in the protocol.

8.1.5 Future Directions

There are a number of privacy issues to examine for each MSN. Some of these have been studied previously in connection with traditional OSNs, but they bear reexamination for MSNs, while other issues are raised due to new features of MSN.

- Availability of user information within MSNs: What pieces of information are supplied by users for each of the MSNs and what are the default privacy settings for their availability to others within an MSN? In particular, how is the availability of a user's location and presence handled by each MSN?
- Interconnection of MSN: To what degree do MSNs have interconnections based on API connections with other MSNs, thereby potentially allowing the leakage of information to users in these other MSNs?
- Leakage to third parties: Beyond leakage of information within or across MSNs, to what extent is information about a user leaked to third parties, and does that differ across the various interfaces of each MSN? Moreover, are there new pieces of personally identifiable information, such as the individual device identifier of mobile devices, unique to the context of MSNs that are being leaked to third parties?

8.1.6 Conclusion

The concept of MSNs is a novel social communication paradigm that exploits opportunistic encounters between human-carried devices and social networks. Like any other emergent archetype of technology, it will take time to ensure that MSNs are totally safe and immune. Because of the social aspects that are included, MSNs encompass more complex and correlated challenging safety problems such as privacy issues and so on. In this section, the basic architecture of MSNs was analytically presented, and an overall view of privacy issues in this young and exciting field along with some privacy protection methods were provided. Finally, several future research directions were outlined.

8.2 Trust and Reputation

8.2.1 Introduction

The purpose of trust and reputation systems in MSNs is to strengthen the quality of markets and communities by providing an incentive for good behavior and quality services and by not sanctioning bad behavior and low-quality services. Therefore, a trust and reputation system (abbreviated TRS hereafter) represents an important decision support tool that can help reduce risk when engaging in MSN transactions and interactions on the Internet. From the individual relying party's viewpoint, a TRS can help reduce the risk associated with any particular interaction. From the service provider's viewpoint, it represents a marketing tool. From the community viewpoint, it represents a mechanism for social moderation and control, as well as a method to improve the quality of online markets and communities.

The same basic principles for creation and propagation of trust and reputation in traditional communities are also used by TRSs in MSNs. The main difference is that trust and reputation formation in traditional communities typically is relatively inefficient and relies on physical communication (e.g., through word of mouth), whereas TRSs in MSNs are supported by extremely efficient networks and computer systems.

The rest of the chapter is organized as follows. Section 2.2 discusses fundamentals of trust and reputation, such as definition and system classifications. Section 2.3 gives some common types of challenges and attacks for TRSs. Section 2.4 presents several existing mechanisms in TRSs, and Section 2.5 concludes this part of the chapter.

8.2.2 Fundamentals of Trust and Reputation

In this section, we will discuss the definition of trust in diverse disciplines and give a clear notion in MSNs. A brief review of trust definition is presented, and we review the literature for three different categories: trust information collection models, trust evaluation models, and trust dissemination models.

8.2.2.1 Definition of Trust in MSNs

Trust has been studied in many disciplines, including sociology, economics, philosophy, psychology, organizational management, and autonomic computing. Each of these disciplines has defined and considered trust from different perspectives, and their definitions may not be directly applicable to MSNs. The concept of trust also has been attractive to MSN protocol designers where trust relationships among participating nodes are critical in building cooperative and collaborative environments to optimize system objectives in terms of scalability, reconfigurability, reliability (i.e., survivability), dependability, or security (as shown in Figure 8.6).

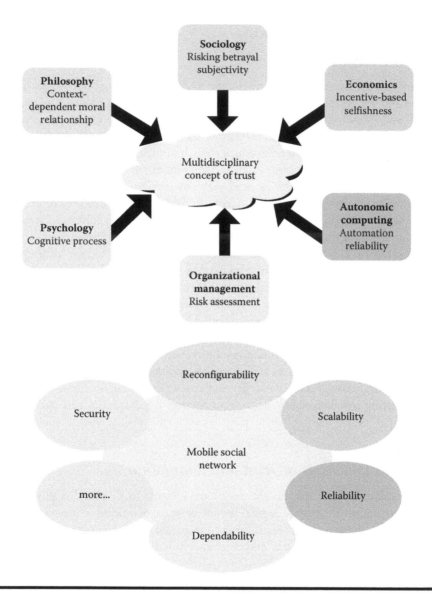

Figure 8.6 **The multidisciplinary concept of trust and its application in mobile social networks.**

8.2.2.2 Trust System Classification

In general, trust literature can be categorized based on three criteria: (i) trust information collection, (ii) trust value assessment, and (iii) trust value dissemination [44], as shown in Figure 8.7.

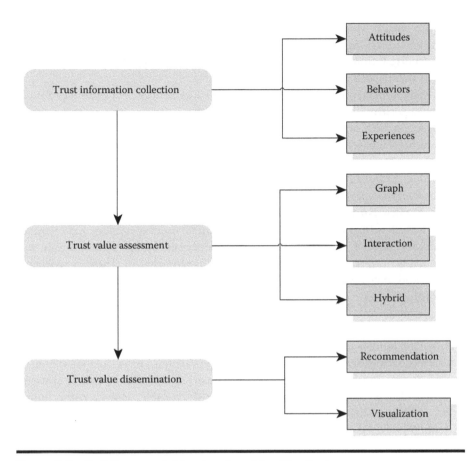

Figure 8.7 Building a social trust system (classification).

8.2.2.2.1 Trust Information Collection Models

In MSNs, trust information can be collected from three main sources: (i) attitudes, (ii) behaviors, and (iii) experiences.

8.2.2.2.1.1 Attitudes — Attitude represents an individual's degree of like or dislike for something. Attitudes are generally positive or negative views of a person, place, thing, or event. People can be conflicted or ambivalent toward an object, meaning they can simultaneously possess positive and negative attitudes toward the item in question. Attitudes are judgments developed on the affect, behavior, and cognition model. An effective response is an emotional response that expresses an individual's degree of preference for an entity. A behavioral intention is a typical behavioral tendency of an individual. A cognitive response is a cognitive evaluation of the entity that constitutes an individual's beliefs about the object. Most attitudes are the result of either direct experience or of observational learning from the environment. It is reported that these attitudes are transmitted from parents to children, indicating that these attitudes can represent the dispositional aspect of trust.

Attitude information is derived from a user's interactions. For example, a user expresses a positive/negative view toward an activity, a service, or a process during an interaction with another member in the community. Relatively little research effort has been focused on using attitude information to derive trust.

8.2.2.2.1.2 Experiences — Experiences describe the perception of the members in their interactions with each other. Experience information can be implicit or explicit. Explicit experiences are direct firsthand interactions with another member. Feedback mechanisms are tools for reflection on direct user experiences. Experiences may affect attitudes or behaviors. Positive experiences encourage members to interact frequently in the community, leading to a change in behavior. These experiences may also lead to changes in attitudes and make the member more receptive toward similar information.

In computer science, a number of experience-based trust management frameworks have been proposed. Reviews such as Sabater and Sierra, and Josang et al., discuss trust models ranging from Bayesian networks to belief models, all based on user feedback/experiences. Paradesi et al. [45] present an integrated trust model for Web service composition. This model progressively updates the beliefs of users of a particular Web service when the feedback is available. However, the foundation of this model is users' feedback/experience rather than their behavior. All reputation-based trust management models use user experience as the main source of trust information. Experiences provide one aspect of trust information in social networks and need to be considered along with other factors, namely attitudes and behaviors.

8.2.2.2.1.3 Behaviors — Behaviors are identified by patterns of interactions. For example, if a member is a highly active participant and suddenly stops participating, this change in behavior is noticeable and might imply that this member's trust in the community or in the member(s) with whom he or she had been frequently interacting has decreased. Findings and observations on behavioral aspects of users have been widely reported in psychology, social science, behavior science, and design systems. The behavior of individuals in a society reflects much more about them than simply their past experiences. Therefore, user behaviors provide an interesting aspect of trust in online communities.

Caverlee et al. [46] propose tracking user behavior over time to encourage long-term good behavior and to penalize bad/misbehavior in MSNs. Adali et al. [47] propose algorithmically quantifiable measures of trust based on communication behavior of members in an MSN, with the premise that trust results in likely communication behaviors that are statistically different from random communications.

Identifying such trust-like behaviors then enables the system to detect who trusts whom in the network.

User behavior is an important aspect of trust in social networks. Current literature has focused on behaviors separately from the other aspects, namely attitudes and experiences. Future research needs to focus on a combination of all aspects for holistic analysis of trust in social networks.

8.2.2.2.2 Social Trust Evaluation Models

Approaches to trust computation can be roughly categorized as network-based trust models, interaction-based trust models, and hybrid trust models. The majority of the literature leverages on the social network structure.

8.2.2.2.2.1 Network Structure/Graph-Based Trust Models — A trust network is usually created for each member. It represents the other members in the person's social network as nodes and the amount of trust he or she has for each of them as the edges. Various approaches are then used to traverse the network and determine trust between any two nodes. These approaches exploit the propagative nature of trust for trust evaluation. Some of the major works are reviewed as follows.

Hang and Singh [48] employ a graph-based approach for measuring trust, with the aim of recommending a node in a social network using the trust network. The model uses the similarity between graphs to make recommendations. Zuo et al. [49] propose an approach for computing trust in social networks using a set of trust chains and a trust graph. The model uses a trust certificate graph and calculates trust along a trust chain. In addition, it exploits the composability of trust in the form of fusion of relevant trust chains to form a base trust chain set.

Caverlee et al. [46] propose a social trust model that exploits social relationships and feedback to evaluate trust. Members provide feedback ratings after they have interacted with another member. The trust manager combines these feedback ratings to compute the social trust of the members. A member's feedback is weighted by their link quality (high link quality indicates more links with members having high trust ratings). Kuter et al. [50] focus on the trust on information obtained from different social channels/chains. They propose a Bayesian trust inference model for estimating the confidence in the trust information obtained from specific sources. In contrast to most "Web of trust" mechanisms that derive data from users' feedback, Kim et al. [51] explores building a Web of trust without using explicit user ratings. The model considers a user's reputation (expertise) and affinity for certain contexts/topics as the main factors for deriving trust connectivity and trust value. For example, user A is interested in reading science fiction novels and watching drama movies. User B writes reviews of movies. User C writes reviews of books. User A trusts user B in the context of movies and C in the context of books.

The context can be derived from the user A, B, and C's activities in the online community. This means a trust value is derived based on the reputation of a provider such as B or C and on the affinity of the user for different topics such as movies and books. This provides a much denser "Web of trust." The approach has three steps: (i) calculating users' expertise on a certain topic, which involves calculating the quality of reviews using the reputation of raters and then the reputation of writers, (ii) calculating the users' affinity for the category: the user's affinity is derived from the average number ratings and reviews provided by them in each category, and (iii) deriving degree of trust: the trust is derived from the user's affinity for the topic and another user's expertise on the topic.

Maheswaran et al. [52] propose a gravity-based model for estimating trust. The trust model has two stages: first, the strengths of the friendships are recomputed along with the extent of the trusted social neighborhood for each user. This is based on the user's annotations of the connections he or she has with others with trust values or constraints. Second, the social neighborhood is used to compute the effective trust flow for users not in the social neighborhood. The model is based on the presumption that social relationships change over time and social relations may impose constraints on the trust relationships. The model also includes provenance of trust in social networks, where a user can ask questions about the trust value given by the underlying system.

Approaches that leverage network structures to compute trust capture one aspect of trust computation, namely how members are related to each other and how trust flows through their network. They fail, however, to capture actual interactions between members. The volume, frequency, and even the nature of interaction are important indicators of trust in social networks. We now discuss trust models that use interactions as the basis for trust evaluation.

8.2.2.2.2.2 Interaction-Based Trust Models — In contrast to the models presented in the previous section, some trust models in the literature only use interactions within the network to compute social trust. Liu et al. [53] propose an approach for predicting trust in online communities using the interaction patterns/behaviors of the users. They identify two types of taxonomies representing user actions and possible interactions between pairs in the community: (i) user action taxonomy for shared data such as reviews, posted comments, rating, and so on with metrics such as number/frequency of reviews, number/frequency of ratings, and average length/number of comments given to reviews; and (ii) pair interaction taxonomy for different possible interactions/connections that could happen between two users (e.g., connections between writers and raters, writers and writers, and raters and raters). The model also considers the time difference between two users' respective actions, which forms the connection and is called the "temporal factor." They describe a supervised learning approach that automatically predicts trust between a pair of users using evidence derived from actions of individual users (user factors) as well as from interactions between pairs of users (interaction factors). These factors are then used to derive the corresponding features for training classifiers that predict trust between pairs of users.

Nepal et al. [54] propose STrust, a social trust model based only on interactions within the social network. The model consists of two types of trust: popularity trust refers to the acceptance and approval of a member in the community, representing the trustworthiness of the member from the perspective of other members in the community, and engagement trust refers to the involvement of the member in the community, representing the trust the member has toward the community. Popularity trust is derived from metrics such as how many members follow, read, and provide positive feedback on the member's posts. Engagement trust is derived from metrics such as how frequently the member visits the site/network, how many members he or she follows, and how many posts he or she reads and comments on. A combination of popularity trust and engagement trust forms the basis for determining the social trust in the community. The model aims to increase the social capital of the community by encouraging positive interactions within the community and, as a result, to increase the social trust in the community.

Similarly, Adali et al. [47] evaluate trust based on communication behavior of members in a social network. Behavioral trust is calculated based on two types of trust: conversation trust and propagation trust. Conversation trust specifies how long and/or how frequently two members communicate with each other. Longer and/or more frequent communication indicates more trust between the two parties. Similarly, propagation of information obtained from one member to other members in the network indicates that a high degree of trust is placed on the information and implicitly on the member that created the information.

Interaction-based social trust models consider interactions in the community to compute trust but ignore the social network structure. Social network structure provides important information about how members in a community relate to each other and is a significant source of information for social trust computation. Therefore, social trust models should consider both graph structures and interactions within the social networks to compute social trust. These models can be considered to be hybrid models. We discuss these models next.

8.2.2.2.2.3 Hybrid Trust Models — Hybrid trust models use both interactions and social network structure to compute social trust. Trifunovic et al. [55] propose such a social trust model for opportunistic networks. Opportunistic networks enable users to participate in various social interactions with applications such as content distribution and microblogs. The model leverages the social network structure and its dynamics (conscious secure pairing and wireless contacts) and proposes two complementary approaches for social trust establishment: explicit social trust and implicit social trust. Explicit social trust is based on consciously established social ties. Each time two users interact, they exchange their friend lists and save them as friendship graphs. Trust is calculated on the friendship graph with a direct link/friend having the highest trust

value of 1. As the number of links between two users grows, trust decreases proportionately. Implicit trust is based on frequency and duration of contact between two users. It uses two metrics: familiarity and similarity of the nodes. Familiarity describes the length of the interactions/contacts between the two nodes. Similarity describes the degree of coincidence of the two nodes' familiar circles. In this model, explicit social trust evaluation is based on network structure, whereas implicit trust evaluation is based on user's interactions in the network. This work considers only the duration and frequency of interactions. However, the nature of interactions is an important indicator of trust between two members. If two members interact frequently but the interaction is negative (e.g., they are having an argument), then this does not indicate trust between the members. In such cases, this model would consider trust to be low. The literature on hybrid social trust models is limited. This presents an interesting area for further research.

8.2.2.2.3 Trust Dissemination Models

There could be many ways of disseminating trust information. Recommendation is one approach for trust value dissemination within a social network, and visualization is another. Both are discussed in the following subsections.

8.2.2.2.3.1 Trust-Based Recommendation Models — Trust-based recommendation usually involves constructing a trust network where nodes are users and edges represent the trust placed on them. The goal of a trust-based recommendation system is to generate personalized recommendations by aggregating the opinions of other users in the trust network.

Recommendation techniques that analyze trust networks were found to provide very accurate and highly personalized results. As described earlier, Hang and Singh [48] use a graph-based approach to recommend a node in a social network using similarity in trust networks. Andersen et al. [56] explore an axiomatic approach for trust-based recommendation and propose several recommendation models, some of which are incentive compatible (i.e., malicious members cannot entice other members to provide false/misleading trust information and trust links because it is always in the interest of the member to provide factual information).

Hess [57] extends trust-based recommendations for single items, such as movies, to linked resources. For this purpose, she builds a second type of network, called a document reference network. Recommendations for documents are typically made by reference-based visibility measures that consider a document to be more important if it is often referenced by important documents. Document and trust networks, as well as networks such as organization networks, are integrated in a multilayer network. This architecture allows for combining classical visibility measures with trust-based recommendations, giving trust-enhanced visibility measures.

Trust-based recommendation techniques provide a way of disseminating trust information within a social network. However, the social network providers would prefer to have a bird's-eye view of trust in the social network at any point in time. Trust visualization provides this mechanism for trust dissemination at the network level.

8.2.2.2.3.2 Visualization Models — Visualization of trust connections as a graph is another means for disseminating trust information. Graphs show the strength of connection between two nodes (i.e., the connection routes between two nodes)—a higher number of connections means a closer relationship. Many social network visualization tools are available on the Internet, such as social network visualization* (SocNetV), network visualization† (NetVis), and graph visualization‡ (Graphviz), among others.

O'Donovan et al. [58] propose a model that extracts negative information from the feedback comments on eBay, computes personalized and feature-based trust, and presents this information graphically. The graph shows the trust value and the trust strength calculated based on the number of transactions/comments between two users. Guerriero et al. [59] propose a trust-based visualization of cooperation context between members.

Trust visualization approaches are very useful for the social network providers to analyze and determine the level of trust in the community. It helps them identify the most and least trustworthy members. In addition, the providers can take preemptive action such as introducing new interesting and relevant material/information if the trust level in the community decreases below a certain threshold. Trust visualization allows the provider to control the social network to encourage positive behavior and to discourage disruptive behavior [44].

8.2.2.3 Definition of Reputation in MSNs

Reputation systems represent a significant trend in decision support for Internet-mediated service provision. The *Feedback Forum* on eBay is the most prominent example of online reputation systems. The basic idea is to let parties rate each other (e.g., after the completion of a transaction) and to use the aggregated ratings about a given party to derive a reputation score, which can assist other parties in deciding whether to transact with that party in the future. A natural effect is that it also provides an incentive for good behavior and therefore tends to have a positive impact on market quality. Reputation is generally defined as the opinion or view of one about the character of somebody or of an entity. Here, an entity could be an agent, a product, or a service. Reputation is frequently used as the basis of a judgment to

---

* http://socnetv.sourceforge.net
† http://www.netvis.org/index.php
‡ http://www.graphviz.org/

trust an individual or organization particularly in the absence of previous direct experience or contact with them. Mui et al. [60] define reputation as "a perception that an agent creates through past actions about its intentions and norms." A similar definition is given by Abdul-Rahman et al. [61]: "a reputation is an expectation about an agent's behavior based on information about or observations of its past behavior." The *Concise Oxford Dictionary* definition of reputation supports the view of MSN researchers. Reputation is what is generally said or believed about a person's or thing's character or standing.

8.2.2.4 Reputation System Classification

In this subsection, representative reputation systems will be reviewed, such as those used by Amazon, YouTube, Digg, and CitySearch.

1. Evidence collection: A reputation system can obtain three types of evidence. The first type is direct observation, usually based on the experiences of the employees of a business (e.g., ConsumerReports.org). The second type is opinions from experts who have verifiable expertise and provide feedback either voluntarily or for a fee. Both types of evidence are considered reliable, but they are costly to collect for a large number of objects. The third type is feedback provided by users, which has been the main source of evidence in most of today's popular reputation systems, such as the product rating system on Amazon.com, restaurant ratings on Yelp.com, and customer reviews at the Apple App Store. However, user feedback is also the least reliable source of evidence because it can be easily manipulated.

2. Reputation aggregation: Reputation aggregation algorithms calculate the reputation scores of objects based on the collected evidence. A good reputation aggregation scheme should be able to compute reputation scores that accurately describe the true quality of objects, even if there is dishonest feedback.

3. Reputation dissemination: Reputation systems not only make the reputation scores publicly available but also release extra information to help users understand the meaning of them. For example, Amazon shows all feedback given by each reviewer. YouTube has started to provide visualization of viewing history for video clips, accompanied with some statistical features [62].

8.2.3 Challenges and Attacks for Trust and Reputation Systems

Attempts to misrepresent reliability and to manipulate reputation are common in traditional communities. Con artists employ methods to appear trustworthy (e.g., through skillful acting or through the fabrication and presentation of false credentials). Similar types of attacks would also apply to online communities.

In case some form of TRS is being used to moderate an online community or market, vulnerabilities in the TRS itself can open up additional attack vectors. It is therefore crucial that TRSs are robust against attacks that could lead to misleading trust and reputation scores. In the worst case, a vulnerable TRS could be turned around and used as an attack tool to maliciously manipulate the computation and dissemination of scores. The consequence of this could be a total loss of community trust caused by the inability to sanction and avoid low quality and deceptive services.

When attacks against a TRS occur, it does not normally mean that a server hosting TRS functions is being hacked. Attacks on TRSs typically consist of those playing the role of relying parties and/or service entities and of manipulating the TRS through specific behavior that is contrary to an assumed faithful behavior. For example, a relying party that colludes or that is identical to the service entity could provide fake or unfair positive ratings to the TRS with the purpose of inflating the service entity's score, which in turn would increase the probability of that service entity being selected by other relying parties.

Many other attack scenarios can be imagined that, if successful, would give unfair advantages to the attackers. All such attacks have in common the fact that they result in the erosion of community trust, which in turn would be damaging to services and applications in the affected market or community. Types of threats against TRSs are presented in the next section [63].

8.2.3.1 Playbooks

A playbook consists of a sequence of actions that maximizes profit or fitness of a participant according to certain criteria. A typical playbook example is to act honestly and provide quality services over a period to gain a high reputation score and then to subsequently profit from the high reputation score by providing low-quality services (at a low production cost). There will be an infinite set of possible playbook sequences, and the actual profit resulting from any particular sequence will be influenced by the actions (and playbooks) of other participants in the community. It should be noted that using playbooks is not necessarily unethical because explicitly generating oscillation in a brand's reputation is commonly used by commercial players (e.g., by brands of consumer goods). In addition, participant agents monitor each other's performance in numerous natural and artificial agent communities similarly to the way a TRS does and try to define the optimal strategy for own survival or to maximize own fitness. This topic is being studied extensively in economics and artificial agent literature. It is only when the norms of a community dictate that specific types of playbooks are unethical that this can be considered as an attack.

8.2.3.2 Unfair Ratings

This attack consists of providing ratings that do not reflect the genuine opinion of the rater. This behavior would be considered unethical in most communities and therefore represents an attack. However, it can be extremely difficult to determine when this attack occurs because agents in a community do not have direct access to each other's genuine opinions; they only see what other agents choose to express. A strategy for detecting possible unfair ratings often proposed in the literature is to compare ratings about the same service entity provided by different agents and to use ratings from a priori trusted agents as a benchmark. However, this method can lead to wrong conclusions in specific scenarios, such as in case of discrimination (described in the next section). It is relatively easy to detect unfair ratings in situations where the quality of the rated or recommended service can be objectively assessed, meaning that the rating can be directly compared to objective quality criteria.

8.2.3.3 Discrimination

Discrimination means that a service entity provides high-quality services to one group of relying parties and low-quality services to another group of relying parties. This behavior can have very different effects on the service entity's score depending on the specific TRS being used. For example, if a TRS uses a method to detect unfair ratings based on comparing ratings from unknown agents with ratings from a priori trusted agents, it is sufficient for the attacker to provide high-quality services to the a priori trusted agents. The method for detecting unfair ratings will have the paradoxical effect that genuine negative ratings will be rejected by the TRS, so that the attacker will not be sanctioned for providing low-quality services.

8.2.3.4 Collusion

Collusion means that a group of agents coordinate their behavior, which can consist of running playbooks, providing unfair recommendations, or discrimination. Clever collusion can have a significant influence on scores and thus increase the profit or fitness of certain agents. Collusion is not necessarily unethical (e.g., if it consists of controlling the quality of services provided by a group in a coordinated fashion). However, if the collusion consists of coordinating unfair ratings, it would be clearly unethical.

8.2.3.5 Proliferation

When faced with multiple sellers, a relying party will choose randomly between equal sellers. By offering the same service through many different channels, a single agent will be able to increase the probability of being chosen by a relying party. Proliferation is not necessarily unethical. For example, in some markets, it is common that the same product is being marketed by multiple representatives. However, proliferation can be considered unethical in cases where multiple representations of the same service entity pretend to represent different and independent service entities.

8.2.3.6 Reputation Lag Exploitation

There usually is a time lag between an instance of service provision and the corresponding rating's effect on that service entity's score. Exploiting this time lag to provide a large number of low-quality services over a short period before the rating suffers any significant degradation would normally be considered unethical.

8.2.3.7 Reentry

Reentry means that an agent with a low score leaves a community and subsequently reenters the community under a different identity. The effect is that the agent can start fresh and thereby avoid the consequences of the low score associated with the previous identity. This would be considered unethical in most situations. Reentry is a vulnerability caused by weak identity and the inability to detect that two different identities represent the same entity.

Theoretically, an entity can have multiple identities within the same domain or in different domains. Each identity normally consists of a unique identifier (relative to a domain) and possibly other attributes. There are two cases to consider. If a reliable mapping is known between an entity's identity in the community and the same entity's identities in other domains (external identities), then it is technically possible to detect reentry. For example, when the real name of an online user is known in the online community, detection of reentry will be possible. On the contrary, if no mapping is known between an entity's identity within the community and external identities, then reentry will be hard to detect. Even though such mappings may not be publicly known within a community, they may be known by certain parties such as identity providers. These parties could assist in detecting reentry.

8.2.3.8 Value Imbalance Exploitation

Ratings provided to a TRS typically do not reflect the value of the corresponding transaction. The effect of providing a large number of high-quality, low-value services and a small number of deceptive high-value services would then result in a high profit, resulting from high value deception without any significant loss in scores. This behavior is only unethical to the degree that providing deceptive services is unethical in a particular market. If the service entity simply provides low-quality, high-value services that cannot be considered deceptive, then the behavior could be considered ethical. This threat is related to the problem of mismatch between trust scopes. Weighing ratings as a function of service value is a simple method to remedy this vulnerability.

8.2.4 Typical Trust and Reputation Systems

This section presents information about current trust and reputation management systems. To date, there are numerous trust solutions have been proposed for each environment (i.e., peer-to-peer, multiagent system, e-commerce, etc.) [64].

8.2.4.1 Peer-to-Peer Multidimensional Trust Model

Ion et al. [65] propose a multidimensional trust model that encompasses several dimensions of trust values, such as users, knowledge, services or nodes, and social inter-institutional. In this multidimensional model, every entity keeps a list of opinions of other entities, data, services, and nodes. A distributed knowledge base (DKB) is used to search and update these lists. Each entity has a contact list showing all trusted entities in a specific context and trust value. The trust value for an unknown entity is computed based on the reputation of that entity based on the trusting entity's contacts. In the list of opinions, each entity would keep his or her experiences with other entities. Each opinion consists of subject, object, keywords, and value. Subject is an entity that provides the opinion about the object or target entity; keywords provide the contextual information about the opinion; and value contains the credibility value of this opinion. The trust rating is represented as probabilistic values ranging from 0 (no trust) to 1 (complete trust).

This model allows the referral method in which a contact is able to retrieve the values of a target entity from other contacts. Further, in order to keep track of an entity's actions in a system, this model uses the concept of credential provider (CP) that is commonly used in identity management. CP is used to authenticate the entities in each service provider and also to establish trust relationships with another CPs. In an event where an entity uses the certificate from his or her certificate authority (CA), the trust value of this user is automatically computed based on the CA's value in a particular CP. A DKB is used to register the institutions and to keep the metadata certificate and social status of the institutions. Therefore, it can be inferred that this model is considered a centralized model where central servers provide the management of trust.

8.2.4.2 DEco Arch

DEco Arch [66] is a trust and reputation service brokering architecture that is tailored for digital ecosystems. DEco Arch addresses some issues faced by the Universal Description, Discovery and Integration (UDDI) directory service as a central registry for service brokering (such as issues in centrally managed registry service, issues of nontransparent trustworthiness value and reputation rankings, etc.). This model is applied on semantic peer-to-peer architecture, such as Juxtapose (JXTA), and it stores the trustworthiness and reputation values in the Distributed Hash Database (DHT). DEco Arch provides the service discovery by matching the semantic attribute of service consumers with the services or product descriptions in a resource description format (RDF).

DEco Arch introduces the group alliance concept in which services are grouped together based on their semantic similarities. Each group alliance has reputation value ranging from –1 (unknown) to 0 (very bad reputation) to 5 (very good reputation). This reputation value is associated with the trend value (i.e., decreasing, neutral, and increasing) and the confidence value (0, 1) that depend on the number of contributing entities. The reputation value of a group alliance is computed as the weighted average of the aggregated reputation values of its members, while the individual reputation value is computed based on the membership percentage. The inclusivity of a new service provider into a group alliance is decided based on its semantic matching degree and on its individual reputation value. The group alliance rejects the service provider whose reputation value is low because it would affect the reputation value of the alliance. Further, it puts the unknown or new service provider in a sandbox until it gains a better reputation value. In case a dishonest entity provides a wrong opinion, authors argue that the credibility value of this entity is only included in the trustworthiness computation during the selection process.

8.2.4.3 TRAVOS

TRAVOS [67] is a trust model that has been proposed for managing trust in a multiagent environment, such as a grid environment. It uses the probabilistic and beta distribution (Baycsian) method to measure a probability that a provider agent will be trustworthy in fulfilling its obligation. TRAVOS measures the trustworthiness of a provider agent based on two criteria: (i) the consumer agent's past experience (direct interactions with the provider agent) and (ii) the provider's reputation (ratings) that are perceived by other agents (raters). For the first criteria, TRAVOS uses probability density function (pdf) to model the probability of a random variable in which the beta family of the pdf is used to measure the probability that a provider's agent will fulfill its obligation.

To compute the provider's reputation as perceived by other agents, TRAVOS first estimates the accuracy (credibility) of the ratings based on all accurate and inaccurate advice provided by the raters in the past. The accuracy of ratings is measured through beta distributions similar to the computation of a consumer's past experience. Once the accuracy of ratings is computed, it then adjusts these ratings according to their accuracy. TRAVOS includes only the accurate ratings for its computation and discards those ratings that are inaccurate. The approach that is taken by TRAVOS in measuring the accuracy of ratings differs significantly from other similar Bayesian approaches in the literature, such as the Beta Reputation System (BRS). In the BRS, an agent's rating that differs significantly from the majority of ratings is deemed inaccurate and is discarded from computation.

8.2.4.4 Measuring Trustworthiness Using a Personalized Approach

Zhang and Cohen [68] proposed a personalized method for measuring the trustworthiness of sellers in e-marketplaces. This trust model is classified as a binary model that takes into account the buyer's past interactions with a seller and also the reputations of a seller. In addition, a novel approach for estimating the credibility of ratings provided by the raters is presented. Authors used two methods for measuring the credibility of ratings: (i) a private method—a buyer estimates the reputation (credibility) of a rater based on the ratings on the same set of sellers and (ii) a public method—all ratings of the sellers that a rater has ever provided are considered. In the private method, a buyer will retrieve all sellers' ratings that each rater has provided. The buyer then identifies a set of sellers that he or she and the rater have both rated. He or she then compares his or her ratings and the ratings provided by the other rater on the same set of sellers within a given context and time. Such comparison measures the credibility of the rater.

The public method is introduced to manage the unavailability of private knowledge that the buyer has about the rater (e.g., buyer never request ratings from the rater). The public method requires central servers to store all ratings provided by all raters in the environment. These central servers are responsible for measuring the credibility of raters by comparing the similarity measures between the ratings provided by each rater and the overall ratings provided by the majority of raters on a same set of sellers. Note that the centralized structure suffers from single point control and failure.

8.3 Conclusion

Trust and reputation systems represent a significant evolution in support for Internet services, especially in helping users decide among a growing number of choices—from which movies to rent to which data sources to trust. However, there are also some threats against trust and reputation systems in MSNs. Most trust and reputation systems have serious vulnerabilities, and it is questionable whether any trust and reputation systems can be considered to be robust. In this section, the basic review of trust and reputation were analytically presented, including their notions and classifications. Challenges and attacks for trust and reputation systems in MSNs were discussed. Finally, several existing trust and reputation schemes were briefly introduced.

References

1. Shen, X. S. Security and privacy in mobile social network. *IEEE Network* 2013; 27(5): 2–3.
2. Madden, M. and A. Smith. Reputation management and social media. 2010. http://www.pewinternet.org/2010/05/26/reputation-management-and-social-media/

3. Hoofnagle, C. J., K. Jennifer, L. Su, and T. Joseph. How different are young adults from older adults when it comes to information privacy attitudes and policies? Available from: http://ssrn.com/abstract=1589864 [cited April 14, 2010].

4. Kemp, J. and F. Reynolds. Mobile social networking two great tastes. In: Proceedings W3C Workshop on the Future of Social Networking, Barcelona, Spain, January 15–16, 2009.

5. Gupta, A., A. Kalra, D. Boston, and C. Borcea. MobiSoC: A middleware for mobile social computing applications. *Mobile Network and Applications* 2009; 14(1): 35–52.

6. Ma, L., Z. Jia, and J. Liu. MoViShare: Building location-aware mobile social networks for video sharing. Burnaby: School of Computing Science, Simon Fraser University, 2009.

7. Karam, A. and N. Mohamed. Middleware for mobile social networks: A survey. In: Proceedings of the 45th Hawaii International Conference on System Science (HICSS), Maui, HI, January 4–7, 2012.

8. Miluzzo, E., N. D. Lane, K. Fodor, R. Peterson, H. Lu, M. Musolesi, et al. Sensing meets mobile social networks: The design, implementation and evaluation of the CenceMe application. In: Proceedings of the 6th ACM Conference on Embedded Network Sensor Systems (SenSys), Raleigh, NC, November 5–7, 2008.

9. Chang, Y. J., H. H. Liu, L. D. Chou, Y. W. Cheng, and H. Y. Shin. A general architecture of mobile social network services. In: Proceedings of the IEEE International Conference on Convergence Information Technology, Gyeongju, Korea, November 21–23, 2007.

10. Pietiläinen, A. K., E. Oliver, J. LeBrun, G. Varghese and C. Diot. MobiClique: Middleware for mobile social networking. In: Proceedings of the 2nd ACM Workshop on Online Social Networks, Barcelona, Spain, August 17, 2009.

11. Sarigöl, E., O. Riva, P. Stuedi, and G. Alonso. Enabling social networking in ad hoc networks of mobile phone. *Proceedings of the VLDB Endowment* 2009; 2(2): 1634–1637.

12. Pelusi, L., A. Passarella, and M. Conti. Opportunistic networking: Data forwarding in disconnected mobile ad hoc networks. *IEEE Communications Magazine* 2006; 44(11): 134–141.

13. Han, B., P. Hui, V. S. A. Kumar, M. V. Marathe, J. H. Shao, and A. Srinivasan. Mobile data offloading through opportunistic communications and social participation. *IEEE Transactions on Mobile Computing* 2012; 11(5): 821–834.

14. Vastardis, N. and K. Yang. Mobile social networks: Architectures, social properties and key research challenges. *IEEE Communications Surveys & Tutorials* 2012; 15(3): 1355–1371.

15. Krishnamurthy, B. and C. E. Wills. Privacy leakage in mobile online social networks. In: Proceedings of the 3rd Conference on Online Social Networks (WOSN), Boston, MA, June 22, 2010.

16. Guha, S., K. Tang, and P. Francis. NOYB: Privacy in online social networks. In: Proceedings of the 1st Conference on Online Social Networks (WOSN), Seattle, WA, August 18, 2008.

17. Dóra, L. and T. Holczer. Hide-and-Lie: Enhancing application-level privacy in opportunistic networks. In: Proceedings of MobiOpp, Pisa, Italy, February 22–23, 2010.

18. Narayanan, A. and V. Shmatikov. De-anonymizing social networks. In: Proceedings of the 30th IEEE Symposium on Security and Privacy, Oakland, CA, May 17–20, 2009.

19. Wondracek, G., T. Holz, E. Kirda, and C. Kruegel. A practical attack to de-anonymize social network users. In: Proceedings of IEEE Symposium on Security and Privacy, Oakland, CA, May 16–19, 2010.

20. Parris, I., G. Bigwood, and T. Henderson. Privacy-enhanced social network routing in opportunistic networks. In: Proceedings of IEEE PERCOM, Mannheim, Germany, March 29–April 2, 2010.
21. Xuan, D., Z. Lan, W. Zhiguo, and G. Ming. De-anonymizing dynamic social networks. In: Proceedings of IEEE GLOBECOM, Houston, TX, December 5–9, 2011.
22. Xie, Q. and U. Hengartner. Privacy-preserving matchmaking for mobile social networking secure against malicious users. In: Proceedings of the IEEE Ninth Annual International Conference on Privacy, Security and Trust (PST), Montreal, QC, July 19–21, 2011.
23. Freedman, M., K. Nissim, and B. Pinkas. Efficient private matching and set intersection. In: Proceedings of the EUROCRYPT, Interlaken, Switzerland, May 2–6, 2004.
24. Narayanan, G. S., T. Aishwarya, A. Agrawal, A. Patra, A. Choudhary, and C. P. Rangan. Multi party distributed private matching, set disjointness and cardinality of set intersection with information theoretic security. In: Proceedings of CANS, Kanazawa, Japan, December 12–14, 2009.
25. Li, R. and C. Wu. An unconditionally secure protocol for multi-party set intersection. In: Proceedings of ACNS, Zhuhai, China, June 5–8, 2007.
26. Li, M., N. Cao, S. Yu, et al. FindU: Privacy-preserving personal profile matching in mobile social networks. In: Proceedings of IEEE INFOCOM, Shanghai, China, April 10–15, 2011.
27. Arb, M. V., M. Bader, M. Kuhn, and R. Wattenhofer. VENETA: Serverless friend-of-friend detection in mobile social networking. In: Proceedings of IEEE WIMOB, Avignon, France, October 12–15, 2008.
28. Wei, D., V. Dave, Q. Lili, and Z. Yin. Secure friend discovery in mobile social networks. In: Proceedings of IEEE INFOCOM, Shanghai, China, April 10–15, 2011.
29. Zhang, R., Y. Zhang, J. Sun, and G. Yan. Fine-grained private matching for proximity-based mobile social networking. In: Proceedings of IEEE INFOCOM, Orlando, FL, March 25–30, 2012.
30. Duckham, M. and L. Kulik. A formal model of obfuscation and negotiation for location privacy. In: Proceedings of PERVASIVE, Munich, Germany, May 8–13, 2005.
31. Ardagna, C. A., M. Cremonini, S. De Capitani di Vimercati, and P. Samarati. An obfuscation-based approach for protecting location privacy. *IEEE Transactions on Dependable and Secure Computing* 2011; 8(1): 13–27.
32. Zakhary, S. and M. Radenkovic. Utilizing social links for location privacy in opportunistic delay-tolerant networks. In: Proceedings of IEEE ICC, Ottawa, ON, June 10–15, 2012.
33. Vicente, C. R., D. Freni, C. Bettini, and C. S. Jensen. Location-related privacy in geo-social networks. *IEEE Internet Computing* 2011; 15(3): 20–27.
34. Li, H., H. Hu, and J. Xu. Nearby friend alert: Location anonymity in mobile geo-social networks. *IEEE Pervasive Computing* 2013; 12(4): 62–70.
35. Wei, C., W. Jie, and C. C. Tan. Enhancing mobile social network privacy. In: Proceedings of IEEE GLOBECOM, Houston, TX, December 5–9, 2011.
36. Liang, X., X. Li, T. H. Luan, R. Lu, X. Lin, and X. Shen. Morality-driven data forwarding with privacy preservation in mobile social networks. *IEEE Transactions on Vehicular Technology* 2012; 61(7): 3209–3222.
37. Magkos, E., P. Kotzanikolaou, S. Sioutas, et al. A distributed privacy-preserving scheme for location-based queries. In: Proceedings of the World of Wireless Mobile and Multimedia Networks (WoWMoM), Montreal, QC, June 14–17, 2010.

38. Zhichao, Z. and C. Guohong. APPLAUS: A privacy-preserving location proof updating system for location-based services. In: Proceedings of IEEE INFOCOM, Shanghai, China, April 10–15, 2011.

39. Gedik, B. and L. Ling. Protecting location privacy with personalized k-anonymity: Architecture and algorithms. *IEEE Transactions on Mobile Computing* 2008; 7(1): 1–18.

40. Beach, A., M. Gartrell, and R. Han. Social-K: Real-time K-anonymity guarantees for social network applications. In: Proceedings of IEEE PERCOM, Mannheim, Germany, March 29–April 2, 2010, pp. 600–606.

41. Lu, R., X. Lin, X. Liang, and X. Shen. A dynamic privacy-preserving key management scheme for location-based services in VANETs. *IEEE Transactions on Intelligent Transportation Systems* 2012; 13(1): 127–139.

42. Najaflou, Y., B. Jedari, F. Xia, et al. Safety challenges and solutions in mobile social networks. *IEEE Systems Journal* 2013; 99: 1–21.

43. Ardagna, C. A., S. Jajodia, P. Samarati, and A. Stavrou. Providing users' anonymity in mobile hybrid networks. *ACM Transactions on Internet Technology* 2013; 12(3): 1–33.

44. Sherchan, W., S. Nepal, and C. Paris. A survey of trust in social networks. *ACM Computing Surveys* 2013; 45(4): 1–33.

45. Paradesi, S., P. Doshi, and S. Swaika. Integrating behavioral trust in web service compositions. In: Proceedings of the IEEE International Conference on Web Services, Los Angeles, CA, July 6–10, 2009.

46. Caverlee, J., L. Liu, and S. Webb. Socialtrust: Tamper-resilient trust establishment in online communities. In: Proceedings of the 8th ACM/IEEE-CS Joint Conference on Digital Libraries, Pittsburgh, PA, June 16–20, 2008.

47. Adali, S., R. Escriva, M. K. Goldberg, et al. Measuring behavioral trust in social networks. In: Proceedings of the IEEE International Conference on Intelligence and Security Informatics (ISI), Vancouver, BC, May 23–26, 2010.

48. Hang, C. W. and M. P. Singh. Trust-based recommendation based on graph similarity. In: Proceedings of the 13th International Workshop on Trust in Agent Societies (TRUST), Toronto, CA, May 10, 2010.

49. Zuo, Y., W. Hu, and T. O'Keefe. Trust computing for social networking. In: Proceedings of the IEEE Sixth International Conference on Information Technology, Las Vegas, NE, April 27–29, 2009.

50. Kuter, U., and Golbeck, J. Sunny: a new algorithm for trust inference in social networks using probabilistic confidence models. In: Proceedings of the 22nd Conference on Artificial Intelligence, Vancouver, CA, July 22–26, 2007; pp.1377–1382.

51. Kim, Y. A., M. T. Le, H. W. Lauw, et al. Building a web of trust without explicit trust ratings. In: Proceedings of the IEEE 24th International Conference on Data Engineering Workshop, Cancún, México, April 7–12, 2008.

52. Maheswaran, M., H. C. Tang, and A. Ghunaim. Towards a gravity based trust model for social networking systems. In: Proceedings of the International Conference on Distributed Computing Systems Workshops, Toronto, CA, June 25–29, 2007.

53. Liu, H., E. P. Lim, H. W. Lauw, et al. Predicting trusts among users of online communities: An epinions case study. In: Proceedings of the 9th ACM Conference on Electronic Commerce, Chicago, IL, June 8–12, 2008.

54. Nepal, S., W. Sherchan, and C. Paris. STrust: A trust model for social networks. In: Proceedings of the 10th International Conference on Trust, Security and Privacy in Computing and Communications (TrustCom), Changsha, China, November 16–18, 2011.

55. Trifunovic, S., F. Legendre, and C. Anastasiades. Social trust in opportunistic networks. In: Proceedings of the IEEE INFOCOM, San Diego, CA, March 15–19, 2010.

56. Andersen, R., C. Borgs, J. Chayes, et al. Trust-based recommendation systems: An axiomatic approach. In: Proceedings of the 17th International Conference on World Wide Web, 2008.

57. Hess, C. *Trust-Based Recommendations in Multi-layer Networks*. IOS Press, The Netherlands, 2008.

58. O'Donovan, J., B. Smyth, V. Evrim, and D. Mcleod. Extracting and visualizing trust relationships from online auction feedback comments. In: Proceedings of the 20th International Joint Conference on Artificial Intelligence, Hyderabad, India, January 6–12, 2007.

59. Guerriero, A., S. Kubicki, and G. Halin. Trust-oriented multi-visualization of cooperation context. In: Proceedings of the second IEEE International Conference in Visualisation, Barcelona, Spain, July 15–17, 2009.

60. Mui, L., M. Mohtashemi, and A. Halberstadt, A Computational Model of Trust and Reputation. In: Proceedings of the 35th Hawaii International Conference on System Science (HICSS), Island of Hawaii, USA, January 7–10, 2002.

61. Abdul-Rahman, A., and S. Hailes, Supporting trust in virtual communities. In Proceedings of the 33rd Hawaii International Conference on System Sciences (HICSS), Island of Hawaii, USA, January 4–7, 2000.

62. Sun, Y. and Y. Liu. Security of online reputation systems: The evolution of attacks and defenses. *IEEE Signal Processing Magazine* 2012; 29(2): 87–97.

63. Jøsang, A. and J. Golbeck. Challenges for robust trust and reputation systems. In: Proceedings of the 5th International Workshop on Security and Trust Management (SMT), Saint Malo, France, September 24–25, 2009.

64. Pranata, I., G. Skinner, and R. Athauda. A holistic review on trust and reputation management systems for digital environments. *International Journal of Computer and Information Technology* 2012; 1(1): 44–53.

65. Ion, M., A. Danzi, H. Koshutanski, and L. Telesca. A peer-to-peer multidimensional trust model for digital ecosystems. In: Proceedings of the Second IEEE International Conference on Digital Ecosystems and Technologies (DEST), Phitsanuloke, Thailand, February 26–29, 2008.

66. Schmidt, S., R. Steele, and T. Dillon. DEco arch: Trust and reputation aware service brokering architecture in digital ecosystems. In: Proceedings of the Inaugural IEEE International Conference on Digital Ecosystems and Technologies (IEEE DEST), Cairns, Australia, February 21–23, 2007.

67. Teacy, W. T. L., J. Patel, N. R. Jennings, and M. Luck. TRAVOS: Trust and reputation in the context of inaccurate information sources. *Journal of Autonomous Agents and Multi-Agent Systems* 2006; 12(2): 183–198.

68. Zhang, J. and R. Cohen. Evaluating the trustworthiness of advice about seller agents in e-marketplaces: A personalized approach. *Journal of Electronic Commerce Research and Applications* 2008; 7: 330–340.

Chapter 9

Economic and Business Models in Mobile Social Networks

9.1 Introduction

Mobile social networking is a form of social networking wherein individuals with similar interests or communalities converse and connect with one another using mobile devices. Like Web-based social networking, mobile social networking occurs in virtual communities.

Similar to other emerging industries, mobile social networking is characterized by a continuously changing and complex environment that creates important uncertainties at the levels of technology, consumer demand, and business strategy. At the technological level, uncertainties are typically caused by the rapid technological development and the battle for establishing standards, which are common in the beginning stages of the life cycle of any industry. Regarding demand, despite a generalized consensus about the huge potential of mobile business services, nobody actually knows how to exploit the new possibilities brought about by technology to create valuable services for which the customers are willing to pay. Finally, strategic uncertainties in business are common situations in emerging industries, where the essential characteristic from the viewpoint of formulating strategies is that there are no established rules of the game. Consequently, mobile social network (MSN) providers should be experienced in various strategic business approaches and must constantly reposition themselves to find the most competitive position in the industry [1].

Despite all the technological possibilities, the number of successful MSNs in the application market is still limited. Generally speaking, the adoption of innovative technologies depends heavily on the viability of the underlying business models, because successful innovations demand business models as much as they demand innovative technology. Literature suggests several factors that may explain the lack of success of MSN services, including a dearth of value-added applications, unwillingness on the part of the user to pay for mobile services, a mismatch between the applications being launched and the everyday needs of the intended users, legal aspects related to privacy, and ineffective business models in general.

While these factors may explain the lack of success, they do not point at practical business model design issues on which organizations should focus when they are addressing the factors in question [2].

This chapter is meant to give a better understanding of the economic and business models in MSN. To understand an MSN in relation to business models, how it works, and its implications, we must first understand what business models are, how they work, and what they imply. Thus, a number of unique economic characteristics are presented in Section 9.2. Then, a brief review of the literature concerning the business model and a related framework are discussed in Section 9.3. Subsequently, present MSN business models and their specific issues are proposed. In Section 9.4, these business models are analyzed based on two-sided market theory. Finally, section 9.5 concludes this chapter.

9.2 Economic Characteristics of MSNs

MSNs have evolved to meet the requirements of data exchange, sharing, and delivery services. MSNs can be broadly classified into two types—Web-based MSNs and decentralized MSNs. Services available to MSN users follow several trends, including social gaming, business, and media [3].

Research regarding MSN services focuses primarily on MSN service architecture, user modeling, privacy resolving, and building reference applications. However, there is relatively little knowledge about the actual added value of different economic characteristics in MSN services and about the factors that enhance user adoption of these services. For example, the tedious balancing of costs and benefits needs mentioning. Tangible costs for consumers are obstacles in this regard, but other sacrifices and uncertainties are important as well (e.g., related to privacy, trust in the service, and the required cognitive effort) [2].

The MSN industry presents a number of unique economic characteristics such as mobility, network effects, and long tail. Developers ought to take these characteristics into consideration when they formulate their strategic approaches. Indeed, successful business models are likely to be those that best comply with these particularities.

9.2.1 Mobility

It is commonly supposed that mobility is the most important characteristic of MSN because it represents its principal distinctive advantage upon which mobile services can build their value proposition. In fact, mobility brings several unique benefits that can be related to a number of attributes such as freedom of movement (services can be used while on the move), ubiquity (the possibility of using services anytime and anywhere, independent of the user's location), localization (user's location information can be exploited to offer location-based services), reachability (users can be reached anywhere, anytime, by selected persons and contexts), convenience (always at hand), instant connectivity (always on), and personalization (personal device, apt to store personal information).

However, mobility also involves some drawbacks, and MSN services are actually inferior to their wire-line counterparts in other dimensions. In particular, mobile applications suffer from limited and more expensive bandwidth and device limitations. Bandwidth limitations are a consequence of the radio spectrum being a fixed and rare resource and its control being restricted to licensed owners. Device limitations are due to the portability requirement of mobile handsets that they have to be small and lightweight, with limited space used for screen, batteries, and input/output interfaces. Even if valuable services can be entirely built around mobility—exploiting the distinctive features of the mobile channel—combining them with the advantages of e-commerce services and other channels would lead to substantially more value delivered to the customer [1].

9.2.2 Network Externalities

Formally speaking, MSNs are composed of a set of components, connected together by links. These structures exhibit a characteristic economic phenomenon known as network externality. An externality occurs when a transaction between two actors affects, as a side effect, a third party that is external to the transaction. A product presents a network externality if the utility that a user derives from consumption of the product increases with the number of other agents consuming the product.

The key reason for the appearance of network externalities is the complementarities among the components of a network, which are inherent in the structure of a network because many components of a network are required for the provision of a typical service. Links on a network are potentially complementary, but it is compatibility that makes complementarities actual. Network externalities can be direct or indirect.

Network externalities play an important role in the MSN market, where the utility of joining a MSN is positively related to the number of its members. In this case, network effects take the form of being able to maintain social connectivity and interaction with a large number of other users (direct externality occurs in same side, i.e., consumer side). MSNs also show signs of indirect externalities, where users benefit indirectly from network size. As an example, in MSNs, an extra customer potentially increases the number of services available to other customers (indirect externality occurring in both the consumer and the third-party service provider sides) because by increasing the demand for services, service provisioning becomes more profitable, and more third-party service providers would be willing to offer additional service based on the MSN platform.

In brief, in MSN, network externalities influence consumers (e.g., when they decide whether to participate in and put effort on an MSN) and producers (including advertisers, business/service/content providers), for example, when they decide whether to standardize their products and services to allow compatibility with other producers, when they set the product quality, and when they chose the pricing policy [1].

9.2.3 Long Tail

Sociological research has shown that in many situations, such as advice-seeking or job search, we usually do not benefit much from people with whom we have strong social bonds, rather we benefit from people who we do not know directly or who we know only very superficially, the so-called weak ties. By granting access to these weak ties, MSNs offer a much larger pool of potentially interesting contacts than traditional means of networking can typically provide.

Second, due to time restrictions, traditional networking only allows individuals to stay in touch with a limited number of people. It requires too much effort to permanently update all contact data in a traditional address book or on an Excel spreadsheet because contacts do not regularly inform the individual about changes in their contact data, such as address, telephone number, job position, or e-mail address. Hence, contact data are not always up-to-date, and the individual might lose track of these people, even if he or she would, in theory, be willing to retain the contact. Thus, relationships expire over time due to a lack of interaction. On MSNs, however, terminating a relationship requires the user's active intervention; otherwise, a contact will be retained in a user's contact list. Thus, it becomes possible to manage a constantly growing number of contacts without any additional efforts. Actively used MSNs grant users access to valid contact data at all times, with the profiles acting as a de facto self-actualizing address book.

The combination of these two factors (i.e., the impact of weak ties and the improved contact management) creates a vast potential for mobile social networking called the long tail of mobile social networking (Figure 9.1). This tail visually conceptualizes the MSN of an individual network member. The X-axis, which depicts the number of contacts the member maintains, is sorted by

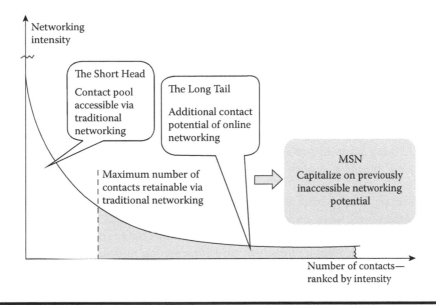

Figure 9.1   The long tail of mobile social networking.

networking intensity. The Y-axis depicts networking intensity, which is a function of the contact frequency and the amount and type of information that is exchanged between the member and his or her contacts. Depending on the number of contacts and the intensity of interaction with other members of the network, the shape of this tail will differ for each individual member. Those users who have many contacts have longer tails than those with only few contacts, and users who interact heavily with others have fatter tails than those who do not. Overall, the long tail curve reflects the fact that we tend to have a few people with whom we have very close relationships (the very top left of the graph), whereas there are more people we only know superficially and whom we contact infrequently.

In summary, MSNs make a larger contact pool available to their members and allow them to, in theory, easily manage and maintain virtually unlimited numbers of contacts by granting access to the long tail of mobile social networking—an additional pool of contacts that is inaccessible via traditional networking. Enders et al. [4] describe the revenue models of social networking sites and classifies them as advertising models, subscription models, and transaction models. They show how each can be used as a different strategy for social networking sites to increase revenues, either by lengthening the "long tail" of revenue (advertising), by "fattening" the tail (subscriptions), or by "driving demand down the tail" (transactions). Advertising relies on high user amounts, subscriptions rely on a set level of willingness to pay, and transactions rely on being able to provide value to another party on the platform.

9.3 Present Business Models of MSNs

9.3.1 Theories about Mobile Business Models

9.3.1.1 Business Model Components

A business model can be defined as the description of an organization or network of organizations involved in creating and capturing value from technological innovation. In recent years, the field of business models has developed from defining business models, via exploring business model components and classifying business models into categories, toward developing descriptive models.

There are several basic components that constitute a business model [2]. Many researchers focus on business model elements such as service and product innovation, the actors involved, the relationships between the actors, information and application architecture, and information and value exchange. Alt and Zimmermann [5] suggest a few common elements that emerge in business model definitions: mission (i.e., overall vision, strategic objectives, and value proposition, as well as the basic features of a product or service), structure (i.e., the actors involved and the roles they play within a specific business environment as well as the specific market segments, customers, and products), process (i.e., the concrete translation of the mission and the structure of the business model into more operational terms), and revenues (i.e., the investments needed in the medium and long term, cost structures, and the revenues being generated). Afuah and Tucci [6] regard business models as a system of components (customer value, scope, pricing, revenue sources, connected activities, implementation, capabilities, and sustainability) and the relationships among these components. Osterwalder and Pigneur [7] are far more systematic in their approach to the concept of business models. Based on what a company has to offer, who it targets, how the proposition can be realized, and how much can be earned, they distinguish four basic elements: (1) product innovation (i.e., the value proposition, the target customer, and the capabilities needed to offer the value); (2) customer relationship (i.e., the information strategy, delivery channels, and trust and loyalty); (3) infrastructure management (i.e., the configuration of the company and its partner network and resources); and (4) financial (i.e., the revenue model, cost model, and profit model. In a literature meta-study, Morris et al. [8] have identified 24 different business model components, of which value offering, economic model, customer interface/relationship, partner network/roles, internal infrastructure/connected activities, and target markets are mentioned the most frequently. In a similar study, Shafer et al. [9] have identified 42 different business model components that can be clustered into four generic components—strategic choices, value network, value creation, and value capturing.

When various business model definitions are compared together, the following common components can be distinguished [2].

- Service component: A description of the value proposition (added value of a service offering) and the market segment at which the offering is aimed
- Technological component: A description of the technical functionality required to realize the service offering
- Organizational component: A description of the structure of the multi-actor value network required to create and distribute the service offering and to describe the focal firm's position within the value network
- Financial component: A description of the way a value network intends to generate revenues from a particular service offering and of the way risks, investments, and revenues are divided among the various actors in a value network.

9.3.1.2 Design Issues

Often, design choices in business model components cannot be considered in isolation but should be balanced in order to develop a viable business model. Knowledge on how to balance requirements and strategic interests effectively within and between the different domains is to a large extent missing in business model literature. To develop insight into the way organizations can design "balanced" business models, designers need to understand the design issues in business models and their interdependencies. A design issue is defined as a design variable that is perceived (by a practitioner and/or researcher) to be of eminent importance to the viability and sustainability of the business model under investigation. Generic design issues within the four mobile business model components for MSNs are specified in Table 9.1.

1. Service domain: Targeting refers to the target group of a service (e.g., niche market or mass market, consumers, or business users). Closely connected to choosing a target group is formulating the value-creating elements that should result in a compelling value proposition for end users (e.g., fun, efficiency, accuracy, speed, and/or personalization). Branding is an important issue when it comes to reaching the targeted customers, influencing the perceived value of service offerings. Customer retention refers to marketing strategies aimed at keeping customers satisfied and loyal to the product or service.

2. Technology domain: Trust on the part of end users and customers with regard to a service offering is partly determined by the way security is implemented in the technological architecture. Often, security issues have to be balanced with ease-of-use. The performance of the technological architecture in delivering the technological functionalities has a profound impact on the quality of service and on perceived value, making it necessary to strike a balance between the quality of a service and the costs. The adoption of a new service is in part determined by the

Table 9.1   Summary of Generic MSN Business Model Design Issues

Component in MSN Business Model	Key Design Issues
Service domain	Targeting
	Value-creating elements
	Branding
	Customer retention
Technology domain	Security
	Quality of service
	System integration
	Accessibility for customers
	Management of user profiles
Organizational domain	Partner selection
	Network openness
	Governance
Financial domain	Pricing
	Division of investment, cost, and revenues
	Valuing the contributions and benefits

extent to which it can be integrated into the existing technological infrastructure (i.e., system integration). Accessibility for customers is obviously influenced by the choice of platform, devices, and architecture. To allow for personalization of a service, a user profile that contains user interests, preferences, and behavior must be created and maintained (i.e., management of user profiles).

3. Organizational domain: Partner selection is important in acquiring access to the resources and capabilities needed to realize a service offering. Network openness indicates the degree to which new business actors can join the value network and are allowed to provide services to customers. Generally speaking, there are two organizational arrangements: the closed model, in which there is a relatively fixed consortium of partners, and the walled garden model, in which new partners are only allowed to join the value network if they comply with certain rules. Governance is relevant because there is often a dominant actor who has access to the customers and end users or who has developed the service offering.

4. Financial domain: For a service to be adopted and actually used, the perceived customer value must at least balance, and preferably exceed, the price of a service. Because developing and introducing a new service involves financial risks, division of investments is another design issue. Division of costs and revenues may follow different rationales (e.g., cost based or value based). For fair and viable revenue sharing arrangements, valuing the contributions and benefits of each partner to the service offering is important (e.g., based on actors' access to resources and strategic interests).

9.3.2 Revenue Models of MSNs

Community or MSN sites can create and capture value using different types of business models. But the most common revenue models of MSNs are advertising, subscription, and transaction models, as shown in Table 9.2. And Table 9.3 summarizes the adopted revenue streams of several existing MSNs.

Ads are a natural, traditional, and easy way to fund services. Popular services are also attractive in the eyes of advertisers. In addition, the analyzed services have other ways to generate revenue. Some of the analyzed services were partially or fully subscription based. Some of these sold something—either virtual or concrete products. The best services combine these methods, and the service has a functional business model from the very beginning [10]. Those three typical revenue streams are presented as follows.

Table 9.2 Basic MSN Revenue Models

Advertising	Revenue is derived from advertising
Subscription	Purchasing mobile value-added services
Transaction	Fees charged within constituent part of value chain

Table 9.3 Overview of Used Revenue Streams of Typical MSN Sites

Name	Web Sites	Revenue Model	Context
Kakao Talk	https://www.kakao.com/	Advertising/Subscription/Transaction	Social
Foursquare	https://foursquare.com/	Advertising/Transaction	Social/Business
Facebook	https://www.facebook.com/	Advertising	Social
Yelp	https://www.yelp.com/	Advertising/Transaction	Business
Orbitz	https://www.orbitz.com/	Advertising/Transaction	Business

9.3.2.1 Advertising

Advertising is one possible way to generate revenue for an MSN. MSN services usually gather a lot of personal information about each user, which enables personalized advertisements. Mobile phones are regarded as personal devices that are with the user most of the time thus creating an ideal target for advertising if the advertisements are highly informative or entertaining [11].

Kakao Talk's advertising works such as WeChat's social media marketing allows companies and brands to communicate with the app's users should people add these brands as their friends. Kakao Talk now has more than 200 "plus friend" advertising partners and more than 66 million such social marketing friendships/follows made, with each brand averaging around 350,000 followers. Overall, there are a total of 15 million unique users of this particular feature [12].

Advertising in an MSN has its unique advantages over online advertising that result from the MSN's economic characteristics of mobility: accessible, personal, and location aware. Accessible means ads can be accessed by a mobile device anywhere and at any time. Personal represents MSN enables advertisers to deliver personalized ads result from mobile device can carry mobile user's assigned identity. Location aware means that a mobile device can help advertisers with delivering location-oriented ads based on the physical location of the device.

Different types of ads that are applied in MSNs can be broadly classified as follows [13].

- Location-based ads are delivered to mobile users according to the mobile device's location. Dao et al. [14] propose a new model of recommender system for location-based advertising (LBA). LBA is a new advertising form that is attracting attention. In LBA, service providers track the location information of target users via their mobile devices and provide the users with location-specific advertising.
- Mobile coupon ads can use different forms such as text messages or images stored on the phone's memory. Text message coupons have been somewhat successful. A new method for mobile coupons has also been discussed [11]. For instance, the mobile coupon could have stamps based on coffee purchase events. After gathering enough stamps, the user would get a free cup of coffee. An MSN enables interesting new viral marketing methods for coupon use. Sharing a coupon with the user's contact list, ranking coupons by popularity, and giving gifts to other contacts via mobile coupons are all possibilities that could open up business opportunities. Some problems exist, though, with redeeming mobile coupons.
- Brand-building ads could increase the awareness of a specific business or product among potential consumers.
- Game-based ads let users play games using mobile devices while the ads pop-up between games.

■ Interstitial ads refer to the advertisements that appear on mobile devices during mobile users' interstitial time—for example, page loading, wireless Web surfing, or waiting for wireless connections. They can increase interactivity by click-through links and buttons, and they provide audio features.

9.3.2.2 Subscription

The subscription-based business (freemium) model is widely used in Internet software and applications. Several notable companies are using it today, including Spotify, the music streaming service offering unlimited music streaming; the publicly traded LogMeIn, which provides remote access to computers over the Internet and is valued at more than $1 billion USD; and Kakao Talk, which earns money by selling digital items such as premium emoticons and stickers to its users [12].

In the subscription model, basic access to a service is often free of charge. This makes it easy to pull in users. Furthermore, the value of a service increases with increased volume of users and content. If users want additional features for a service, such as networks that offer premium memberships to find business partners or former classmates, subscribers can use all services (i.e., they get any information about their account via short notifications or newsletters, receive and send e-mails, get job offers: Xing, LinkedIn, Stayfriends, etc.); these features are often available for a monthly or annual fee [10].

The word freemium is a combination of "free" and "premium." The concept was popularized by Wilson [15]: Give your service away for free, possibly ad-supported but maybe not; acquire a lot of customers very efficiently through word-of-mouth, referral networks, organic search marketing, and so forth; then offer premium-priced value-added services or an enhanced version of your service to your customer base. Thus, freemium is a concept whereby a company offers something for free, with no requirements to purchase now or later, in order to build a large user base, of which at least some purchase the premium-priced offer. This is then a change from what economists call tying because the customer is not conditioned to buy the premium version; it is not indispensable.

Bekkelund [16] found that low marginal costs are required, at least on the free version, as a precondition for choosing freemium. Otherwise, the premium version would be too expensive for the paying customers. Freemium also depends on the company having a large addressable market because it entails possibly giving away a company's product or service for free into perpetuity for some users. The key metric is then the customer lifetime value—especially because freemium is more opaque than more traditional models where each user pays directly. Because of this opaqueness, tracking and measuring the costs and revenues associated with the users are found to be vital in order for fermium to succeed.

In traditional economic theory, there is an inverse relationship between price and quantity demanded, but what happens when the price becomes zero?

According to Shampanier et al. [17], people perceive benefits associated with free products as of higher value. Thus, decreasing the price to zero increases perceived value considerably more; people tend to overreact to free products. Shampanier et al. also saw a decrease in the demand for the more expensive product when including a free offer. They termed these two findings the zero-price effect. Driving this effect may be the fact that choosing a free product is a much simpler decision.

Haruvy and Prasad [18] found that, in order not to cannibalize the premium offer, the quality of the free offer must be sufficiently low and the price of the premium over must not be too high—but, importantly, the free offer must not be of too low quality; it must induce customers to use it.

According to Shampanier et al., much additional work is still needed to properly understand the complexities of gratis products. However, the research thus far on free services/products can be broadly categorized into three categories:

- Network externalities, in which free adopters increase future adopters' valuation of a product
- Demonstration effects, in which users can try the software before buying
- Word-of-mouth effects, in which free spyware adopters help speed up the diffusion of a new product

9.3.2.2.1 Network Externalities

When a network externality is present, the value of a product or service increases as more people use it. The classic example is telephony because having a telephone is only valuable if there are other people with compatible telephones. The value can be seen in two eponymous laws: (1) Metcalfe's law, which states that the value of a network is proportional to the square of the number of connected users of the system and (2) Reed's law, which states that the utility of large networks, particularly social networks, can scale exponentially with the size of the network. He calls these group-forming networks, wherein network members can create and maintain communicating groups.

One of the problems with distributing free versions of a product is that it can cannibalize sales of the premium version, thereby lowering the profits for the firm. On the contrary, the price users are willing to pay, in part and ceteris paribus, is determined by the number of users in the network to which the product belongs; establishing this initial network is simpler when giving away the product for free.

Because it is advantageous to achieve a significant share of the market quickly, the initial purchase should be made as easy as possible. This is especially important because the network benefits are lower for the early adopters. Lee et al. [19] divided customers into two types: power users and light users. Power users are less sensitive to compatibility (e.g., lock-in) than light users and are also assumed to be keener to try new technologies. Thus, segmenting the customers such that the initial customers are power users can make their initial purchase simpler.

9.3.2.2.2 Demonstration Effect

Compared to network effects, product demonstrations play on the intrinsic rather than extrinsic features of the product. According to Faugère and Tayi [20], a primary purpose of offering a free sample is to increase sales by providing firsthand experience for potential buyers. Being able to try a service/product before buying it has been shown to play a significant role in the adoption of information technologies. Gallaugher and Wang [21] found that, especially in the study of the market for Web server software, trial versions were associated with price premiums. As for network externalities, Gallaugher and Wang found that being able to try the product before buying it helped seed the initial market.

9.3.2.2.3 Word-of-Mouth Effect

Jiang and Sarkar [22] found that even if other benefits do not exist (i.e., network externalities or demonstrations), a company can still benefit from giving away fully functioning service/software because the free adopters help speed up the diffusion process (e.g., giving away service/software the first month when it is on sale). The different methods that free adopters can use to help speed up the adoption of the service/product are verbal (e.g., spreading information about its availability and quality) or nonverbal (e.g., through others seeing the product in use, peer pressure, or through social influence). Together, these are referred to as a word-of-mouth effect. The influence of word-of-mouth means that companies can increase their profit when they have a free offer. However, an important consideration with regard to Jiang and Sarkar is that their study focuses on free software that is offered without any restrictions, while for freemium we would either see time or functionality limitations. No study was found that discussed this more specifically for freemiums [16].

9.3.2.3 Transaction-Based Fee

A transaction fee–based revenue model means a company receives commissions based on volume for enabling or executing transactions. The revenue is generated through transactions by the customer paying a fee for a transaction to the operator of a platform. The company is a marketplace operator providing the customer with a platform on which to place his or her transactions. During this process, the customer may be presented as a buyer as well as a seller. Customers must register to actively participate in this e-market, so both parties in a transaction are identified. From a business perspective, the offer is determined by others because customers offer their goods online and are acting as sellers. The amount of the transaction fee can be fixed or calculated as a percentage.

The number of third-party developers will increase significantly as the improved revenue arrangements and increasing user base attract more developers to the platform. Some MSN services generate this revenue model by selling products.

The products can be either virtual or physical. The third-party developers make up the seller side and the users are the customer side.

For example, companies and brands can cooperate with Kakao Talk to produce their own special series of emoticons as a brand marketing tool. The revenue will then be shared between the brand and the messaging start-up. Kakao Talk's other business model is mobile commerce, which allows users to purchase online goods and then send them to a friend as a gift via the app. The friend will then be able to redeem the gift through the corresponding merchant. There are now more than 320 brands that have cooperated with Kakao Talk for this feature, with 7700 items available for purchase [12].

Usually, the third-party developers are creators of mobile applications. There are several kinds of third-party developers [23]:

■ Hobbyists: Those developing mobile applications in their spare time for recreation and profit
■ Professionals: Those developing mobile applications as a main source of income, either alone or as part of a business centered on mobile application development
■ Contractors: Those developing mobile applications on behalf of another entity or individual

Third-party developers can create revenue from MSN applications in several ways [23]:

■ Purchase of the mobile application by a consumer
■ In-app purchases (game money, for example)
■ In-app advertising
■ Subscription services (a limited number of mobile applications are available via subscription)
■ Sale of the mobile application as a product

9.3.3 Specific Issues

9.3.3.1 Activeness

MSN is based on the concept that users not only consume but also produce contents. It is a vital question for many services as to whether they can make people excited about the new service. However, a common hypothesis of the MSN user base is: "1% produce content, 9% provide comments, and the rest are passive consumers."* This is called participation inequality [10].

Active users share the following characteristics:

■ An ambition and desire to express oneself
■ "Ego-casting" or emphasizing one's own competence to peers or headhunters

---

* http://www.useit.com/alertbox/participation_inequality.html

- Membership in a larger community
- A particular hobby creating a desire to produce and comment on content associated with it
- Less technically oriented people becoming content producers, as tools have become easier to use

Passive users can be characterized as follows:

- Passive consumption of information without an ambition toward critical evaluation or enrichment.
- The view that almost all necessary information is already available on the Internet and that it is sufficient to pick the subjects that interest you.
- One's own expression is not at its best using recordable media methods (the user would rather speak than write, draw, take photos, etc.).
- The tools differ from the conventional.
- Social media is not considered important, and there is no desire to learn about it.
- Sufficient information and networks can be gained through conventional media (news, letters to editors, hobby clubs, etc.).
- Available time is limited.
- Individuality and anonymity issues (the desire to protect one's real-world identity from the online identity and make a clear distinction between them).

However, it is noteworthy that a user can change from being passive to more active once MSN applications become more familiar. Therefore, it is important that the MSN service provides different possibilities of participation in order to take small steps toward an active role. A transition from the 90% group of passive consumers to the 9% group of minor contributors is significant as such [10].

9.3.3.2 Identity

With regard to privacy, MSN services require either complete identification, provide for various types of pseudonymity (such as pen names in newsgroups), or allow complete anonymity.

Identity is a crucial concept. Generally, it can be said that protecting your identity is easier on the Web than in the "real-world." A user can create different "online identities" for different services, and for outsiders, these can seem completely detached from each other and need not return to the user's true identity (person). On the flip side, such freedom brings a variety of problems. All kinds of malpractice are possible; using an online identity, a user can act in ways that he or she would not in real life. On the contrary, identity theft (i.e., acting in someone else's name) is a dangerous phenomenon.

The incoherence and diversity of online identities also pose disadvantages for the user. Practical problems include remembering user IDs and passwords. However, the problem is more serious on the psychological level. The more time a user spends in a virtual world and the more different the virtual identities are from real-world identities, the greater the risk of damage to one's own mind.

9.3.3.3 Copyright

In the social network field, it is easy for a user to distribute not only his or her own material but also content that was produced by others in his/her own name without asking the original producer for permission. This phenomenon raises issues of copyright. If the original content producer is a private individual, this opens the way for all kinds of cases hampering privacy as well as defamation. On the contrary, if the content producer is a commercial party, the potential issues include brand deterioration and the distribution of commercially significant material through free channels.

Specific issues related to brands are the distribution of original material through a suspicious hosting service, alteration of material before distribution, or imitation of material. So-called spoof commercials imitate the style and/or storyline of original commercials to a certain point but ultimately try to shock the viewer in various ways. The producer of a spoof commercial may have several reasons to create such content; it can be based on humor, globalization criticism, or on a bad experience with the company in question. However, all of these motives affect the brand of the party targeted by the spoof commercial. On the contrary, there is some speculation that such spoof commercials may actually be produced by the companies themselves, even though they strongly deny this in public. Is the saying "any publicity is good publicity" also true in this respect?

9.3.3.4 Mobility

Mobility and context (situational and environmental) awareness can be taken into account in various ways when using MSNs. For example, the location of a user can serve as a filter when offering content. For example, it can be assumed that a user looking for a place to eat will be interested in opinions about a particular restaurant if the opinions are those of people located fairly nearby. A specific social media application related to geographic information systems (GISs) is geo caching, in which users can search for various kinds of physical objects cached by others with the help of Global Positioning System devices.

In addition to location, a user's social environment—that is, the people in his or her company (either physically or through another kind of connection) at each time—can affect the filtering and offering of essential material to the user.

Currently, most MSN services provide users with the possibility of sharing their "presence information" with other users, which can also be connected to traditional OSNs so that the "presence flow" allows access to photos taken and shared by the users in OSNs.

Emphasizing mobility in connection with social media imposes special requirements on the usability of applications. The input and output features of mobile devices are more restricted compared to conventional PCs. Typically, this is also true of the data communication connections between the servers hosting MSN sites and the user applications on mobile devices. These special issues must be taken into account in the design of MSN applications and services [10].

9.3.3.5 Trust

Trust is a subjective relationship between a trustor and a trustee. Because trust is subjective, no general metrics associated with the level of trust can be defined. However, trust is affected by trustworthiness, which is an objective value or set of values. Trustworthiness can be measured and compared with a threshold that contributes to the trust decision of a rational trustor. When establishing trust, the trustor examines the different features of the trustee, trying to establish an aggregate on the basis of which to make a decision regarding the trustee. The decision may lead to starting to use a service, buying a product, or to engaging in other interaction. In some cases, the factors contributing to the assessment may be context-linked.

The establishment of trust in the MSN is very essential. If the user of an MSN application or service is satisfied with the content or service received, it is more probable that he or she will later come back to use the same provider's services. This can be interpreted as an expression of trust.

Context issues and their effect on the establishment of trust in OSNs are particularly applicable to MSNs. Naturally, desktop activities on the Web can also be made more efficient by taking the context into account. For example, a user's navigation through different links can be saved, and conclusions can be made on the user's potential interests and the content that should be offered to him or her. However, the mobile world brings new dimensions to the situation through changes in the user's physical and social environment. A user can be offered social network content associated with geographically nearby objects, or in a popular blog, messages from people within Bluetooth range can be filtered and highlighted.

Traceability is an important factor affecting the establishment and upkeep of trust that probably becomes a challenge in the MSN field. Traceability means that the consumer of content/a service can at his or her option find out who the content producer or service provider is. The challenge is caused by the open nature of the MSN on the one hand and the possibility for anonymity on the other.

The huge and constantly increasing volume of available material makes it more difficult to find trusted parties, requiring the application of filtering mechanisms based on trust. Traceability is linked to MSN identity.

The large and constantly increasing volume of MSNs creates a need for filtering essential and meaningful content. With respect to this, there are many approaches, with reputation, recommendations, and context awareness described earlier. Recommendations can be either explicit or implicit. At the simplest, implicit recommendations refer to the most viewed content, measured by the number of hits on a Web page or video. In many cases, limiting the scope to smaller communities and determining the relevance of information through collaborative filtering applied to these communities' works more precisely.

Explicit recommendations may be comments on Web content or services actively provided by users. Comments can be either positive or negative. The ability to provide negative comments distinguishes this approach from implicit recommendations, such as page hits, and therefore provides additional dynamics. Extensive stabilization of explicit recommendations would require the provision of feedback to become a natural part of user activities. It can naturally be supported by various external incentives (prizes, participation in competitions, etc.), but a more sustainable alternative would be to motivate the users for the sake of the activity itself. Users should find it important to be a referee and to belong to a community of referees.

Generally speaking, at least the following factors can be said to affect the establishment of trust [10]:

- The service provider's brand and reputation
- Recommendations from peers
- Recommendations from well-known/recognized parties
- The general reputation of the service (based on previous experience)
- The country (the service's or the server's)
- Privacy protection
- Information security practices and policy
- Information security for the network connection (encryption in particular)
- The party implementing an application or the method of implementation
- A terminal device's support for an application
- The operation and usability of an application
- The appearance of an application
- Application maintenance (number of updates and effort required)
- A user group's properties or assessments of a user group
- Recognition of an application (in traditional media and online)
- The age/history of an application
- The type of content (fact, entertainment, advertisement)
- Consequences of use (spam, viruses, etc.)

9.4 Analyzing MSN Business Models Based on Two-Sided Market Theory

9.4.1 Definitions of Two-Sided Market

Two-sided markets, also called two-sided networks, are economic platforms having two distinct user groups that provide each other with network benefits. The organization that creates value primarily by enabling direct interactions between two (or more) distinct types of affiliated customers is called a multisided platform (MSP). Two-sided networks can be found in many industries, sharing the space with traditional product and service offerings. Example markets include credit cards, composed of cardholders and merchants; HMOs (patients and doctors); operating systems (end users and developers); night clubs (men seeking women, and often buying them drinks); yellow pages (advertisers and consumers); video game consoles (gamers and game developers); recruitment sites (job seekers and recruiters); search engines (advertisers and users); and communication networks, such as the Internet. Examples of well-known companies employing two-sided markets include such organizations as American Express, eBay, Facebook, Mall of America, Match.com, Monster.com, Sony, Skype, Google, and others. Benefits to each group exhibit demand economies of scale [24].

There has been a recent surge of interest in two-sided market platforms, and the two papers by Roson [25] and Rochet and Tirole [26] do an admirable job in surveying much of the extant literature. Armstrong and Wright [27] have considered different scenarios of competition in two-sided markets. In particular, Gupta et al. [28] discuss the difference between electronic retailers and in-store retailers, which are examples of two-sided markets. Muylle and Basu [29] define and discuss electronic intermediary in general.

In multisided markets (two-sided markets), several (two) distinct groups of agents interact through an intermediary called a platform. Because the participation of each group gives value to the other group, two-sided markets (2SM) are characterized by a specific class of network externalities called cross externalities. Many industries exhibit such characteristics. Video game platforms, credit card payment systems, newspapers, and dating agencies provide well-known examples of 2SM [30].

Two-sided markets represent a refinement of the concept of network effects. There are same-side network effects and cross-side network effects. Each network effect can be either positive or negative. Quite often in a two-sided network, members of at least one group exhibit a preference regarding the number of users in the other group; these are called cross-side network effects. Each group's members may also have preferences regarding the number of users in their own group; these are called same-side network effects. Cross-side network effects are usually positive but can be negative (as with consumer reactions to advertising). Same-side network effects may be either positive (e.g., the benefit from swapping video games

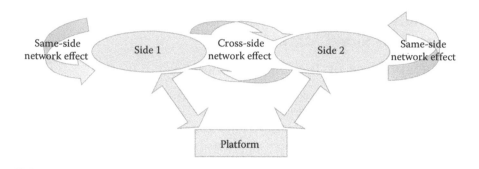

Figure 9.2   Cross-side and same-side network effects in a two-sided network.

with more peers) or negative (e.g., the desire to exclude direct rivals from an online business-to-business marketplace) [24]. Figure 9.2 depicts these relationships.

Rochet and Tirole observed that there are two types of indirect externalities—usage externality and membership externality.

A usage externality exists when two economic agents need to act together to use the platform to create value. We see that with OpenTable. There is a person who wants to dine at a restaurant at a particular time and a restaurant that would benefit from serving that individual at that time. They can enter into a value-increasing exchange only if they can get together. In practice, that means the person and the restaurant finding each other and entering into a transaction. OpenTable and similar businesses help generate these usage externalities by making it easier for restaurants and diners to enter into this transaction. They also increase the value of externaliza-tion by increasing the quality of the matches: they make it easier for people to find the best restaurant for the particular occasion involved. It is possible that the usage externalities are positive for one type of economic agent but negative for another type of economic agent. So long as the net value of these externalities is positive, there is a benefit to facilitating interaction, some of which the platform may be able to capture. Some advertising-supported media are examples. An advertiser benefits from being able to communicate with a possible customer, but consumers may place a negative value on seeing ads. The platform enables a value-increasing interaction by subsidiz-ing the consumer so that he or she is willing to see the ad. Most advertising-supported media do this by bundling subsidized content with advertising.

There is a membership externality when the value received by agents on one side increases with the number of agents or some related measure of their aggregate value—participating on the other side. In the case of OpenTable, the more restau-rants that participate in the service, the more valuable it is to consumers. Smart mobile phone software platforms provide another example. Developers of applica-tions value a platform more if there are more potential users; users value a platform more if there are more applications. This phenomenon results in the well-known positive feedback loop. More agents on one side attract more agents on the other side, thereby fueling growth.

The platform plays a key role in creating these indirect network effects. In fact, as described in Evans and Schmalensee [31], the major challenge for aspiring platforms is to get enough participants on each side to secure enough critical mass to propel indirect network effects. Platforms generate indirect network effects (and thus value for the economic agents they aspire to serve) through pricing, product design, marketing, and other efforts to attract agents on each side. Jullien [32], for instance, has stressed the value of "divide and conquer" strategies for a start-up or an entrant challenging an established platform; subsidize agents in the most price-sensitive group and then use their participation to attract agents in the other group.

The presence of these indirect network externalities has an important implication for economic analysis of multisided platforms. For traditional one-sided industries, economists ordinarily assume that demand depends on the price of the product as well as the prices of complements and substitutes. For multisided platforms, the demand by one group of economic agents also depends on the number (or other measures of the size and quality) of each of the other groups of economic agents that the platform serves. Loosely speaking, the sides complement each other in demand. Failure to account for these demand interdependencies in an economic model renders the results of that model suspect, if not completely unreliable, when such interdependencies are important. We will return to this point when we discuss economic models commonly used in antitrust [33].

In particular, an MSN has several features that are common among two-sided platforms.

First, it has two sorts of indirect network externalities. There is a usage externality: both users and merchants benefit when the system is used to make a transaction. And there is a membership externality: the more users will attract more advertisers/merchants/third-party developers. And the more services provided by the other side, the more valuable the platform is to users.

Second, the price structure for the two different types of economic agents, which determines the relative incremental profit earned from the two types, is an important tool in solving the coordination problem between the two sides in order to capture value from the externalities that link them. Subsidizing end users is more likely to increase the value of the platform to the other side, resulting in an MSN getting more advertisers/merchants/third-party developers, which in turn makes the MSN more attractive to users. This sort of dramatic asymmetry, with one group of agents paying prices below marginal cost, is also not uncommon with multisided platforms [33].

9.4.2 The Model and Analysis

9.4.2.1 Platform Monopoly

We start with a platform monopoly model of a two-sided market [34]. Figure 9.3 illustrates the basic analytical 2SM model of monopoly MSN platform, notations used in this model are summarized in Table 9.4.

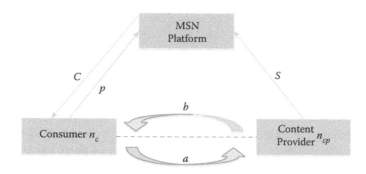

Figure 9.3   Two-sided market model of monopoly MSN platform.

Table 9.4   Summary of Notations

Notation	Meaning of Notation
p	Subscription price
s	Fee from content provider
c	The cost of providing the platform service per consumer
n_{cp}	The number of active content providers
n_c	The number of consumers paying the platform for access to content providers
a	The value for a content provider of an additional consumer connected to the MSN platform
b	The marginal value that a consumer places on an additional content provider on the MSN platform
v	The benefit that a platform offers to consumers
x_i	The customer's utility for using the context
t	Transportation cost per customer's unit of distance "traveled"
y_j	The index of the content provider's location on the unit interval
f	Transportation cost per content provider's unit of distance "traveled"

A platform sells access to consumers at a subscription price p and possibly collects a fee s from each advertiser/merchant/third-party developer (content provider) to allow the content to reach the consumer. Finally, we assume that the cost of providing the platform service is c per consumer.

Consumers are interested in accessing the MSN to reach games, commerce, media, and services. A user i's location x_i indexes his or her utility for using the context. Consumers pay a transportation cost equal to t per unit of distance "traveled." Consumers' locations are uniformly distributed on the interval zero to one with the platform located at $x = 0$. This specification allows for an easy extension to a duopoly setting.

Consumer i's utility is specified as

$$u_i = v + bn_{cp} - tx_i - P \tag{9.1}$$

where $v > c$ is an intrinsic value that a consumer receives from connecting to the MSN irrespective of the amount of content, b is the marginal value that a consumer places on an additional content provider on the MSN platform, and n_{cp} is the number of active content providers.

Content providers rely on advertising/transaction revenue per consumer a to generate revenue. We assume content providers are uniformly distributed on the unit interval and have a unit mass. We make the simplifying assumption that content providers are independent monopolists, each in its own market, and therefore do not compete with each other. Each content provider then earns an_c, where n_c is the number of consumers paying the platform for access to content providers. Thus, a is the value for a content provider of an additional consumer connected to the MSN. Content providers are heterogeneous in terms of the fixed costs of coming up with a business idea and setting up their business.

A content provider indexed by j faces a fixed cost of fy_j, where y_j is the index of the content provider's location on the unit interval. The marginal costs for serving advertisements to consumers are taken to be zero. Thus, a content provider j's profit is

$$\pi_j = an_c - s - fy_j \tag{9.2}$$

In this two-sided market, the demand for content depends on the expected amount of content provided since more consumers will connect to the network if more expected content is available. In addition, the provision of content depends on the expected number of consumers. When the expected number of consumers is n_c^e and the expected number of content providers is n_{cp}^e, the marginal consumer x_i who is indifferent between subscribing to the Internet not subscribing is located at:

$$x_i = n_c = \frac{v + bn_{cp}^e - p}{t}. \tag{9.3}$$

The marginal content provider y_j indifferent between being active and exiting the market is located at:

$$y_j = n_{cp} = \frac{an_c^e - s}{f}.$$ (9.4)

We look for a fulfilled expectations equilibrium where each side's expectations are fulfilled: $n_c^e = n_c$ and $n_c^e = n_{cp}$. The number of consumers and active content providers are then given by the solution to the simultaneous equation systems (9.3) and (9.4), shown as Equations (9.5) and (9.6).

$$n_c(p,s) = \frac{f(v-p) - bs}{ft - ab}$$ (9.5)

and

$$n_{cp}(p,s) = \frac{a(v-p) - ts}{ft - ab}.$$ (9.6)

Positivity of the demands requires $ft > ab$ and v to be sufficiently large: $v > P + bs/f$ and $v > p + ts/a$.

We now study the monopoly platform optimum—the optimum with network neutrality regulation and the social optimum.

9.4.2.1.1 Monopoly Platform Optimum

Consider first the monopoly platform private optimum under which the platform is free to set both the subscription price p and the fee s to content providers. The platform faces the problem of choosing p and s to maximize.

$$\Pi(p,s) = (p-c)n_c(p,s) + sn_{cp}(p,s)$$ (9.7)

Because the two markets provide complementary products, the monopolist finds an inverse relationship between p and s, that is, maximizing with respect to p results in a smaller p when s is larger, and maximizing with respect to s results in a smaller s when p is larger. Specifically, the optimal p for the monopolist given, s, defined by $\partial\Pi/\partial p = 0$, is given by

$$p(s) = \frac{f(v+c) - (a+b)s}{2f}$$ (9.8)

and the optimal s for the monopolist given p, defined by $\partial\Pi/\partial s = 0$, is

$$s(p) = \frac{av + bc - (a+b)p}{2t}$$ (9.9)

To ensure that the second order conditions are fulfilled here and in the analysis that follows, it is assumed that there is sufficient differentiation among consumers and content providers, or equivalently, that the network effects are not too strong, and moreover, that the market is never entirely covered on the consumer or the content provider side.

9.4.2.1.1.1 Assumption 1 — (i) Cross-side externalities (network effects) are not too strong, or equivalently, consumers and content providers are sufficiently differentiated: $ft - (a + b)^2 > 0$. (ii) The market is never entirely covered on the consumer or the content provider side: $ft - (a + b)^2 > \max\{ f(v - c), (a + b)(v - c)\}$.

We make the assumption strong enough to cover not only the monopolist's maximization problem but the social planner's as well. Hence, the conditions in Assumption 1 come from the second order conditions and equilibrium participation levels when determining socially optimal prices. The conditions for obtaining interior solutions for the privately optimal monopoly prices are weaker, but we impose this stricter assumption here because we want to compare the privately optimal price balance to the socially optimal price balance under the same assumptions.

Solving the two aforementioned equations simultaneously gives the consumers' subscription price and the fee charged to the content providers that maximize the platform's profits.

$$p^M = \frac{(2ft - ab)(v + c) - b^2c - a^2v}{4ft - (a + b)^2} \tag{9.10}$$

and

$$s^M = \frac{(a - b)f(v - c)}{4ft - (a + b)^2} \tag{9.11}$$

The participation levels are

$$n_c^M = \frac{2f(v - c)}{4ft - (a + b)^2} \tag{9.12}$$

and

$$n_{cp}^M = \frac{(a + b)(v - c)}{4ft - (a + b)^2} \tag{9.13}$$

and the profits of the monopoly platform are

$$\Pi^M = f(v - c)^2/(4ft - (a + b)^2). \tag{9.14}$$

Superscript M indicates the fully private optimum where both p and s are chosen by the monopoly platform. The price consumers pay is above the marginal cost if $2ft - a(a + b) > 0$, which holds under Assumption 1. The monopoly platform service provider sets a positive fee to content providers for accessing users ($s^M > 0$) only if $a/b > 1$. This means that if content providers value additional consumers more highly than consumers value additional content providers, the platform will charge content providers a positive price for accessing consumers.

9.4.2.2 Platform Duopoly

Here, the 2SM-based model is extended to duopoly competition between two platforms with multihoming content providers. Consumers are assumed to be single-home; content providers, however, are assumed to be multihome (i.e., they sell through both platforms, paying the fees charged by the platforms). The two platforms, although providing similar functionality, appeal to different types of users, who are spatially distributed in a Hotelling sense [35] with the two platforms situated at the two ends of the linear model, as shown in Figure 9.4. As in monopoly, we assume that platforms only offer linear subscription prices and content provider fees.

There are two platforms (1 and 2) located at $x = 0$ and $x = 1$. Each platform offers the same intrinsic benefit v to consumers. Given an expected number of content providers n^e_{cpk} in each platform k, $k \in \{1,2\}$ the marginal consumer, indifferent between buying from platform 1 or 2, is located at x_i that obeys:

$$v + bn^e_{cp1} - tx_i - p_1 = v + bn^e_{cp2} - t(1 - x_i) - p_2. \tag{9.15}$$

Assuming full market coverage, the sales of the two platforms are

$$n_{c1} = \frac{1}{2} - \frac{b\left(n^e_{cp2} - n^e_{cp1}\right) - \left(p_2 - p_1\right)}{2t} \tag{9.16}$$

and

$$n_{c2} = 1 - n_{c1}. \tag{9.17}$$

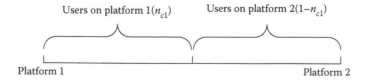

Users on platform 1(n_{c1})   Users on platform 2(1−n_{c1})

Platform 1   Platform 2

Figure 9.4 User distribution for the competing platform.

Content providers are defined as in the monopoly model shown here, that is, they are heterogeneous with respect to the fixed costs of setting up MSN services. The expected number of consumers that are able to reach each content provider is n_{ck}^e, if the content provider buys access from platform k, $k \in \{1,2\}$. The total revenue for each content provider is an_{ck}^e.

Platform k collects a fee s_k from each content provider to allow access to its users. Thus, a content provider j's profit from selling through platform k is

$$\pi_{jk} = an_{ck}^e - s_k - fy_j \tag{9.18}$$

Each content provider with $\pi_{jk} \geq 0$ sets up its business, pays platform k for access to its consumers, and makes nonnegative profits from sales to those consumers. Thus, the marginal content firm that is indifferent about being active or staying out of the market is $n_{cpk} = (an_{ck}^e - s_k)/f$ with $k \in \{1,2\}$. Because consumers are single-home, content providers can only reach the consumers of each platform by buying access from that platform. Essentially, as Armstrong [27] points out, a "competitive bottleneck" arises as there is no competition for content providers because they make a decision to join one platform independently of the decision to join the other.

As in the monopoly setup, we look for a fulfilled expectations equilibrium. Each side's expectations are fulfilled; therefore $n_{ck}^e = n_{ck}$ and $n_{cpk}^e = n_{cpk}$, $k \in \{1,2\}$. Solving the four equations given by (9.16), (9.17) and $n_{cpk} = (an_{ck}^e - s_k)/f$, the number of consumers and active content providers are

$$n_{c1} = \frac{1}{2} + \frac{b(s_2 - s_1) + f(p_2 - p_1)}{2(ft - ab)} \tag{9.19}$$

$$n_{c2} = \frac{1}{2} + \frac{b(s_2 - s_1) + f(p_2 - p_1)}{2(ft - ab)} \tag{9.20}$$

$$n_{cp1} = \frac{a(b(s_1 + s_2) + f(t + p_2 - p_1)) - (a^2 b + 2fts_1)}{2f(ft - ab)} \tag{9.21}$$

and

$$n_{cp2} = \frac{a(b(s_1 + s_2) + f(t + p_1 - p_2)) - (a^2 b + 2fts_2)}{2f(ft - ab)} \tag{9.22}$$

When the duopoly platforms are free to set prices to both consumers and content providers, platform k maximizes $\pi_k(p_1,p_2,s_1,s_2) = (P_k - c)n_{ck} + s_k n_{cpk}$, with $k \in \{1,2\}$ resulting in equilibrium prices $p_1^D = p_2^D = t + c - (a^2 + 3ab)/4f$ and

$s_1^D = s_2^D = (a-b)/4$. The monopolists split the market on the consumer side and profits are $\pi_1^D = \pi_2^D = \left(4ft - (a+b)^2 + 4(ft - ab)\right)/16f$.

9.5 Conclusion

Mobile social networking is an evolving paradigm that could change the way people communicate and exchange data. An MSN offers user-centric services in which mobile users who are parts of the MSN not only use but also improve the services of that MSN. In MSNs, information about social relationships among users can be used to optimize data exchange, sharing, and delivery to meet the requirements of the services and applications.

At the same time, the MSN industry presents a number of peculiar economic characteristics such as mobility, network effects, and long tail, which should be taken into consideration in the design of a MSN's business models. For example, mobility adds a great deal of flexibility (and complexity) to development and deployment of MSN applications and services, and MSNs make a larger contact pool available to their members and allow them to easily manage and maintain virtually unlimited numbers of contacts by granting access to the long tail of mobile social networking. Direct and indirect network effects have an impact on both sides of MSN platforms—consumers as well as advertisers/merchants/third-party developers. This can be either positive (customers vs. third-party service providers) or negative (customers vs. advertisers). This chapter specifies generic design issues within the four MSN business model components (STOF) to develop insight into the way organizations can design "balanced" business models. Today, the most common revenue models for MSNs are advertising, subscription, and transaction models. The best service provisioning should properly combine these business models. Finally, this chapter preliminarily discusses theoretical analysis of 2SM-based MSN business models, aiming to maximize the profit of MSN platform operators.

References

1. Camponovo, G. and Y. Pigneur. Business model analysis applied to mobile business. In: Proceedings of the 5th International Conference on Enterprise Information Systems (ICEIS), Angers, France, April, 23–26, 2003; pp. 173–183.
2. De Reuver, M. and T. Haaker. Designing viable business models for context-aware mobile services. *Journal of Telematics and Informatics* 2009; 26(3): 240–248.
3. Jabeur, N., S. Zeadally, and B. Sayed. Mobile social networking applications. *Journal of Communications of the ACM* 2013; 56(3): 71–79.
4. Enders, A., H. Hungenberg, H. P. Denker, and S. Mauch. The long tail of social networking: Revenue models of social networking sites. *European Management Journal* 2008; 26(3): 199–211.

5. Alt, R. and H.-D. Zimmermann. Preface: Introduction to special section-business models. *Electronic Markets* 2001; 11(1): 3–9.
6. Afuah, A. and C. Tucci. *Internet Business Models and Strategies*. McGraw-Hill, Boston, 2003.
7. Osterwalder, A. and Y. Pigneur. An e-business model ontology for modeling e-business. In: Proceedings of the 15th Bled Electronic Commerce Conference, Bled, Slovenia, June 17–19, 2002.
8. Morris, M., Schindehutte, M. and J. Allen. The entrepreneur's business model: Toward a unified perspective. *Journal of Business Research* 2005; 58: 726–735.
9. Shafer, S. M., H. J. Smith, and J. C. Linder. The power of business models. *Business Horizons* 2005; 48: 199–207.
10. Kangas, P., S. Toivonen, and A. Back, (eds). Ads by Google and other social media business models, VTT Tiedotteita-Research Notes 2384, 2007.
11. Mäkinen, O. Mobile social media business models. Master's Thesis. Helsinki University of Technology, 2009.
12. KakaoTalk now profitable. Available from: http://www.gamesinasia.com/kakotalk-profitable-aim-indonesia/
13. Gao, Z. and A. Ji. Smartmobile-ad: An intelligent mobile advertising system. In: The 3rd International Conference on Grid and Pervasive Computing Workshops of IEEE, 2008, pp. 164–171.
14. Dao, H., S. R. Jeong, and H. Ahn. A novel recommendation model of location-based advertising: Context-aware collaborative filtering using GA approach. *Journal of Expert Systems with Applications* 2012; 39(3): 3731–3739.
15. Wilson, F. The Freemium business model. Retrieved Oct. 14, 2010, Available online: http://www.avc.com/a_vc/2006/03/ the_freemium_bu.html.
16. Bekkelund, K. J. Succeeding with Freemium: Exploring why companies have succeeded & failed with Freemium, Master thesis, NTNU, Trondheim, Norway, 2011.
17. Shampanier, K., N. Mazar, and D. Ariely. Zero as a special price: The true value of free products. *Marketing Science* 2007; 26(6): 742–757.
18. Haruvy, E. and A. Prasad. Optimal freeware quality in the presence of network externalities: an evolutionary game theoretical approach. *Journal of Evolutionary Economics* 2001; 11(2): 231–248.
19. Lee, J. and H. Lee. Exploration and exploitation in the presence of network externalities. *Management Science* 2003; 49(4): 553–570.
20. Faugère, C. and G. K. Tayi. Designing free software samples: A game theoretic approach. *Information Technology and Management* 2007; 8(4): 263–278.
21. Gallaugher, J. M. and Y. M. Wang. Understanding network effects in software markets: Evidence from web server pricing. *MIS Quarterly* 2002; 26(4): 303–327.
22. Jiang, Z. and S. Sarkar. Speed matters: The role of free software offer in software diffusion. *Journal of Management Information Systems* 2009; 26(3): 207–240.
23. Emerging business models in the digital economy—The mobile applications, market. Available from: http://www.acma.gov.au/webwr/_assets/main/lib310665/emerging_business_models.pdf
24. Two-sided market. Available from: http://en.wikipedia.org/wiki/Two-sided_market
25. Roson, R. Two-sided markets: A tentative survey. *Review of Network Economics* 2005; 4(2): 142–160.
26. Rochet, J. C. and J. Tirole. Platform competition in two-sided markets. *Journal of the European Economic Association* 2003; 1(4): 990–1029.

27. Armstrong, M. and J. Wright. Two-sided markets, competitive bottlenecks and exclusive contracts. *Economic Theory* 2007; 32(2): 353–380.
28. Gupta, A., B. Su, and Z. Walter. Risk profile and consumer shopping behavior in electronic and traditional channels. *Decision Support Systems* 2004; 38(3): 347–367.
29. Muylle, S., and A. Basu. Online support for business processes by electronic intermediaries. *Decision Support Systems* 2008; 45(4): 845–857.
30. Gazé, P. and A. G. Vaubourg. Electronic platforms and two-sided markets: A side-switching analysis. *Journal of High Technology Management Research* 2011; 22(2): 158–165.
31. Evans, D. S. and R. Schmalensee. Failure to launch: Critical mass in platform businesses. *Review of Network Economics* 2010; 9(4): 1–26.
32. Jullien, B. Competition in multi-sided networks: Divide-and-conquer. *American Economic Journal: Microeconomics* 2011; 3(4): 1–35.
33. Evans, D. S. and R. Schmalensee. The antitrust analysis of multi-sided platform businesses. Research of National Bureau of Economic Research, Cambridge, MA, 2013.
34. Economides, N. and J. Tåg. Network neutrality on the Internet: A two-sided market analysis. *Journal of Information Economics and Policy* 2012; 24(2): 91–104.
35. Li, S., Y. Liu, and S. Bandyopadhyay. Network effects in online two-sided market platforms: A research note. *Journal of Decision Support Systems* 2010; 49(2): 245–249.

APPLICATIONS OF MSN

Chapter 10

Socially Inspired
Mobile Networking

10.1 Introduction

The increasing pervasiveness of mobile devices with short-range networking capabilities offers novel communication opportunities. Nodes can directly communicate when they come within radio range of each other with no need for any preinstalled networking infrastructure. Recently, the diffusion of mobile personal devices has exploded. The mobile handset is, by its own nature, a social artifact—an object carried by people to connect with people. This is the reason why the next big development in mobility and mobile service involves social behaviors in some form or fashion—to enable better ways to find, communicate, and share with friends and family, to learn about nearby places, and to consume information, all while on the go. The handsets are used by ordinary people—not only technology geeks—to communicate, to use applications once run only by desktops, and to organize their life. Typically, these devices can communicate with each other over short distances by using wireless technologies such as Bluetooth and Wi-Fi. In this way, a new kind of network emerges where nodes are carried by people, and links appear and disappear as people move and get in contact. This kind of network can be key technology to provide innovative services to the users without the need of any fixed infrastructure. The goal of socially inspired mobile networking is to harness adaptive human social structures (people and their social practices) as critical resources to design successful information and communication systems. Contrary to conventional Internet design, socially inspired mobile networking technologies do not seek to enforce or maintain end-to-end connectivity; instead they seek to create an underlying structure among devices where opportunistic interactions could be harnessed to convey relevant content. There is a huge amount of information about socially inspired networking technologies, including opportunistic communications, epidemic information propagation, mobile search and ranking, and so on.

The rest of this chapter is structured as follows. In Section 10.2, we review the real human mobility traces and models. The real human mobility traces include Cambridge trace and MIT reality data, which are usually used to evaluate and analyze the performance of some forwarding protocols. We survey the utilizations of social characteristics in the design of social-based, delay-tolerant network routing protocols and give a comparison of these routing protocols in Section 10.3. Finally, we briefly conclude the chapter in Section 10.4.

10.2 Human Mobility Traces and Models

A thorough study of individual movements became possible only recently with the availability of real-life mobility traces collected by cell phone operators, academic experiments, and Internet communities. The most accurate data come from systems that are directly designed to track location, such as by exploring satellites (e.g., Global Positioning System [GPS]). GPS provides accurate measurements of position and speed in outdoor locations (fine granularity of the location data), but signal quality is reduced or completely lost in indoor environments. Another approach is based on using nodes of communication systems such as Global System for Mobile Communications (GSM) base stations or Wireless Local Area Network (WLAN) access points (APs). The coordinates of the base station (e.g., Wi-Fi router) can be approximately considered as the location of the users currently connected to it. Some useful characteristics of human connectivity can be derived from tracing Bluetooth or Wi-Fi one-hop communications between hand-held devices. Finally, the routes of items exchanged between humans, such as bank notes, can also be considered as a proxy for human mobility. Human mobility traces gave significant impulse to studying statistical characteristics and patterns of human movements and have attracted researchers from different domains. As highlighted in Figure 10.1, the ultimate goal of the analysis of human mobility, as far as mobile social network in proximity (MSNP, which will be introduced in Chapter 12) is concerned, is the definition of a mobility model that allows us to evaluate the performance of forwarding protocols in realistic settings by means of either simulation or mathematical analysis.

10.2.1 Data Collection Techniques

One of the basic components in human mobility studies is data collection technique because it indicates the accuracy of the positioning system. Human mobility

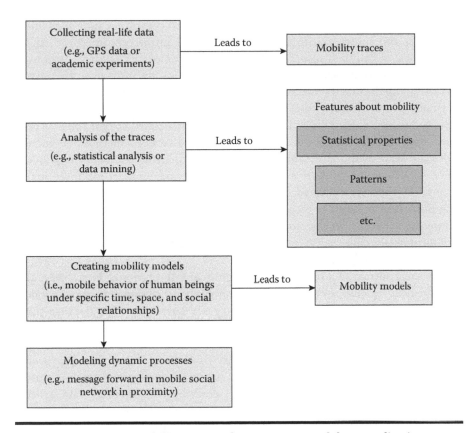

Figure 10.1 Human mobility—From data traces to models to applications.

researchers have traditionally relied on expensive data collection methods, such as surveys and direct observation, to get a glimpse into the way people are moving. These high-cost methods typically result in infrequent data collection or in small sample sizes. For example, a national census produces a wealth of information on where millions of people live and work, but it is carried out only once every ten years. Brockmann et al. [1] used the data of bank notes to study human traveling behavior. Later on, many other studies used GPS to track individuals or any moving objects. GPS provides accurate measurements of position and speed in outdoor locations (fine granularity of the location data), but signal quality is reduced or completely lost in indoor environments. Moreover, phone users tend to keep GPS turned off when not in use to avoid battery drain. When the GPS signal is available, however, it tends to be a very good candidate for differentiating between dwelling and mobility.

Continuous scanning for Wi-Fi APs has been used in context-aware computing to detect user mobility. This method is attractive because it can be performed online and in real time, both desirable qualities for this class of applications.

In recent years, the emergence of information and communication technologies (ICTs) and substantial investments in wireless infrastructures have led to extensive use of call data records (CDRs) in human mobility studies. Although CDR may have some bias on human mobility studies, up until now, they have been providing the best data sets to study the human mobility. Each CDR contains the time a phone placed a voice call or received a text message and the identity of the cellular antenna with which the phone was associated at that time. When joined with information about the locations and directions of those antennas, CDRs can serve as sporadic samples of the approximate locations of the phone's owner. CDRs are an attractive source of location information for three main reasons: (1) They are collected for all active cellular phones, which number in the hundreds of millions of records. (2) They are already being collected to help operate the networks, so that additional uses incur little marginal cost. (3) They are continuously collected as each voice call and text message completes, thus enabling timely analysis.

On the contrary, CDRs have two significant limitations. First, they are sparse in time because they are generated only when a phone engages in a voice call or text message exchange. Second, they are coarse in space because they record location only at the granularity of a cellular antenna (with an average error of 175 meters). It is not obvious a priori whether CDRs provide enough information to characterize human mobility in any useful way. The first problem could be solved if we modify the data collection system and track the user in fixed time intervals. Smoreda et al. [2] describe two different data collection methods from a cellular phone network: active and passive localization. Active localization provides a tool for recording positioning data on a survey sample over a long period of time. Passive localization, on the contrary, is based on phone network data that are automatically recorded for technical or billing purposes (CDRs).

In Ref. [3], the authors try to explore how visibility and signal strength of GSM cell towers and Wi-Fi beacons, which are already available on standard mobile handsets, can be used to generate mobility profiles. They have proposed the use of GSM data in order to avoid privacy risk of "fine-grained" locations and other practical limitations of continuous GPS sampling such as reduced phone battery life, inconsistent coverage for typical users, and limited availability of integrated GPS in current mobile phones. Nowadays, the social network, temporal dynamics, and the mobile behavior of mobile phone users have often been analyzed independently from each other using mobile phone data sets. Table 10.1 summarizes the properties of different data collection methods. It is important to notice that the active localization method, by provoking cell localization of the mobile device, solves the problem of time scarcity and provides a valuable resource of human trajectories. Considering the limitation of this type of data (spatial accuracy, when compared with other data types), it is important to choose a proper level of detail on which to apply analysis.

Table 10.1 Comparative Summary of Different Data Collection Techniques

Methods	Advantages	Disadvantages
• Wi-Fi localization	• Accuracy • Energy usage ~50% GPS	• Low coverage area • Providing access point is expensive
• GPS localization	• Highly precise ~5 m error • Can distinguish between transportation modes	• High battery (energy) usage • Expensive • No (low-quality) signal in indoor environment
• Cellular network • Localization (passive)	• Automatically generated	• Sparse in time • Needs more filtering • Less accuracy (~175 m error)
• Cellular network • Localization (active)	• More accuracy than passive localization • Less expensive than previous methods	• More costly than passive form • Issue of large database arises

10.2.2 Real Human Mobility Traces

10.2.2.1 Reality Mining Trace

The Reality Mining trace [4] was collected by the Reality Mining project group from MIT Media Labs. It is an experimental study involving about 97 people for the duration of nine months. Each person was given a Nokia 6600 cell phone with software that continuously logs data about the location and contacts of the cell phone. The logged data from all the cell phones total around 350 K hours of monitoring time and fit into a database of 1 GB size.

10.2.2.2 Cambridge Trace

The Cambridge trace [5] was collected by Scott et al. at Cambridge University. This trace includes the contacts about 36 people for three days. Each person was asked to carry the mobile devices (i.e., iMotes) with them at all times for the duration of the experiment. In addition, a number of stationary nodes were deployed in various locations that many people were expected to visit such as grocery stores, pubs, marketplaces, and shopping centers in and around the city of Cambridge, UK. A stationary iMote was also placed at the computer lab in which most of the experiment participants are students.

10.2.3 Human Mobility Models

In recent years, studying human mobility has been one of the major focuses of different disciplines. The main findings can be classified along the three axes of spatial, temporal, and connectivity properties. Brockmann et al. [1], who analyzed a huge data set of records of bank notes, showed that travel distances Δr (frequently called jump size) of individuals follow a power-law distribution. As for the temporal properties of human movements, Gonzalez et al. [6] detected the tendency of people to return to a previously visited location with a frequency proportional to the ranking in popularity of the location with respect to other locations. The authors computed the return time probability distribution (probability of returning at time t to a selected place) and concluded that prominent peaks (at 24, 48, and 72 hours) capture the tendency of humans to return regularly to the location they visited before. Song et al. [7] extended the experiment to a larger data set and measured the distribution of the visiting time (i.e., the time interval Δt a user spends at one location). The resulting curve is well approximated by a truncated power law with an exponent $\beta = 0.8 \pm 0.1$ and a cutoff of $\Delta t = 17$ h, which the authors connected with the typical awake period of humans. Connectivity properties have been extensively studied in the context of Mobile ad hoc network (MANET) research. Chaintreau et al. [8] showed that the distribution of intercontact time has a power-law nature over a wide range of values from a few minutes to half a day. Duration of contact times was also shown by Hui et al. [9] to follow an approximate power-law distribution.

Several metrics that depict regularities in people's movements have been observed by researchers. These metrics, however, cannot capture some aspects of human mobility such as the distinction between periodic and frequent (but not periodic) trips. The concept of human mobility patterns (Figure 10.2) has been introduced to fill this gap in recent studies. The motivation comes from the results of Song et al. [10], which provide a quantitative evaluation of the limits in predictability of human walks. The authors define some entropy measures (the metric used in information theory to evaluate uncertainty of random variables) ranging from one that depends only on frequencies of visits to one that considers the probability of finding particular time-ordered subsequences in the trajectory. The study shows that if we rely only on heterogeneous spatial distribution, as does the first entropy measure, the predictability across the whole population is insignificant and varies widely from person to person. Instead, if we also take into account a historical record of the daily mobility patterns, the potential predictability reaches 93% and does not vary a lot across the population (i.e., users who usually travel over long distances are as predictable as those who navigate within a narrow vicinity of their homes). This means that, knowing the history of a person's movements, we can potentially foresee his or her current location with an extremely high probability of success.

When talking about predictability or uncertainty, we usually connect it with some regularity or patterns in a structure or in the behavior of the phenomena.

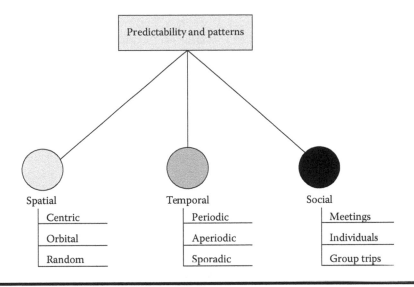

Figure 10.2 Predictability and patterns in human mobility traces.

If we think of regularity in human mobility, the first association that springs into mind is the periodic (e.g., daily, weekly, monthly) nature of movements in time. For example, each single day in a working week includes a trip from our home location to the office, conducted at more or less a similar time of the day. On a higher scale, we can think of repeated trips once every few weeks from a hometown to other cities to visit relatives or friends. These two examples, however, are different. Although the first case presents a stronger periodicity, the second one also gives us a flavor of predictability but in a weaker sense. On the contrary, there are always some completely random trips that might not ever be repeated (e.g., a vacation spent on a desert island). All these types of movements are connected with the frequency of location visits and thus with some temporal patterns in individual walks. More specifically, the first class represents periodic returns to the same location, the second characterizes frequent returns without strict regularity (called aperiodic), and the last one depicts unrepeated walks to new sites (sporadic trips).

We can also consider predictability in the context of patterns observed in the structure of trajectories. As mentioned by Song et al. [10], there is still a lack of knowledge in this area, and further development in that direction requires application of advanced data mining and knowledge discovery techniques. However, let us imagine the structure of mobility traces of an individual over a long time (i.e., years) and take the spatial grain at the level of a town. In other words, let us approximate the city (and human movements inside it) as a point and only consider trips between different localities (e.g., towns, cities). We can predict that, for an average person, mobile trajectories would have an approximately star-like structure with a distinctly emphasized center (hometown) and rays (depicting two-way trips) connecting the home with other cities. If we look, however, at a smaller temporal and spatial scale (e.g., a work week of a student on a campus), we can imagine the appearance of orbits in movements (e.g., from home to study room in the morning, from study room to gym in the evening and back home at night). These regularities in the structure of trajectories are called spatial patterns.

Among the driving forces that define human mobility behavior, considering the social aspect of predictability is also very important. For example, a substantial part of our free time is usually spent with our friends. Although this obviously gives a sense of predictability, it does not depend on some specific location (i.e., we might meet in different places) or other factors but is only driven by the necessity to interact with our social contacts. In other words, knowing our social surroundings helps predict our location. These social patterns of movements were classified into three categories: meetings (when the trip is taken to meet social contacts), group trips (i.e., organized movements of a group of socially connected humans), and individual trips.

The last aspect to consider is the scale of mobility. The scale can be categorized into three levels (building-wide, city-wide, and global views), which characterize the mobility behavior of individuals. This is, however, only a spatial scale, which can also be extended with two other dimensions—temporal and social. The building-wide view characterizes our movements only during a short period of time (i.e., from a few hours up to one day) when we are located inside a building. The scope of our social interactions is limited to a community of people staying with us simultaneously at the same location (e.g., colleagues from office during the workday). On a higher city-wide scale, considerably longer periods of time (between two trips to other cities) must be spent. Finally, as for the worldwide (i.e., global) views, months-long or years-long tracks and interactions with the whole social network were considered. The scale in modeling mobility is crucial due to the deviations in patterns and properties that can be observed at each of the layers. The driving forces behind our movements can also differ on different scales. For example, at the global level, our movements significantly depend on country borders, travel cost (increasing with the distance), and so on. An interesting observation is that the duration of the visit to a location may significantly depend on the trip taken in order to reach it (e.g., if we pay the cost of travel from Europe to the United States, we probably want to stay there at least for a few days, as opposed to a one-day business trip to a nearby town). If we, in contrast, consider a daily circulation of a student on the university campus between library, study rooms, gym, and so on, probably the movements are defined by the function of the buildings, neglecting the cost of the walks. Now, if we think from a top-down perspective, we can imagine a hierarchical mobility model divided into layers. Thus, we can start by modeling a worldwide view and approximate all trajectories of movements inside the city with one point on the map, considering the whole time spent as a sojourn on a global scale and the cluster of our contacts inside it as a point. If you are interested in the details of human motion inside the city or in a particular building, you can zoom in on the map as well as the social and temporal scales.

10.3 Social-Based Routing Protocols in Delay-Tolerant Networks

Mobile delay-tolerant networks (DTNs) are composed of a finite set of mobile devices, such as cell phones and personal digital assistants that can communicate with each other via short-range wireless protocols (e.g., Bluetooth). The DTN paradigm has attracted increased attention in the research and industrial community in recent years. Intermittent connectivity in DTNs results in the lack of instantaneous end-to-end paths, large transmission delays, and unstable network topology. These characteristics make the classical ad hoc routing protocols not applicable for DTNs because these protocols rely on establishment of a complete end-to-end route from the source to the destination.

All of the current DTN routing methods share a similar paradigm, the "store and forward" fashion. If there is no connection available at a particular time, a DTN node can store and carry the data until it encounters other nodes. When the node has such a forwarding opportunity, all encountered nodes could be the candidates to relay the data. Thus, relaying selection and forwarding decisions need to be made by the current node based on a certain routing strategy. Various DTN routing approaches adopt different strategies based on different metrics. Examples of such metrics include estimated delivery probability to the destination node, network resources available (including bandwidth, storage, and energy), estimated delay, and current network congestion level. However, the unpredictable mobility and restricted resource in DTNs significantly obstruct us from designing an ideal forwarding mechanism.

Lately, the consideration of social characteristics provides a new angle of view in the design of DTN routing protocols. In most of the DTN applications, a multitude of mobile devices are used and carried by people whose behaviors are better described by social models. This opens new possibilities of social-based DTN routing, in which the knowledge of social characteristics is used to make better forwarding decisions in DTN routing. In general, social relations and behaviors among mobile users are long-term characteristics and less volatile than node mobility. Based on this observation and taking the recent advances in social network analysis, several social-based DTN routing methods have been proposed recently to exploit various social characteristics in DTNs (such as community and centrality) to assist the relay selections. Many of the social properties related to DTN routing have been recently studied in social network analysis. They can be categorized into positive properties that benefit the relay selection (e.g., community, centrality, similarity, and friendship) and negative properties that hurt the network performance (e.g., selfishness).

10.3.1 A Survey of Social-Based Routing Protocols

Several social-based routing methods that take advantage of social characteristics in DTN networks have been investigated in recent years. This section surveys the utilization of social characteristics in the design of social-based DTN routing protocols.

10.3.1.1 Label Routing

Hui and Crowcroft [11] introduced a routing method (called "label routing" here) based on community labels in pocket-switched networks (PSNs). A PSN is a type of DTN where mobile devices are carried by people, and the devices communicate with each other when people meet. To reduce the amount of traffic created by forwarding messages in PSNs, the proposed routing method uses a labeling strategy to select forwarding relay. Because people in the same community are likely to meet regularly, they are appropriate forwarders for messages destined for the members of their community. In their solution, Hui and Crowcroft assumed that each node has a small label telling others about its affiliation/group (i.e., its social community), just like name badges used in a conference. Based on the labels, label routing chooses to forward messages to destinations directly or to next-hop nodes that belong to the same group (label) with the destinations.

Label routing takes advantage of the knowledge of the social community. It assumes that people from the same community tend to meet more often than people from different communities; therefore, they can be good forwarders to relay messages destined to the other members in the same community (with the same label). Label routing requires very little information about each individual (only its group/affiliation). This is easy to implement in PSN applications by tapping a mobile device and writing down the affiliation of the owner. In other words, the community (or group) information relies on user inputs in label routing. However, user-defined communities may not always reflect the position/contact relationship among nodes. For example, two DTN nodes in the same community may be physically far away and could never meet with each other. In this scenario, using one node to be the forwarder for the other may not be a good choice. In addition, in label routing, the message forwarding from the source to the destination is purely via the members within the destination's same community. This may significantly increase the delay or may even mean the message is not delivered. For instance, message delivery will fail when the source does not meet any member from the destination's community even though there are possible relay nodes from other communities.

10.3.1.2 SimBet Routing

Daly and Haahr [12] proposed a social-based routing protocol (SimBet routing) that uses betweenness centrality and similarity metrics to identify some "bridge" nodes (with high values of these metrics) in networks. To avoid exchanging information of the entire network topology, they only estimated the betweenness centrality Bet_n for each node n in its local neighborhood. For similarity metric, they considered the similarity $Sim_n(d)$ and the number of common neighbors of the current node n with the destination node d. Both of the social metrics are maintained and updated dynamically in DTNs. Therefore, the proposed SimBet routing makes a forwarding decision by considering not only the pre-estimated betweenness centrality metric but also the locally determined social similarity. Nodes with high betweenness centralities are those nodes that can act as bridges in their neighborhood, while nodes with high similarities with the destination are more likely to find a common neighbor with the destination that can act as the forwarder.

In SimBet routing, when a DTN node n meets another DTN node m and holds a message with destination d, n calculates its relative betweenness utility and similarity utility to node m, shown as

$$SimUtil_n = \frac{Sim_n(d)}{Sim_n(d) + Sim_m(d)} \tag{10.1}$$

$$BetUtil_n = \frac{Bet_n}{Bet_n + Bet_m} \tag{10.2}$$

Then node n can compute its SimBet utility, which is a weighted combination of betweenness utility and similarity utility:

$$SimBetUtil_n(d) = \alpha SimUtil_n(d) + (1-\alpha) BetUtil_n. \tag{10.3}$$

Here, α is a tunable parameter that can adjust the relative importance of the two utilities. For the message with d as its destination, if $SimBetUtil_m(d) > SimBetUtil_n(d)$, node n forwards the message to node m. Otherwise, it continues to hold the message. The message may eventually reach d via possible multihop relays.

In summary, SimBet routing uses two social metrics (centrality and similarity) to estimate or predict the probability that potential relay nodes may meet the destination. It is obvious that both metrics are effective at identifying suitable relays in different scenarios. Take an example graph, as shown in Figure 10.3, where a few low-degree bridges (i.e., a, b, and c) connect two well-connected components C_1 and C_2.

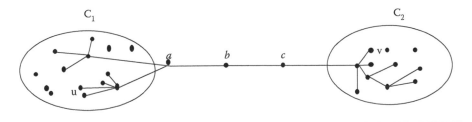

Figure 10.3   Illustration of problems in SimBet routing.

Assume that node u wants to send a message to node v. When node u encounters node a, it compares its SimBet utility with that of node a. Both u and a have zero similarity to v, but u's global betweenness centrality is less than a's because a sits on more of the shortest paths. Thus, u will transfer the message to a based on SimBet routing. In this case, the centrality metric helps to pick the better relay node. On the contrary, if node a wants to send a message to v and it encounters node b, the similarity metric will play a role because the global betweenness centralities of a and b are the same. Therefore, a has a smaller similarity (zero common neighbor) to v than b has (one common neighbor with v). Combining multiple social metrics may make the social-based protocol more effective in broad situations. However, due to the uncertainty of future encounters and the underlying social graph, it is still possible that the node with high SimBet utility fails to deliver the message to the destination.

To avoid global information exchanges, SimBet routing provides a distributed method to calculate social metrics locally, which is desirable in a DTN environment. However, estimating centrality based solely on local information may lead to inaccurate "bridge" identification. For instance, in the example shown in Figure 10.3, it is assumed that u wants to send a message to v. When u encounters node a, based on the two-hop information, u's local betweenness Bet_u is much larger than a's, Bet_a. Because both u and a have zero similarities to v, the overall $SimBetUtil_u(v) > SimBetUtil_a(v)$. Then node u will not pass this message to node a and thus will miss the opportunity to deliver the message. Nonetheless, considering global betweenness, each of the nodes of a, b, and c has the highest betweenness in the entire network (because they form the only path connecting components C_1 and C_2) and can then be correctly identified. A possible way to increase the chance of correct "bridge" identification is by using larger neighborhood information, although this may increase the communication cost. Similarly, to increase the chance of delivery, multiple relay nodes could be used. The trade-off is always between delivery performance and communication cost.

10.3.1.3 BUBBLE Rap Forwarding

The strategy, BUBBLE Rap forwarding, proposed by Hui et al. [13] also relied on two social characteristics (community and centrality). They assumed that each node belongs to at least one community and that its node centrality (either betweenness or degree centrality) in the community describes the popularity of the node within this community. Each node has a global centrality across the whole network (the global community) and a local centrality within its local community. A node may also belong to multiple communities and hence have multiple local centralities. Taking advantages of these social characteristics, BUBBLE Rap forwarding basically includes two phases: a bubble-up phase based on global centrality and a bubble-up phase based on local centrality. In both phases, the bubble-up forwarding strategy is used to forward messages to nodes that are more popular than the current node (i.e., with higher centrality). When a node s has a message with a destination of d, it first bubbles the message up based on the global centrality until the message reaches a node that is in the same local community C_d as the destination d. This procedure is shown as light grey arrows in Figure 10.4. After the message reaches d's community at node u, BUBBLE Rap forwarding switches to the second phase, which uses members of C_d as relays. This forwarding strategy continues to bubble the message up through the local community based on local centrality until the destination is reached. This later procedure is shown as dark grey arrows in Figure 10.4. In order to reduce cost, whenever a message is delivered to the community, the original carrier must delete this message from its buffer to prevent further dissemination.

BUBBLE Rap forwarding uses the concept of community in addition to node centrality to help with the forwarding decision. The introduction of local centrality inside a community is more beneficial than local centrality around a local neighborhood (i.e., k-hop) [12]. The bubble-up operations allow fast transfer of a message toward the destination or its community. However, such a strategy may fail when the destination belongs only to the communities whose members are all have low global centrality values. In this case, the bubble-up process in the first phase of BUBBLE Rap forwarding cannot find the relay node that is in the same local community as the destination node. A possible solution for this problem is to have a timeout timer for the bubble-up process and to change to another backup strategy for data delivery after time out. Hui et al. [13] used a flat community (instead of a hierarchical one) to demonstrate the efficiency of BUBBLE Rap forwarding. However, they did not provide details about how to handle hierarchical communities where the destination d may belong to multiple overlapping communities. In that scenario, they

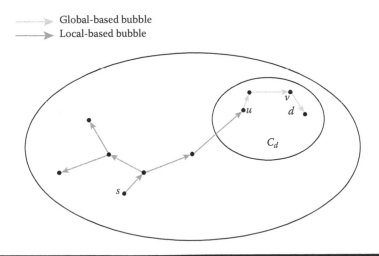

Figure 10.4 An illustration of BUBBLE Rap forwarding from source s to destination d.

may face problems in the second phase of BUBBLE Rap forwarding. For example, if the current encountering node u shares multiple communities with d, a problem arises regarding which one of d's local communities should be chosen to bubble up. A simple solution to this problem is picking the local community with which d has the highest centrality. This solution also matches the spirit of BUBBLE Rap forwarding, which keeps looking for nodes with high centralities.

10.3.1.4 Social-Based Multicasting

All of the social-based routing methods discussed here are unicast routing protocols for DTNs. Social-based approaches can be applied to multicast routing protocols for DTNs as well. Gao et al. [14] have proposed a set of multicast routing methods that use both centrality metric and community for relay selection.

Instead of using traditional centrality metrics such as betweenness, Gao et al. introduced a new metric (called cumulative contact probability) based on the Poisson modeling of social networks. Betweenness is purely defined based on the topology of the contact graph and may not be sufficient to represent the probabilities for a node to contact others. Thus, a weighted social network model is used to differentiate the contact frequencies of different node pairs. In such a model, the contact process of each node pair (i,j) is formulated as a Poisson process with an average contact rate of λ_{ij}. Then the cumulative contact probability of node i can be defined as

$$C_i = 1 - \frac{1}{N-1} \sum_{j=1, j \neq i}^{N} e^{-\lambda_{ij}T} \tag{10.4}$$

Here, N is the total number of nodes in the network and T is the total time period. In other words, C_i indicates the average probability that node i meets a random node within time T. This centrality metric (or its variation) is used in the proposed multicast methods to select a relay node with higher centrality.

Two multicast problems are considered by Gao et al. that are single-data multicast and multiple-data multicast. Their goals are to deliver a data item or a set of data items to a set of destinations within the time constraint T. The additional optimization objective is to minimize the number of relays used to achieve the average delivery ratio P.

For single-data multicast, the authors assumed the destinations are uniformly distributed; thus, they tried to ensure that all nodes are contacted by the data source or its selected relays within T. Based on cumulative contact probabilities (i.e., centrality) of nodes, a minimal number of relay nodes are selected among the contacted neighbors of the data source to guarantee the average delivery ratio is larger than p. This selection problem can be solved as a unified knapsack problem. The centrality metric is also refined for the case where the data source does not meet its contacted neighbors.

For multidata multicast, the authors proposed a community-based approach that only requires nodes to maintain the probabilities of forwarding each data item to other nodes in the same community. When the destinations are in other communities, data forwarding is conducted through some gateway nodes that belong to multiple communities. The data source selects relays among its contacted neighbors based on the local centrality metric and places appropriate data items on each relay. They used a two-stage relay selection scheme where both data item selection and relay selection are modeled as knapsack problems. Their method can ensure that the average delivery probability is larger than p.

10.3.1.5 Friendship-Based Routing

Bulut and Szymanski [15] also used friendship to aid in the delivery of packets in DTNs. They introduced a new metric, the social pressures metric (SPM), to accurately detect the quality of friendship. Different from social-based multicasting routing methods [14], where friendship is defined by users based on their social relationships, this approach considered friends as nodes that contact to each other frequently and that have long-lasting and regular contacts. Therefore, the SPM between nodes i and j can be estimated from the encounter histories of these nodes (recorded by the nodes) as:

$$SPM_{i,j} = \frac{\int_0^T f(t)dt}{T} \tag{10.5}$$

where $f(t)$ denotes the remaining time to the first encounter of these nodes after time t and T is the total time period. SPM describes the average forwarding delay if node i has a message destined to j at each time unit. Then the link quality $w_{i,j}$ between each pair of nodes, (i, j), is defined as

$$w_{i,j} = \frac{1}{SMP_{i,j}} \tag{10.6}$$

The authors assumed that the bigger value of $w_{i,j}$ represents the closer friendship between i and j. Using the value of $w_{i,j}$, each node can construct its friendship community for each period T as a set of nodes whose link quality with itself is larger than a threshold. When a node i, having a message destined to d, meets with node j, it forwards the message to j if and only if (1) j and d are in the same friendship community (in the current period) and (2) j is a stronger friend of d than i.

In summary, this friendship-based routing (FBR) method uses the node contact information in each period to calculate the friendship metric (i.e., SPM), and it constructs the friendship community. These social metrics can indeed help with making smarter forwarding decisions. However, the calculation of these metrics needs the whole contact information during each period, which may not be realistic in most DTNs. To obtain $f(t)$ in the current period, node i needs to know the time of its first encounter with node *j* after time *t* in this period, which is an event in future. Therefore, either the values in contact history from previous periods are used for this calculation at the current period or the estimated future contacts in this period are available for this calculation. This is clearly a drawback of this proposed method. In addition, this FBR uses a similar forwarding scheme to label routing, which may lead to the same problem. If the source node fails to meet with any node in the same friendship community with the destination node, the delivery fails. Therefore, more felicitous forwarding strategies should be studied for this FBR.

10.3.1.6 Give2Get

Mei and Stefa [16] proposed a reputation-based incentive scheme (Give2Get) for DTN routing that can detect misbehaving nodes and remove them from DTN routing. They proved that their proposed scheme achieves a Nash equilibrium; that is, no rational node has any incentive to deviate (i.e., a selfish node cannot find a better choice other than truthfully following the protocol). They provided two versions of their proposed scheme for epidemic forwarding (where messages are forwarded to the modes encountered first) or delegation forwarding (where messages are forwarded to nodes with higher forwarding qualities). The basic ideas of both versions are the same: (1) hiding the content of the message (including its source and destination) from the candidate relay before the relay agrees to serve and (2) requiring proof of relay after the selected relay agrees to serve and receive the encrypted message. A proof of relay is just an encrypted message sent from the relay to the sender. The sender can show this message to the source node later to prove it has forwarded the message.

Here, we only use the example of epidemic forwarding to explain their proposed Give2Get scheme. The nodes that generate the messages hide sender information to every possible relay except for the destination (by using the destination's public key). Once a message *m* is generated, the sender *s* tries to find two other nodes and relay the message to them. When *s* encounters a node *u*, it first negotiates a symmetric key with *u*, and then every communication after this point is encrypted with this key. Node *s* asks node *u* if it has already handled a message *m* by sending *u* the hash H(m) of *m*. Node *u* would not lie because it does not know the content of *m* (including its destination), that is, whether the destination is itself. If *u* lies, it may reject a message destined for itself. If *u* accepts the relay request, *s* encrypts the message with a random key *k* and sends it to *u*. Then *u* sends a proof of relay to *s*. Finally, *s* sends the key *k* to *u*. After *u* gets message *m*, *u* follows the same steps as *s* did. It must collect two proofs of relay before a timeout Δ_1 and hold the message until Δ_2 if it cannot collect two proofs within $[0, \Delta_1]$. During $[\Delta_1, \Delta_2]$ node, *s* can test node *u* if it meets *u* again by requesting *u* to show either two proofs of relay or a proof of still having the message in its memory. Otherwise, *s* will broadcast a proof of misbehavior to the whole network that, in turn, will remove node *u* from the network unless *u* can prove that *s* is wrong. Node *u* can stop looking for relays only when two proofs are collected or Δ_1 expires and can discard the message after Δ_2. The timeout

Δ_1 and Δ_2 should be carefully chosen to achieve a high success forwarding rate and positive probability that node s will meet node u again before Δ_2 expires.

The Give2Get algorithm uses encryption techniques to avoid selfish nodes that lie to elude their duties or that forge proof of relay to obtain a good reputation. The Nash equilibria of the proposed scheme make sure that no rational node has any incentive to deviate. It is worth noticing that the choices of the timeouts Δ_1 and Δ_2 could significantly affect the protocol's performance. The longer Δ_1 is, the higher success rate could be achieved. However, as Δ_1 and $\Delta_2-\Delta_1$ increase, the buffer of nodes will be occupied for a longer time and the packet loss ratio will increase as well. Therefore, the choice of time out could significantly depend on application requirements and on the underlying network mobility. The choice of time out over their datasets was also discussed by Mei and Stefa [16].

10.3.1.7 Tit-for-Tat

Shevade et al. [17] proposed the use of pair-wise Tit-for-Tat (TFT) as a simple and robust incentive mechanism for DTN routing. This incentive-aware routing protocol allows selfish nodes to maximize their own performance without any significant degradation of system-wide performance. Rather than attempting to detect misbehavior, it focuses on detecting good behaviors by using TFT. Differing from traditional TFT, the method proposed by Shevade et al. [17] incorporates generosity and contrition. Generosity enables bootstrapping and absorbs transient asymmetries, while contrition prevents mistakes from causing endless retaliation. Bootstrapping happens when two nodes meet for the first time. Because no packets have ever been successfully relayed by both nodes, the basic TFT prevents the start of any relay. Generosity allows a node to send ε number of packets more than it has earned. This can absorb up to ε amount of traffic imbalance and can stimulate the beginning of cooperation. However, this also allows selfish nodes to do ε less work than others. Although generosity absorbs a small amount of traffic imbalance, any imbalance exceeding ε could lead to lengthy retaliation between two neighbors. Therefore, contrition is introduced in their solution to prevent mistakes from causing endless retaliation and to provide a way to return to stability after perturbation—by stopping a node from reacting to a valid retaliation to its own mistake.

The proposed TFT-based algorithm can be easily implemented in DTNs because the verification of relaying only occurs between neighboring nodes. However, the generosity in this new TFT-based method may be exploited by selfish nodes. These nodes may not provide generosity to their neighbors while enjoying others' generosity. Therefore, this TFT-based algorithm cannot fully avoid selfish behaviors.

10.3.1.8 Social Selfishness Aware Routing

In all of the aforementioned methods, a selfish node is considered as having the same level of selfishness with every other node. This selfishness can be called individual selfishness. However, in a more realistic scenario, a selfish node may have different levels of selfishness with different groups of nodes. Hui et al. [18] found that, in PSNs, each node tends to help with forwarding more for people inside their community and less for people in other communities. This kind of selfish behavior is called social selfishness (to distinguish it from individual self-ishness). Li et al. [19] considered social selfishness in DTNs by allowing a user to define a different willingness (i.e., level of selfishness) with other users. The authors proposed Social Selfishness Aware Routing (SSAR), which enforces users' social selfishness in routing while maintaining an acceptable routing performance. When considering social selfishness, SSAR allocates resources such as buffers and band-widths based on packet priority, which is related to the social relationship among nodes (i.e., willingness defined by the nodes themselves). To maintain routing per-formance, SSAR quantifies a relay's willingness to evaluate its forwarding capa-bility and thus reduces packet dropping rate. Furthermore, SSAR formulates the forwarding process as a multiple knapsack problem with assignment restrictions and uses a greedy algorithm to select the forwarding packet set.

10.3.2 Comparisons of Social-Based Routing Protocols

We survey various social-based routing protocols that use social characteristics to assist packet forwarding in DTNs. Table 10.2 summarizes the social characteristics used by these routing protocols.

Even though social-based routing in DTNs has recently received much atten-tion in the wireless network community as a relatively new area, there are still quite a few challenges left. First, most of the current social-based approaches use only sim-ple definitions of one or two social characteristics (such as k-clique for community, node degree for centrality, contact frequency for friendship). Thus, it is interesting to see whether there are other more realistic and accurate social characteristics that can be used to further improve the performance of DTN routing (even if it would be more complex). Second, social-based approaches significantly rely on accurate modeling of the social characteristics used by them. However, due to the lack of continuous connectivity and time-varying topology, it is hard to accurately esti-mate certain social characteristics without global or future information—even with simple definitions. Therefore, another challenging task is how to model and extract accurate social characteristics in dynamic DTNs. Last, there are not many social-based solutions that consider both positive and negative social characteristics. This is a potential direction worthy of investigation in the future. For example, it could be more harmful for a node with higher social importance (with higher centrality or similarity) than a node with lower importance to be selfish. Considering positive social metrics together with negative ones, different approaches can be applied to avoid selfish behaviors from nodes with different social roles.

In this section, we mainly focus on social-based DTN routing protocols. But there are also other interesting directions within DTNs in which social studies

Table 10.2 Comparison of Social-Based Routing Protocols

Routing Protocols/Social Characteristics	Community	Centrality	Similarity	Friendship	Individual	Social
Label	✓					
SimBet		✓	✓			
BUBBLE Rap forwarding	✓	✓				
Social-based multicasting	✓	✓				
Friendship-based routing	✓			✓		
Give2Get					✓	
TFT					✓	
SSAR						✓

Note: TFT, Tit-for-Tat; SSAR, Social Selfishness Aware Routing.

may be used for protocol design, such as privacy protection, cooperative caching, and content-based sharing. We believe that social-based approaches will be applied to wider research topics in communication networks far beyond DTN routing protocols.

10.3.3 A Typical Social-Based Routing Protocol

In this section, we introduce a concrete protocol (FBR) in mobile delay-tolerant networks, in which the social characteristics of community and friendship were used. Bulut and Szymanski [15] define a new metric that measures different aspects of friendship behavior recorded in the history of their encounters with other nodes and consider both direct and indirect friendship. They also differentiate friendships according to time of day and propose to use different friendship communities in different time periods. But they do not consider users' willingness and implicitly assume that a node is willing to forward packets for all others. This may not work well because some packets may be forwarded to nodes unwilling to relay.

Bulut and Szymanski presented the FBR protocol for delay-tolerant networks. First, they define a new metric to more accurately understand relations between nodes. Second, a local community formation based on this new friendship detection metric was proposed. They use not only direct relations but also indirect ones in a different way. Third, a new approach to handle temporal differentiations of node relations was introduced.

10.3.3.1 Analysis of Node Relations

Because the nodes in an MSN are encountered intermittently, the link quality between each pair of nodes needs to be estimated to learn about the possible forwarding opportunity between them. Then, the temporal encounter information between nodes can be condensed to a single link weight and the neighboring graphs of nodes can be constructed.

Several metrics, including encounter frequency, total or average contact period, and average separation period, were used to extract the quality of links between pairs of nodes. However, all these metrics have some deficiencies in the accurate representation of forwarding opportunities between nodes. For example, consider the six different encounter histories of two nodes, i and j, in Figure 10.5. Shaded boxes show the encounter durations between these nodes in the time interval T. In cases a and b, the encounter frequencies are the same, but the contact between the nodes lasts longer in case b than in case a. Therefore encounter pattern b offers better forwarding opportunities than a does. Comparing cases b and c, we notice that the contact durations are the same, but the encounter frequencies are different. Because frequent encounters enable nodes to exchange messages more frequently, case c is preferable to b for opportunistic forwarding. Among the previously proposed metrics, encounter frequency fails to represent the stronger link in the comparison of

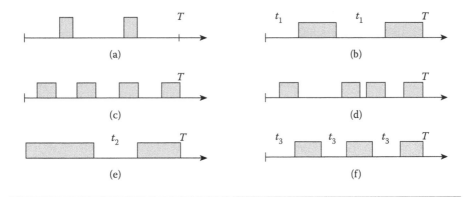

Figure 10.5 **Six different encounter histories between nodes *i* and *j* in the time interval [0, *T*]. (Shaded boxes show the encounter durations during which the nodes are within their communication ranges.)**

cases a and b, and total contact duration fails in the comparison of cases b and c. Although the average separation period can assign correct link weights representing the forwarding opportunities in cases a, b, and c, it fails in other cases. When we compare cases c and d, both the contact durations and encounter frequencies are the same. However, case c is preferred to d due to the even distribution of contacts. Preference of case c is achieved by using irregularities in separation period as a penalty factor. However, deciding on how much it will affect the link quality in different cases is still difficult. Moreover, for the cases such as e and f, the average separation period fails to assign accurate link weights. If $t_1 = t_2$, average separation period cannot differentiate cases b and e, but case e is preferable due to its longer contact duration. (The average separation period can even give preference to case b if t_1 is slightly less than t_2.) Similarly, if $t_1 = t_3$, the average separation period cannot differentiate cases b and f even though case b offers a better forwarding opportunity.

To find a better link metric that reflects the node relations more accurately, Bulut and Szymanski [15] have considered the following three behavioral features of close friendship: high frequency, longevity, and regularity. That is, to be considered friends of each other, two nodes need to contact frequently; their contacts must be long-lasting and regular. Note that frequency and regularity are different. Two nodes may meet infrequently but regularly (e.g., once a week) and still be considered friends. This is, of course, a weaker friendship than the one with contacts that are frequent and regular. Some metrics take into account some of these features but not all of them at the same time. To account these properties in one metric, a new metric called a SPM was introduced. SPM may be interpreted as a measure of a social pressure that motivates friends to meet to share their experiences. This amounts to answering the question: What would be the average message forwarding delay to node *j* if node *i* has a new message destined to node *j* at each time unit? Then, they define the link quality ($w_{i,j}$) between each pair as the inverse of this computed value. More formally:

$$SPM_{i,j} = \frac{\int_0^T f(t)dt}{T} \quad \text{and} \quad w_{i,j} = \frac{1}{SMP_{i,j}} \tag{10.7}$$

where $f(t)$ returns the remaining time to the first encounter of these nodes after time t (if they are currently in contact $f(t) = 0$). The bigger the value of $w_{i,j}$, the closer is the friendship (the higher the forwarding opportunities) between the nodes i and j. Note that, when we evaluate all cases in Figure 5, $w_{i,j}$ gives preference to cases that offer more forwarding opportunities.

10.3.3.2 Friendship Community Formation

Each node can compute its link qualities ($w_{i,j}$) with other nodes from its contact history. Then, it can define its friendship community as a set of nodes having a link quality larger than a threshold (τ) with itself. But this set will include only direct friends. However, two nodes that are not close friends directly (they even may not have contacts at all) still can be close indirect friends. This happens if they have a very close friend in common so that they can contact frequently through this common friend. To find such indirect friendships between nodes, using conditional SPM (CSPM) between nodes was proposed. Consider the sample encounter history in Figure 10.6. Although the upper portion shows the contacts between nodes i and j, the lower one shows the contacts between nodes j and k. $CSPM_{j,k|i}$ is defined as the average time it takes for node j to give node k the message received from node i. That is, for the contact history in Figure 10.6, node j computes

$$CSPM_{j,k|i} = \left(\int_{s=0}^{t_{d1}} s + \int_{s=0}^{t_{d2}} s \right) \Big/ T \quad \text{instead of} \quad SPM_{j,k} = \left(\int_{s=0}^{t_{d3}} s + \int_{s=0}^{t_{d4}} s \right) \Big/ T.$$

Each node can detect its direct friendships from its own history. To detect indirect friendships, a node needs CSPM values of its friends for its noncontact nodes.

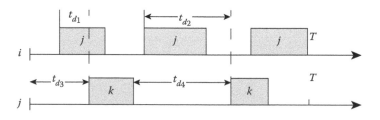

Figure 10.6 Encounter history between node *i* and *j* (upper) and between node *j* and *k* (lower) in the same time interval [0, *T*].

Once such CSPM values are received periodically, each node can form its community using the following definition:

$$F_i = \left\{ j \mid w_{i,j} > \tau \text{ and } i \neq j \right\} \bigcup \left\{ k \mid w_{i,j,k} > \tau \text{ and } w_{i,j} > \tau \text{ and } i \neq j \neq k \right\}$$

where, to simplify the data collection and computation, we approximate the indirect link weight as $w_{i,j,k} = 1/(SPM_{i,j} + CSPM_{j,k|i})$. This equation enables nodes to detect their one-hop direct and two-hop indirect friends. Indirect friendship can easily be generalized to friends more than two hops away. We have not included such an extension because we have demonstrated that nodes in the same community are usually at most two hops away from each other.

Note that the introduced method for detecting the indirect strong links between nodes is different than some approaches (based on transitivity) that basically consider the links between node pairs separately and that assume a virtual link between node i and k if $w_{i,j} \, w_{j,k} > \tau$. However, in this model, the indirect relations were detected more accurately. For example, if node j has a weak link with node k, $w_{i,j} \, w_{j,k}$ may be less than τ. However, if node j usually meets node k in a short time right after its meeting with node i, our metric can still consider node k as a friend of node i.

It is reasonable to expect similar behavior in other mobile social networks. For example, i can be a school, work, or home friend of a node j and the encounter times can differ accordingly. Moreover, i can be both a school and home friend of j so that they stay together during the day. Previously, some aging mechanisms [20] were used to reflect this feature of node relations. However, the most significant drawback of such models is their slow reaction time for periodically changing the quality of a node's links. To reflect the differentiation in the strength of friendships properly, temporal friendship communities were proposed. That is, each node i will compute its F_i for different periods of the day and will have different friendship communities in different periods.

10.3.3.3 Forwarding Strategy

Once each node constructs its friendship community for each period, the forwarding algorithm works as follows. If a node i having a message destined to d meets with node j, it forwards the message to j if and only if node j's current friendship community (in the current period) includes node d and node j is a stronger friend of node d than node i is. It should be noted that even if node j has a better link with node d than node i has, if node j does not include d in its current friendship community, node i will not forward the message to node j.

It is necessary to handle period boundary cases that arise when the encounter of two nodes is close to the end of the current period. In such a case, nodes use their friendship communities in the next period. For example, if we use three-hour periods for community formation and node i meets node j at 2:45 p.m., it would be better if the nodes use their communities in the next three-hour period (3 p.m.–6 p.m.)

to check whether the destination is included. Because the time remaining in the current period is very limited, using the current communities may lead to wrong forwarding decisions being made. In the algorithm, we use threshold t_b and let the nodes use next period's community information if remaining time to the end of current period is less than t_b.

10.4 Conclusion and Future Work

In this chapter, we first introduce two real human mobility traces that are widely used in the evaluation. Afterward, a thorough survey of social-based routing protocols in delay-tolerant networks is discussed, and a detail introduction to FBR protocol is given.

In order to realize the wide-scale deployment of socially inspired mobile networks, more work needs to be accomplished. First, several social properties, such as centrality and community, have been defined and widely used in the design of efficient routing protocols. The performance of these routing protocols has been improved by using these social properties. Most existing protocols assume that these properties do not vary frequently. Thus, investigating the evolution of these properties based on real mobility traces is important. Second, the issue of social selfishness is still an open issue. Many routing algorithms have been proposed for DTNs; most of them do not consider users' willingness and implicitly assume that a node is willing to forward packets for all others. However, nodes tend to behave selfishly due to saving their own storage and power resources. Selfishness of nodes seriously influences the performance of existing routing protocols. According to the investigation, we can see that there is no efficient incentive strategy to prevent social selfishness. Third, researchers recently have found DTN applications cannot catch up with the development of theory research. Therefore, in the future, when devising a DTN routing, we should pay more attention to its deployment and applications. And the target of research should be increasing the efficiency of routing while decreasing the difficulties of deployment.

References

1. Brockmann, D., L. Hufnagel, and T. Geisel. The scaling laws of human travel. *Nature* 2006; 439: 462–5.
2. Smoreda, Z., A. M. Olteanu-Raimond, and T. Couronne. Spatiotemporal data from mobile phones for personal mobile assessment. In: Proceedings of the International Conference on Transport Survey Methods: Scoping the Future While Staying on Track, Termas de Puyehue, Chile, November 14–18, 2013.
3. Mun, M. Y. et al. Parsimonious mobility classification using GSM and Wi-Fi traces. In: Proceedings of the HotEmNets, Charlottesville, VA, June 2–3, 2008.
4. Eagle, N. and A. Pentland. Reality mining: Sensing complex social systems. *Personal and Ubiquitous Computing* 2006; 10(4): 255–68.

5. Scott, J., R. Gass, J. Crowcroft, P. Hui, C. Diot, and A. Chaintreau. CRAWDAD trace cambridge/haggle/imote/content(v.2006-09-15). Available from: http://crawdad. cs.dartmouth.edu/cambridge/haggle/imote/content [cited September, 2006].

6. Gonzalez, M. C., C. A. Hidalgo, and A.-L. Barabasi. Understanding individual human mobility patterns. *Nature* 2008; 453: 779–82.

7. Song, C. et al. Modeling the scaling properties of human mobility. *Nature Physics* 2010; 6: 818–23.

8. Chaintreau, A., P. Hui, J. Crowcroft, C. Diot, R. Gass, and J. Scott, Impact of human mobility on opportunistic forwarding algorithms. *IEEE Transactions on Mobile Computing* 2007; 6: 606–20.

9. Hui, P., A. Chaintreau, J. Scott, R. Gass, J. Crowcroft, and C. Diot, Pocket switched networks and human mobility in conference environments. In: Proceedings of SIGCOMM Workshop on Delay-Tolerant Networking, Philadelphia, PA, August 26, 2005.

10. Chaoming Song, C., Qu, Z., Blumm, N., and Barabasi, A. L. Limits of predictability in human mobility. *Science* 2010; 327: 1018–21.

11. Hui, P. and J. Crowcroft. How small labels create big improvements. In: Proceedings of the International Workshop on Intermittently Connected Mobile Ad hoc Networks in Conjunction with IEEE PerCom, New York, USA, March 19–23, 2007.

12. Daly, E. M. and M. Haahr. Social network analysis for routing in disconnected delay-tolerant MANETs. In: Proceedings of the 8th ACM International Symposium on Mobile Ad Hoc Networking and Computing (MobiHoc), Montreal, QC, September 9–14, 2007.

13. Hui, P., J. Crowcroft, and E. Yonek. BUBBLE RAP: Social-based forwarding in delay tolerant networks. In: Proceedings of the 9th ACM International Symposium on Mobile Ad Hoc Networking and Computing (MobiHoc), Hong Kong, China, May 26–30, 2008.

14. Gao, W., Q. Li, B. Zhao, and G. Cao. Multicasting in delay tolerant networks: A social network perspective networks. In: Proceedings of the 10th ACM International Symposium on Mobile Ad Hoc Networking and Computing (MobiHoc), New Orleans, LA, May 18–21, 2009.

15. Bulut, E. and B. K. Szymanski. Friendship based routing in delay tolerant mobile social networks. In: Proceedings of IEEE Global Telecommunications Conference (GLOBECOM), Miami, FL, December 6–10, 2010.

16. Mei, A. and J. Stefa. Give2get: Forwarding in social mobile wireless networks of selfish individuals. In: Proceedings of the IEEE 30th International Conference on Distributed Computing Systems (ICDCS), Genova, Italy, June 21–25, 2010.

17. Shevade, U. B., H. H. Song, L. Qiu, and Y. Zhang. Incentive-aware routing in DTNs. In: Proceedings of the 16th IEEE International Conference on Network Protocols (ICNP), Orlando, FL, October 19–22, 2008.

18. Hui, P., K. Xu, V. O. Li, J. Crowcroft, V. Latora, and P. Lio. Selfishness, altruism and message spreading in mobile social networks. In: Proceedings of the 1st IEEE International Workshop on Network Science for Communication Networks (NetSciCom), Rio de Janeiro, Brazil, April 24, 2009.

19. Li, Q., S. Zhu, and G. Cao. Routing in socially selfish delay tolerant networks. In: Proceedings of the 29th IEEE International Conference on Computer Communications (INFOCOM), San Diego, CA, March 15–19, 2010.

20. Link, J., N. Viol, A. Goliath, and K. Wehrle. SimBetAge: Utilizing temporal changes in social networks for pocket switched networks. In: Proceedings of ACM Workshop on User-Provided Networking, Rome, Italy, December 1–4, 2009.

Chapter 11

Location-Based Services

11.1 Introduction

With the popularity of smartphones, the function of the Global Positioning System (GPS) is also widely added to those products, and the number of mobile Internet users is growing significantly. Location-based service (LBS) quickly has emerged as a hot issue within this context. Usually, although traditional online social networks (OSNs) tell others who a user is (e.g., WeiBo users may disclose their interests to public), location-based mobile social networks (MSNs) can tell others where a user is (e.g., "check-in" in a typical LBS, such as Foursquare). The application of "check-in" may become the connection between the virtual digital world and physical reality. And moreover, it hopefully will become the "trump card" service for the mobile Internet. Among the top 10 noteworthy mobile applications named by Gartner, the world's most authoritative information technology (IT) research and consulting firm, geographic LBS is ranked as the first [1]. Therefore, LBS has attracted considerable attention due to the potential for transforming mobile communications and the potential for a range of highly personalized and context-aware services.

Since the days of the early location tracking functionalities introduced in Japan and in some U.S. networks in 2001, LBSs have made considerable progress. In addition to the currently emerging satellite-based systems, such as GPS (United States), GLONASS (Russian), GALILEO (EU), and COMPASS (China), that will provide wider coverage to benefit LBSs, some location information can also be derived and used from sensors, Radio Frequency IDentification (RFID), Bluetooth, WiMax, and wireless local area networks (WLANs). These systems can be used as stand-alone or to supplement the coverage for location tracking in indoor environments, where satellite coverage is intermittent or inaccurate [2]. Wi-Fi, especially, could be used as the basis for determining position—like an indoor form of GPS, with access points acting as satellites.

Table 11.1   Various Location-Based Protocols and Standards for Development of Location-Based Services

Protocols and Standards	Description
MLP	• An application-level protocol for obtaining the position of mobile stations (mobile phones, wireless personal digital assistants, and so on) independent of underlying network technology • Serves as the interface between a location server and an LCS • Defines the core set of operations that a location server should be able to perform
OpenLS	• To complement LIF's advanced MLP services, OGC has come up with OpenLS services, which address the geospatial interoperability issues
Geopriv	• A standard for the transmission of location information over the Internet that is being developed by IETF

Note: MLP, Mobile Location Protocol; LIF, Location Interoperability Forum; LCS, location services client Internet Engineering Task Force (IETF).

In the development of LBS standards, there are many organizations that play significant roles, such as the Open Mobile Alliance (OMA), the Open Geospatial Consortium (OGC), and so on. These organizations offer various location-based protocols and standards for the development of LBSs, as shown in Table 11.1. The most important specification that OMA has come up with is the Mobile Location Protocol (MLP). MLP enables LBS applications to interoperate with a wireless network regardless of its interfaces (Global System for Mobile Communications [GSM], CDMA, etc.) and positioning methods. MLP defines a common interface that facilitates exchange of location information between the LBS application and location servers in wireless networks. It also supports the privacy of users by providing access to location information only to those who are authorized users. Hence, the OMA is the key enabler of mobile service specification standards that support the creation of interoperable end-to-end mobile services. The OGC is an international organization responsible for the development of standards for geospatial and LBSs. To complement Location Interoperability Forum's (LIF) advanced MLP services, the OGC has come up with OpenLS services that address geospatial interoperability issues. Key services handled by OpenLS specifications are coordinate transformation, Web mapping, Geography Markup Language (GML), geoprocessing, and Web integration [2].

In general, the availability of LBS applications has transformed electronic social networking from being a people-to-people phenomenon that exists irrespective of geographical place to one in which individuals use technology to maintain a network of strong social ties within a local geographical context such as a city or a university community [3].

First, LBS will change the formations and maintenance of social relationship chains. The traditional relationship chains usually occurring among the individuals are relatively stable. But the location-based relationship is very different. This relationship is based on behavior, hobbies, and colocation. Previously, establishing such a relationship chain was not easy; individuals would usually go to their hobbies forum or participate in various activities on the ground. Now, through the LBS, they can easily have such a relationship. LBS software records each person's footprint; as long as the users have the same frequented places, they can communicate by this common point. This changing chain makes it easier to set up a group, and it is adapted to the "group" habits of human beings. These groups have more in common, such as hobbies, spending habits, and geographical characteristics. Also, it is easier to initiate group activities.

Second, LBS will change the pattern of business marketing. By using LBS software, we can set personal preferences, and it can display the preferential information about nearby shops automatically. At the same time, the organizational activities that are near users can also be shown. Moreover, the LBS platform can also accumulate a large number of user data and consumer behavior data, which can be used to provide targeted advertising. Therefore, the LBS platform is a new marketing platform. It provides many more advantages than the traditional marketing methods, which cost a lot and were without strongly targeted consumers.

Third, as a tool, LBS brings convenience to our daily life, and in a sense, it changes our way of obtaining information (e.g., appropriate location-based information can be pushed to users). For example, LBS can be used in public transport areas. The driver can use a mobile phone to report the location through the LBS service, and we can ask about the driving progress in real time. For emergency service, LBS software located within 500 meters can report location information. This will greatly and accurately increase the speed of assistance.

In brief, as an emerging and promising MSN application, LBS has the most potential for mobile value-added services. This chapter is organized as follows: Section 11.2 collects some positioning technologies, which act as the supporting premise for any LBS applications. Section 11.3 describes special characteristics and types of LBSs. LBSs are the key enablers for a plethora of applications across different domains ranging from tracking and navigation systems to directory services, entertainment to emergency services, and various mobile commerce applications. Various LBS services can be classified under multiple categories: person- or device-oriented, push versus pull, and direct versus indirect profile, among others. Section 11.4 summarizes LBS architecture and LBS service request processing and then presents several typical LBS applications. LBSs facing challenges are introduced in Section 11.5. Finally, Section 11.6 briefly concludes this chapter.

11.2 Positioning Technologies and Assessment

11.2.1 Positioning Technologies

LBSs provide users with information about geographical location; they require specific infrastructure for positioning the mobile terminal. Positioning means the determination of the location of the object in a reference system. The reference system can be a coordinate or an address system, an areal division or a route system. Geocoding is a process used for associating the object to a general coordinate system if some other system, such as an address, is used as a reference.

The technologies offering positioning for mobile terminals in LBSs can be roughly classified as outdoor and indoor technologies, depending on what they are targeted for. Different positioning techniques vary due to their features—such as accuracy, reliability, and time to fix. Each technique has its own place, and they complement each other in certain cases [4]. Their performance assessments are covered in the next section.

11.2.1.1 Outdoor Technologies

For outdoor environments, the technologies used to determine location information can be divided into three categories: network-based technologies, handset-based technologies, and the hybrid systems. Network-based technologies depend on the ability of a mobile device to receive signals from a mobile network covering the area of presence, and these usually perform well in densely populated environments. Handset-based technologies do not need mobile network coverage to work but do require hardware or software computing capability in the handset to locate position. The most common application of handset-based technologies is GPS. Hybrid systems are those that incorporate a combination of network- and handset-based technologies. A summary of outdoor positioning technologies is shown in Table 11.2 [5].

11.2.1.2 Indoor Technologies (Indoor Localization Techniques)

In indoor environments, the GPS signal cannot be used because the signal power is too reduced to enter buildings. An open research problem is to design an indoor location sensor providing accurate spatial information that is also scalable, robust, and not expensive. Location-sensing systems relying on standard wireless networking hardware measure signal intensity and attenuation to determine user location [6]. For short-range positioning technologies, location identification relies on the cooperation between the moving target object and a fixed reference point. Some popular technologies for indoor environments are introduced as follows.

Table 11.2 Summary of Outdoor Positioning Technologies

Technology	Description	Advantages	Disadvantages
Network based			
Cell ID	Information generated by identifying the base station to which the user is currently connected	Use existing network; fast implementation; no handset modification	Low resolution
AOA	Measures angle of signal from mobile device to cell station (minimum of two cell sites required)	Time synchronization not required; no handset modification	Expensive (required network upgrade); line-of-sight required; medium resolution
TDOA[a]	Triangulate at least three stations to measure and compare arrival time of signal from a user	No handset modification	Line-of-sight required; expensive; medium resolution (appropriate for CDMA)
ECID	Software-based solution that compares list of cell sites available to user and checks for overlap	Line-of-sight not required; moderate cost to upgrade	Works only with GSM; some modification required in handset and network
Handset based			
GPS	Based on 24 low-orbit satellites, triangulation by measuring the time it takes to communicate with three satellites	High accuracy, 5–10 meters; not dependent on network	Line-of-sight issues; significant handset requirement; no indoor services

Continued

Table 11.2 (*Continued*)   Summary of Outdoor Positioning Technologies

Technology	Description	Advantages	Disadvantages
Hybrid technology			
E-OTD	Similar to TDOA, but handset calculates the location	Accuracy of 50–125 meters	Suited to GSM only; network handset modification needed; cell coverage necessary
A-GPS	Processing done by network while using satellites	Moderate modification to handset; line-of-sight constant	Significant changes to network; new handset

Note: AOA, Angle of Arrival; TDOA, Time Difference of Arrival; ECID, Enhanced Cell ID; GPS, Global Positioning System; E-OTD, Enhanced Observed Time Difference; A-GPS, Assisted Global Positioning System.

[a] TDOA techniques can be applied to either the uplink signals received by the base stations from the mobile or the downlink signals transmitted from the base stations received by the mobiles. The use of uplink signals for location computation is commonly referred to as time of arrival (TOA). The use of downlink signals is commonly referred to as enhanced observed time difference (E-OTD).

11.2.1.2.1 Ultrasound

Ultrasound transmitters (beacons) send signals to a receiver, allowing the device to calculate its location based on proximity. If using timing differences between the ultrasound and additional radio reference signals, the system can achieve very high accuracy to few centimeters.

11.2.1.2.2 Infrared

This system combines infrared sensors, which are usually placed throughout a building, and an attached device (a badge) on an object to detect the object's position. Infrared requires visual line of sight and does not have very high accuracy.

11.2.1.2.3 Radio Frequency Identification

A low-power radio frequency signal is emitted by a movable target and can be detected by receivers placed at specific places around a building. Additional ID information from the signal is used to identify each target.

11.2.1.2.4 Bluetooth

These are very short-range radio frequency standards used for personal area network (PAN) access. They are mainly conceived as wire replacement (such as between a cell phone and headphones). When a Bluetooth device comes within range of a service point, such systems could be used for proximity-based location services.

11.2.1.2.5 Wireless Local Area Network

Wireless LAN is primarily used to provide Ethernet connections and high-speed Internet access. Many stations have been deployed to accept device access, and each of them has ranges of roughly 50 meters of coverage; gross location information can be obtained by determining which users are being served by a particular base station [5].

11.2.2 Performance Assessment of Outdoor Technologies

The key factor in the realization of an LBS is to let the system determine a user's location in an automatic way. Most position-sensing technologies are based on sensor infrastructure, although some of them have sensors integrated into the mobile device. Based on the nature of the wireless technologies, some are suitable for short-distance applications while others can provide long-distance determinations. Indoor positioning systems are typically based on small radio or infrared cells or, for higher accuracy, on sensor arrays in the environment [5]. For outdoor environments, the positioning process normally uses either a satellite positioning system such as GPS or the mobile telecommunication network to which the mobile user is connected. Satellite positioning systems are independent of the mobile telecommunication networks. Currently, a number of outdoor positioning technologies are used. The network-based ones such as Cell ID, Enhanced Cell ID (ECID), Angle of Arrival (AOA), and Time Difference of Arrival (TDOA) use mobile network data to determine the geographic position of the mobile device. The most widely used handset-based technology is GPS. Hybrid technologies such as Enhanced Observed Time Difference (E-OTD) and Assisted GPS (A-GPS) use both approaches. As seen in Figure 11.1, positioning technologies differ in terms of the accuracy of the geographic location data they provide. The more advanced technologies are more expensive to deploy although some may be restricted in their usefulness because they require line-of-sight (e.g., AOA, TDOA). Handset and hybrid technologies (e.g., ECID, GPS, E-OTD, A-GPS) may also require significant handset modification [7].

The main difficulty in specifying comparative performance variables for positioning methods is that each is highly dependent on the environmental circumstances of the user equipment, the network configuration, or both. Table 11.3 presents a summary of positioning capabilities for the most prominent techniques.

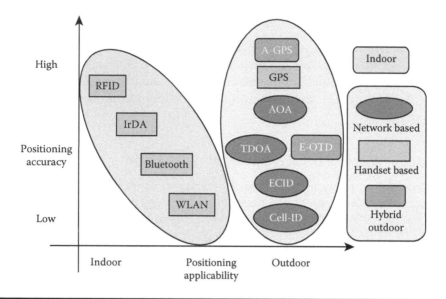

Figure 11.1 Comparison of various positioning technologies on accuracy and coverage in location-based services.

Table 11.3 Summary of Accuracy of Various Positioning Techniques

Method	Accuracy (m)	TTFF	Environment	Power Consumption
GPS	5–20	5–30	Outdoors; limited in dense urban areas and indoors	High for GPS, lower for A-GNSS
D-TDOA	Medium accuracy	5–10	Limited in rural areas with sparse cell networks and linear cell site locations	Low
U-TDOA[a]	100–300	<10	Limited in rural areas with sparse cell networks and linear cell site locations	Low
Wi-Fi	10–50	1–4	Limited in outdoor locations where Wi-Fi coverage is poor	Above average
Cell ID	500–10000	2–6	Very poor accuracy in rural areas with sparse cell networks	Low

Table 11.3 (*Continued*)   Summary of Accuracy of Various Positioning Techniques

Method	Accuracy (m)	TTFF	Environment	Power Consumption
ECID	100–300	2–6	Better in urban areas with greater base station densities; function of enhancement techniques applied	Low

Note: NTTFF, Time-to-First-Fix; GPS, Global Positioning System; ECID, Enhanced Cell ID.

[a] U-TDOA is similar in principle to the D-TDOA method but relies on signals sent from the user equipment to the base stations. D-TDOA uses timing reference signals from the cellular base stations to the user equipment in a trilateration algorithm to produce a location fix. Hardware and firmware support is needed in the mobile device to support the position calculation [8].

11.3 Categories and Characteristics of LBSs

11.3.1 Categories of LBSs

LBSs can be defined as services that depend on and are enhanced by positional information of mobile devices. An LBS is a mobile information service that extends spatial and temporal information processing capability to end users via Internet and wireless communications [2]. From a geographic point-of-view, LBSs can be simplified into indoor and outdoor categories, but this taxonomy is very narrow in scope. LBSs are applications that take a user's location into account in order to deliver a service. From a value chain perspective, LBSs are services that increase location information value to users through specific services. Therefore, an LBS can be further defined as a value-added service offered in a wireless environment that exploits mobile terminal location position information. Based on this definition, location-based applications or services can be classified as follows.

- Emergency services: The ability to locate an individual who is either unaware of his or her exact location or is not able to reveal it because of an emergency situation. This requires the system to automatically determine the mobile user's exact location after receiving an emergency call and to transfer the location information to the emergency assistance agency.
- Navigation services: Navigation services are based on a mobile user's need for directions within their current geographical location. The ability of a mobile network to locate the exact position of a mobile user can be manifested in a series of navigation-based services. In navigation services, the system needs to determine the position by direction prediction and routing calculation, which is based on the destination and relevant information displayed. In addition,

services can be provided to allow mobile users to locate friends, family, workers, or other members of a particular group in order to improve communication.

■ Tracking and navigation services: Applicable on the personal and the corporate level to monitor the exact whereabouts of a person or property. One popular business application is fleet management, which refers to tracking and managing vehicles so that companies know where their goods are at any time so that they can thereby organize their business. These kinds of applications can also be used by companies in order to locate and manage their working team; this is known as field management. Other applications are those used for product tracking within a supply chain.

■ Information/directory services: Information services generally refer to the digital distribution of content to mobile terminal devices based on their location, time specificity, and user behavior. Mobile users can be provided with a wide range of localized information such as landmarks, restaurants, theaters, and public transportation options. The following services can be identified within this category: (a) Travel services that can be provided to tourists moving around in a foreign city, such as guided tours, notification about nearby places of interest, and transportation services. (b) Mobile yellow pages that provide a mobile user, upon request, with information regarding nearby facilities. (c) Infotainment services such as information about location-specific multimedia content and community events.

■ Advertising services: Wireless advertising refers to advertising and marketing activities that deliver advertisements to mobile devices using wireless networks and mobile advertising solutions to promote the sales of goods and services or to build brand awareness. Typical LBS advertising takes the form of mobile banners, short message service (SMS) messages, and proximity-triggered advertisements, but its intrusive nature is a big challenge [5].

LBSs are the key enabler for a plethora of applications across different domains ranging from tracking and navigation systems to directory services, entertainment to emergency services, and various mobile commerce applications. Some LBS examples and their quality of service (QoS) requirements are presented in Table 11.4. QoS requirements of LBSs can be expressed from the following aspects: location accuracy required, response time, and reliability of operation. These attributes could offer some guidance to network designers and operators on the need for applications and the functionalities required in location-aware network infrastructure [2].

11.3.2 Characteristics of LBSs

The key characteristics of LBSs can be derived from the applications mentioned earlier. To implement LBSs properly, the middleware should be designed to include the major characteristics of the LBS applications. Some of these characteristics are shown in Table 11.5. The different LBSs can be classified under multiple types: person- or device-oriented, push versus pull, and direct versus indirect profile, among others [2].

Table 11.4   Location-Based Services

Location-Based Services	Applications	Quality of Service
Emergency services	• Police and fire response • Search and rescue missions • Roadside assistance • Emergency medical/ambulance	Location accuracy of a few meters Response time of a few seconds Need for very high reliability (Goal should be 100%)
Tracking and navigation services	• Tracking of children or locating lost pets • Locating friends in a particular area • Tracking stolen vehicles or asset tracking • Dynamic navigational guidance • Voice-enabled route description	Location accuracy of a few meters Response time of a few seconds Need for very high reliability (Goal should be 100%)
Information/ directory services	• Guided tours, notification about nearby places of interest, transportation services • Dynamic yellow pages that automatically inform user of location of nearest facilities, hospitals, restaurants, etc. • Location multimedia content, community events	Location accuracy of tens of meters Response time of a few seconds Need for high reliability (98%–99%)
Advertising services	• Wireless coupon presentation, targeted and customized ads • Marketing promotions and alerts • Customer notification and identification in the neighborhood store	Location accuracy of meters Response time of a minute Need for high reliability (98%–99%)

Table 11.5   Characteristics of Location-Based Services

Types of LBS	Characteristics
Person-oriented LBS	• Consists of applications where a service is user based • User typically controls how location information is collected and used
Device-oriented LBS	• Applications are external to user • Person or device located is not controlling the service
Push-versus pull-based applications	• Push-based: information delivered to the mobile terminal (end user) automatically when certain event occurs • Pull-based: Mobile terminal (end user) initiates the request
Direct versus indirect profile	• Based on how the user profile is collected—directly from the user during the set up phase, by tracking the user's behavior pattern, or from third parties • Security and privacy issues become critical to maintain user trust and to avoid fraudulent activities
Availability of profile information	• Profile information requested on the fly or already available to the LBS
Mobility and interaction	• Range of mobility scenarios exist based on combinations of mobility of users and network components • Level and type of interactions depend the mobility scenario
State of interaction	• Stateless interaction: Each request is an independent transaction unrelated to previous request • Stateful interaction: LBS preserves the state across service requests (beneficial for forecasting future transactions, requests, and behavior)
Static versus dynamic information source	• Static: Data about historical buildings and landmarks, places of attraction, hotels and restaurants, maps • Dynamic: Information that changes with time (weather, traffic, and road conditions)
Source of location information	• Location information provided by the user or the network infrastructure or by a third party
Accuracy of location information	• Depending on the positioning technology used in the network infrastructure, different accuracy of localization request of mobile terminals may result

11.4 Architecture and Typical Applications of LBSs

11.4.1 Architecture of LBSs

11.4.1.1 Components Needed to Deliver LBSs

Delivering LBSs usually requires cooperation among some fundamental components. Working together, they determine a user's geographic location and provide specific information to the user based on his or her location and service requirements.

When users want to use an LBS, several architecture elements are necessary. These components and their connections are as follows.

- Mobile devices: A tool used for requesting and representing information, acts as an interface for the information and as part of the positioning device. Devices most commonly associated with LBSs are personal digital assistants (PDAs) and cell phones.
- Positioning systems: To process location information and position information, and to determine user position. Geographically, they are classified as either outdoor or indoor positioning. Outdoor positions can be obtained by using the mobile communication network or GPS. Indoor position can be determined by active badges or radio beacons.
- Communication networks: Used to transfer the user's data and service request from the mobile device to the service provider and then to transfer requested information back to the user.
- Service and application provider: Offers a number of different services to users and is responsible for service request processing.
- Data and content provider: Business and industry partners in a particular field that satisfy the user's specific information request or that provide data maintenance [5].

11.4.1.2 Common LBS Architecture

Location information is usually provided in terms of a latitude/longitude pair that is sent to an application in a server or back to the mobile device, which is then transmitted with some additional identification information to an application in a server that may reside inside or outside the wireless network. The location information can be tied to a location database server that could interest users and that will then send a related message to the user about restaurants or hotels [5]. For example, a common LBS architecture, as illustrated in Figure 11.2, consists of four major entities: mobile devices, positioning systems, communication networks, and service providers. Users use their mobile devices (e.g., smartphones) to send queries to LBS servers. The locations in the queries are obtained via positioning systems such as GPS. The queries, as well as their responses from the LBS servers, are transmitted via communication

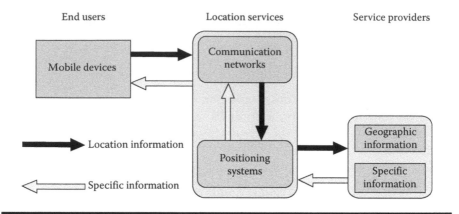

Figure 11.2 A conceptual model of location-based service.

networks, such as third-generation (3G) networks. LBS servers are offered by service providers, replying to queries with well-tailored responses based on the location information in the queries [9].

11.4.2 Requirements of LBS Architecture

Derived from the user actions, different requirements on the LBS system architecture emerge. Further different types of services are offered by companies to satisfy the needs of users.

Unlike geographic information systems (GISs), which are usually desktop or client/server applications with a limited number of users, LBSs have to provide access and information to dozens of users with mobile terminals. The following capabilities of LBSs usually exceed the general requirements of static GISs use:

- High performance: Delivering answers in a sub second if querying information from Internet and databases
- Scalable architecture: Support thousands of concurrent users and terabytes of data
- Reliable: Capable of delivering up to 99.999% uptime
- Current: Support the delivery of real-time, dynamic information
- Mobile: Availability from any device and from any location
- Open: Support common standards and protocols, such as HTTP, Wireless Application Protocol (WAP), Wireless Markup Language (WML), Extensible Markup Language (XML), Multimedia Markup Language (MML)
- Secure: Manage the underlying database locking and security services
- Interoperable: Integrated with e-business applications such as customer relationship management, billing, personalization, and wireless positioning gateways

These requirements lead to a complex LBS architecture involving a number of players. These players include hardware and software vendors, content and online service providers, wireless network and infrastructure providers, wireless handset vendors, and branded portal sites. Only common specifications and agreements among these players can ensure that a user receives a satisfying offering and deployment of services [10].

11.4.3 Typical LBS Applications

There have been many location-sharing systems developed over the past two decades, and only recently have they started to be adopted by consumers. We can broadly categorize location-sharing applications as purpose driven, where people explicitly request another person's current location (e.g., AT&T FamilyMap, Glympse, Verizon Family Locator), and social driven, where people broadcast their location to "friends" in their social networks. Examples of social-driven applications include, for example, Brightkite, Dodgeball (both of them discontinued), Foursquare, Gowalla, and Facebook Places. Although purpose-driven location-sharing applications have not yet achieved critical mass in any system, the same is not true for social-driven applications [11]. In this section, we will introduce several typical LBS applications such as Foursquare, Sindbad, and so on.

11.4.3.1 Foursquare

Foursquare is a Web and mobile application that allows registered users to post their location at a venue ("check-in") and connect with friends. Check-in requires active user selection, and points are awarded at check-in [12]. Foursquare has features that distinguish it from other services (it is not yet clear which factors contribute to its popularity). For example, Foursquare positions itself simultaneously as a mobile game, a way of exploring cities, a way of telling friends where you are, and a way of tracking where friends have been and who they have been colocated with.

11.4.3.1.1 Foursquare Check-In Service

Foursquare describes their service [11] as a "mobile application that makes cities easier to use and more interesting to explore. It is a friend finder, a social city guide, and a game that challenges users to experience new things, and rewards them for doing so. Foursquare lets users 'check-in' to a place when they're there, tell friends where they are, and track the history of where they've been and who they've been there with." Foursquare has smartphone clients, such as iPhone, BlackBerry, Palm, and the Android platform.

Foursquare lets people connect to friends, which is equivalent to the concept of friends on other OSNs. Users can check-in to locations to say that they are currently there. When doing a check-in, Foursquare examines the user's current location and shows a list of nearby places. Users can also register new places.

When a user checks in to a place, a check-in notification is pushed to their Foursquare contacts by default. People can choose to be notified of all check-ins by their contacts. At the time of the check-in, users can also decide if they want to check-in off-the-grid, in which the check-in is recorded by Foursquare but not shared with contacts. These private check-ins still count toward gathering points, badges, or mayorships (these are described in the next paragraph). People can also connect their Foursquare account to other online services, such as Facebook and Twitter, and can have their check-ins announced on these services. Users who have checked-in to a place can also see who else has recently checked-in ("Who's here?"). Users can also allow local businesses to view check-ins to their location.

The game aspect of Foursquare offers virtual and tangible rewards for check-ins. Virtual rewards come in the forms of points, badges, and mayorships visible in one's public profile. Badges are awarded for a variety of reasons (e.g., for starting to use the service, checking-in on a boat, checking-in with 50 people at the same time, or checking-in at a special event). Mayorships are awarded to a single individual for having the most check-ins in a given place in the past 60 days, where only one check-in per day is counted. Some companies offer discounts for mayors; for example, some coffee shops offer discounts on coffee.

Foursquare also enables social recommendations through tips, a small snippet of text associated with a place. Tips are intended to suggest possible activities for that place [11].

11.4.3.1.2 Motivations for Sharing Location

Tang et al. [13] point out that one-to-many sharing with friends or followers is more complex than one-to-one or one-to-few sharing. Rather than deciding whether users want to share with one specific person or a limited group, they now need to decide whether to share with a varied audience with whom they share their locations for different reasons. Especially for services such as Foursquare, they hypothesize that the connection to larger audiences via Twitter or Facebook may lead to more performative uses rather than as a coordination tool. As Brown et al. [14] point out, the value of location technology is not in tracking or communicating location—it is about how this is used, read, viewed, and manipulated. Sharing one's location and knowing the whereabouts of others is not only a practical tool for coordination and communication. Although Tang et al. [13] distinguish between "purpose-driven" sharing (e.g., for coordination) and "social-driven" uses, Brown et al. [14] argue that location sharing is not just about practicality and accurately sharing location or one's activity there; rather, location sharing is an emotional and moral affair. It is used not only to express whereabouts but also moods, lifestyle, and events. It can support social repartee and tell the ongoing story within social groups, while also providing a resource for other interactions and a tangible representation of shared locations, supporting exchange enjoyment and friendship. People can share information for social purposes that is interesting, that enhances self-presentation, and/or that leads to serendipitous interactions. Location sharing can also serve as a reassurance, communicating and knowing that all is well and as it "should be" by bringing a sense of connectedness, togetherness, and identity and a moral position within the group you share your location with [14,15]. Using a set of interviews and two surveys, Lindqvist et al. [11] identified clusters of motivations for sharing one's location using Foursquare, including games and badges (which included playing for fun but also self-presentation and being proud of badges), social connection (keeping in touch, ad hoc meet-ups, seeing where friends have been), place discovery and keeping track of places, and meeting new people, and also simply "something to do." Rather than focusing on these motivations for checking-in and privacy considerations as in Ref. [16], Cramer et al. [17] focused on performative aspects of check-ins, their "audience" and emerging social norms on when to check-in and whom to share with—and their conflicts.

11.4.3.2 Sindbad

Sindbad [18] is a location-based social networking system built with an open source database management system; it distinguishes itself from existing social networking systems (e.g., Facebook and Twitter) because it injects location awareness into every aspect of social interaction and functionality in the system. For example, posted messages in Sindbad have inherent spatial extents (i.e., spatial location and spatial range), and users receive friend news feed based on their locations and the spatial extent of messages posted by their friends. The functionality of Sindbad is fundamentally different from current incarnations of location-based social networks (e.g., Facebook Places, Foursquare). These existing location-based social networks are strictly built for mobile devices, and they only allow users to receive messages about the whereabouts of their friends (e.g., Foursquare "check-ins" that give an alert that "your friend Alice has checked-in at restaurant A"). Sindbad, on the contrary, takes a broader approach that marries functionality of traditional social networks with location-based social scenarios (e.g., friend news posts with spatial extents, location-influenced recommendations). Thus, Sindbad is appropriate for traditional social networking scenarios (e.g., desktop-based applications) as well as location-based scenarios (e.g., mobile-based applications).

Users of Sindbad can select their friend list as well as being listed as friends of other users in way similar to traditional social networking systems. In addition, users can post (spatial) messages and/or rate (spatial) objects (e.g., restaurants), which will be seen by their friends. Once a user logs on to Sindbad, the user will see an incoming location-aware news feed posted by his or her friends.

Sindbad has a location-aware news feed module, GeoFeed. For any user, GeoFeed efficiently retrieves the relevant messages from their friends based on the user location and the message spatial extents. GeoFeed minimizes the total system overhead for delivering the location-aware news feed to the user while guaranteeing a certain response time for each user to obtain the requested location-aware news feed.

Sindbad users can receive location-aware recommendations about spatial items (e.g., restaurants) or nonspatial items (e.g., movies). To this end, Sindbad is equipped with a location-aware recommender module, called the location-aware recommender system (LARS). For any user, LARS efficiently suggests (spatial) items based on user locations, item locations, and previous ratings by user friends. Sindbad produces recommendations by employing user partitioning and travel distance penalty techniques in order to deliver relevant recommendations to its users. In addition to GeoFeed and LARS, Sindbad adopts a location-aware ranking module that efficiently selects the top-k relevant objects produced from either the location-aware news feed or the recommender system modules.

A major part of Sindbad is built inside PostgreSQL, an open source Database management system (DBMS). Hence, Sindbad takes advantage of the scalability provided by the DBMS and is able to employ early pruning techniques inside the DBMS engine, which yields an efficient performance for the news feed, recommendation, and ranking functionalities. Moreover, Sindbad provides a RESTful Web Application Programming Interface (API) so that it would allow a wide variety of applications to easily communicate with Sindbad and make use of its unique features. The system is demonstrated using Web and smartphone (i.e., Android) applications built on top of Sindbad particularly for demonstration purposes. Moreover, the system internals are demonstrated through an administrator-like interface that shows the different system parameters as well as general statistics about Sindbad.

11.4.3.2.1 Sindbad Architecture

Figure 11.3 depicts the Sindbad system architecture that consists of three main modules, namely location-aware news feed (GeoFeed), location-aware ranking, and location-aware recommender (LARS), and three types of stored data, namely spatial messages, user profiles, and spatial ratings. The communication between Sindbad and the outside world is held through RESTful Web API interface (named Sindbad API Functions in Figure 11.3).

The API functions facilitate building a wide variety of applications (e.g., Web applications, smartphone applications) on top of Sindbad. As shown in Figure 11.3, Sindbad API functions can also be used to complement the functionality of existing social networking Web sites (e.g., Facebook) and can make their news feeds and recommendations be location aware. Sindbad can take five different types of input (i.e., through the API interface): profile updates, a new message, a new

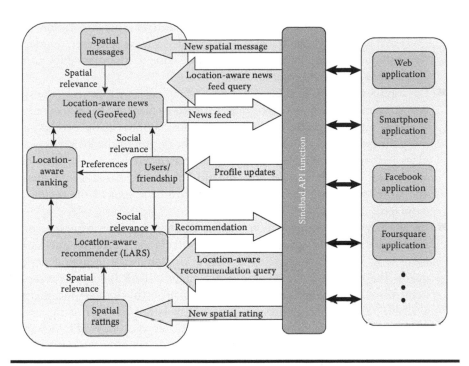

Figure 11.3   Sindbad system architecture.

rating, a location-aware news feed query, and a location-aware recommender query. The actions taken by Sindbad for each input are described as follows:

■ Profile updates: As in typical social networking systems, Sindbad users can update their personal information and their friend list, or accept a friend invitation from others.

■ A new message: Users can post spatial messages to be seen by their friends, if relevant. A spatial message is represented by four-tuple: (MessageID, Content, Timestamp, Spatial), where MessageID and Content represent the message identifier and contents, respectively; Timestamp is the time the message is generated; while Spatial indicates the spatial range for which the message is effective. The message is deemed relevant only to those users who are located within its spatial range.

■ A new rating: Sindbad users can give location-aware (spatial) ratings to various items on a scale from one to five. Location-aware (spatial) ratings can take any of these three forms: (1) spatial ratings for nonspatial items, represented as a four-tuple (user, user location, rating, item), for example, a user located at home rating a book; (2) nonspatial ratings for spatial items, represented as a four-tuple (user, rating, item, item location), for example, a user with an unknown location rating a restaurant with an inherent location; and (3) spatial ratings for spatial items, represented as a five-tuple (user, user location, rating, item, item location), for example, a user at his or her office rating a restaurant with an inherent location.

■ Location-aware news feed queries: Once a Sindbad user logs on to the system, a location-aware news feed query is fired to retrieve the relevant news feed (i.e., messages posted by the user's friends that have spatial extents covering the location of the requesting user). The output of the location-aware news feed module (GeoFeed) will be processed further by the location-aware ranking module to get only the top-k news feed based on the spatial and social relevance, which will be returned to the user as the requested news feed.

■ Location-aware recommendation queries: Sindbad users can request recommendations of either spatial items (e.g., restaurants, stores) or nonspatial items (e.g., movies) by explicitly issuing a location-aware recommendation query. The location-aware recommender module (LARS) suggests a set of items based on: (a) the user location (if available), (b) the item location (if available), and (c) ratings previously posted by either the user or the user's friends. Similar to location-aware news feed queries, the output of LARS goes through the location-aware ranking module to select only the top-k items based on spatial and social relevance [18,19].

11.5 Challenges for LBSs

The potential of LBSs is attractive to industry players and customers. However, results from academic and professional literature show that LBSs also face challenges, which are arising mostly from the technological, business, and social perspectives. Some of the most critical challenges identified in prior work are summarized as follows.

11.5.1 User Privacy

One important issue that has emerged in LBSs is wireless location privacy. This involves keeping a user's location information confidential and preventing users from being illegally tracked. Public concern about potential threats to personal security and use of personal location records for commercial purposes and legal action have resulted in a desire to be able to control who receives location information while still being provided with highly personalized services and applications. Developing strict location information protection is still a challenge for many countries. Currently, LBSs and location data are not regulated in most parts of the world. Protection is provided under other legislation such as general data protection acts [5].

As shown in Figure 11.2, LBS servers might be regarded as malicious observers, and all the other components in the LBS architecture could be considered benign. An adversary can be the owner/maintainer of an LBS server or able to compromise and then seize control of an LBS server. In both cases, the adversary has the ability to access all the information stored on the servers, such as the Internet protocol (IP) address and location information associated with each query. Although we embrace a generic model without imposing additional requirements on the usage of LBSs (e.g., we do not assume that users need to login to use LBSs), an adversary can still make use of side channels (e.g., the IP address of each query) as well as sophisticated object-tracking algorithms to relate consecutive anonymous LBS queries to the same user.

Note that the aforementioned widely accepted threat model intentionally simplifies the privacy protection problem by concentrating only on how to regulate the location information contained inside LBS queries. In reality, the location privacy issue is a complicated and multifaceted problem that needs to be handled with care from several perspectives. For example, users' mobile devices can be compromised, thus becoming malicious and actively revealing users' privacy information (including, but not limited to, location information). How to secure users' mobile devices is an active research topic by itself, and interested readers are referred to the related literature, such as Ref. [20].

In addition, although LBS queries and responses are transmitted via communication networks, they need to be protected against eavesdroppers and man-in-the-middle attacks. Conventional solutions, such as cryptography and hashing, can be used to protect the secrecy, integrity, and freshness of the queries and responses transferred through networks [9].

11.5.2 Interoperability

LBSs usually involve more than one mobile network and operator. There is a need for services that cross different networks. One scenario is when an individual is using services from one operator and trying to locate his or her friend who is using a different operator. This kind of service requires the crossing of networks and operators and is a big challenge for the operators, particularly when the two operators have different mobile network infrastructures.

11.5.3 Accuracy and Reliability

Accurate, high-quality performance and reliable positioning are major priorities in LBSs. The more accurate the position, the more relevant the information that can be provided. Generally, high accuracy depends on technology and infrastructure integration such as the A-GPS technology, which provides reliable accuracy of between 5–30 meters. The time of determination is another issue in LBSs because it usually conflicts with high accuracy, one study showing that determining position within 30 seconds or less is thought satisfactory.

11.5.4 Information Availability

As an important component of LBSs, location-sensitive information can change daily. Keeping this information up-to-date is essential for the quality of the services; updating information objects such as newly opened restaurants and deleting closed ones is necessary to satisfy users. For navigation services in particular, new geographical and road information is crucial for the system to determine direction and routing.

11.5.5 Adaptation

Another challenge concerns design and implementation of adaptive discovery of friends or people sharing the same interests. Adaptation may help minimize the energy consumption of mobile devices, supporting dynamic changes in context and benefiting from historical information. Novel mechanisms may support prediction of friends and identification of those nearby who share the same interests. Indeed, user A frequently detects friend B nearby but not friend C due to the limitations of proximity-aware devices. However, if friend B is able to detect friend C, then friend B is able to notify friend A that friend C is not far away and could be expected to appear soon. A can then be prepared to be in touch with C or alternatively might leave to avoid contact with C [21].

11.5.6 Legal, Ethical, and Social Challenges

Theoretically, when a user employs LBS technologies to look after his or her family members, particularly the elderly or young children, this may involve monitoring and restricting the subject's activities, especially when he or she is alone. This brings up the question of control in LBSs that are used for care or convenience purposes [5].

11.6 Conclusion

With the in-depth development of MSNs, LBSs have found great applications on mobile terminals and have become an indispensable part of people's everyday lives. Delivering LBSs is complex and challenging. The chapter thoroughly provides information about LBS, including positioning technologies, categories and characteristics of LBS, and challenges in developing LBS. Finally, several typical LBS applications, such as Foursquare and Sindbad, are presented.

References

1. Qiu, P., J. Zhang, and J. Zeng. Study on the mobile LBS development model. In: Proceedings of IEEE International Conference on Computer Science & Service System (CSSS), Nanjing, China, August 11–13, 2012.

2. Dhar, S. and U. Varshney. Challenges and business models for mobile location-based services and advertising. *Communications of the ACM* 2011; 54(5): 121–128.

3. Ziv, N. D. and B. Mulloth. An exploration on mobile social networking: Dodgeball as a case in point. In: Proceedings of IEEE International Conference on Mobile Business (ICMB), Copenhagen, Denmark, June 26–27, 2006.

4. Tsalgatidou, A., J. Veijalainen, J. Markkula, A. Katasonov, and S. Hadjiefthymiades. Mobile e-commerce and location-based services: Technology and requirements. In: Proceedings of the 9th Scandinavian Research Conference on Geographical Information Science (ScanGIS), Espoo, Finland, June 4–6, 2003.

5. Wang, B. Mobile location-based services in New Zealand. Master's dissertation, AUT University, 2008.

6. Francese, R., I. Passero, and G. Tortora. Current challenges for mobile location-based pervasive content sharing application. In: *Ubiquitous Computing* (E. Babkin, Ed.). InTech, Croatia, 2011; pp. 199–214.

7. Petrova, K. and B. Wang. Location-based services deployment and demand: A road-map model. *Electronic Commerce Research* 2011; 11(1): 5–29.

8. Ofcom. Assessment of mobile location technology—Update. Available from: http://stakeholders.ofcom.org.uk/binaries/consultations/emergency-mobiles-cfi/annexes/mobile-location-technology.pdf

9. Shin, K. G., X. Ju, Z. Chen, and X. Hu. Privacy protection for users of location-based services. *IEEE Wireless Communications* 2012; 19(1): 30–39.

10. Steiniger, S., M. Neun, and A. Edwardes. Foundations of location based services. Available from: http://www.geo.unizh.ch/publications/cartouche/lbs_lecturenotes_steinigeretal2006.pdf

11. Lindqvist, J., J. Cranshaw, J. Wiese, J. Hong, and J. Zimmerman. I'm the mayor of my house: Examining why people use foursquare—A social-driven location sharing application. In: Proceedings of SIGCHI, Vancouver, CA, May 7–12, 2011.

12. Foursquare. Available from: http://en.wikipedia.org/wiki/Foursquare

13. Tang, K., J. Lin, J. Hong, D. Siewiorek, and N. Sadeh. Rethinking location sharing: Exploring the implications of social-driven vs. purpose-driven location sharing. In: Proceedings of the 12th ACM International Conference on Ubiquitous Computing (Ubicomp), Copenhagen, Denmark, September 26–29, 2010.

14. Brown, B., A. Taylor, S. Izadi, et al. Locating family values: A field trial of the where-abouts clock. In: Proceedings of the 9th International Conference on Ubiquitous Computing (Ubicomp), Innsbruck, Austria, September 16–19, 2007.

15. Toch, E., J. Cranshaw, P. Hankes-Drielsma, et al. Empirical models of privacy in location sharing. In: Proceedings of the 9th International Conference on Ubiquitous Computing (Ubicomp), Copenhagen, Denmark, September 26–29, 2010.

16. Wagner, D., M. Lopez, A. Doria, I. Pavlyshak, V. Kostakos, I. Oakley, et al. Hide and seek: Location sharing practices with social media. In: Proceedings of the 12th International Conference on Human Computer Interaction with Mobile Devices and Services, Lisboa, Portugal, September 7–10, 2010.

17. Cramer, H., M. Rost, and L. E. Holmquist. Performing a check-in: Emerging practices, norms and "conflicts" in location-sharing using foursquare. In: Proceedings of the 13th International Conference on Human Computer Interaction with Mobile Devices and Services, Stockholm, Sweden, August 30–September 2, 2011.

18. Mohamed Sarwat, M., J. Bao, A. Eldawy, J. J. Levandoski, A. Magdy, and M. F. Mokbel. Sindbad: A location-based social networking system. In: Proceedings of

the ACM SIGMOD International Conference on Management of Data, Scottsdale, AZ, May 20–24, 2012.

19. Mohamed S. M., J. Bao, A. Eldawy, J. J. Levandoski, A. Magdy, and M. F. Mokbel. The anatomy of Sindbad: A location-aware social networking system. In: Proceedings of the 5th International Workshop on Location-Based Social Networks, Redondo Beach, CA, November 6, 2012.

20. Bose, A., X. Hu, K. Shin, and T. Park. Behavioral detection of malware on mobile handsets. In: Proceedings of the 6th International Conference on Mobile Systems, Applications, and Services, Breckenridge, CO, June 17–20, 2008.

21. Jabeur, N., S. Zeadally, and B. Sayed. Mobile social networking applications. *Communications of the ACM* 2013; 56(3): 71–79.

Chapter 12

Mobile Social Networking in Proximity

12.1 Introduction

Historically, it is widely recognized that an individual's life is usually characterized by social structure: people's interactions and social relations are highly local, grounded in and organized around shared physical space. It is argued that these physical space–mediated interactions are essential to social relations/community. Recently, the convergence of social and computing disciplines has focused attention on the design of online social networking (OSN) applications, such as Facebook, MySpace, and LinkedIn. They enable online social interactions and lead to a paradigm shift from physical to virtual communities: interpersonal interactions free from geographic constraints. With the development of the Internet in recent years, smart devices are becoming prevalent in our daily lives. Nowadays, Web-based OSN has already extended to mobile terminals, so-called mobile social networks (MSNs). Popular social networking Web sites such as Facebook and MySpace are already providing the possibility of using their products on mobile phones. For example, Facebook announced in its earnings report that, by March 31, 2013, it had 1.11 billion active monthly users with a 23% annual increase; they also had 751 million active users on mobile with a 54% annual increase [1]. MSNs contain the main functions of Web-based social networking services, such as sharing pictures, chatting, sending messages, and looking for new acquaintances. Compared with Web-based social networks, MSNs elevate the freedom of movement; they allow users to share information and stay in touch with their friends on the go.

However, besides exploiting integrated technologies of mobile terminals (the Global Positioning System [GPS], camera, wireless technologies, etc.), most existing MSN applications only regard mobile terminals as entry points to existing social networks. We argue that their drawbacks are two-fold.

■ First, the centralized architecture for data storage and processing is always adopted. For example, location-based service (LBS)–based MSN applications usually ask a user to register first and infer his or her location automatically or artificially; that information is then stored in centralized servers. This implies that the omniscient OSN provider has become a "big brother." The provider collects and stores all the user's data (messages, profiles, location, relations, etc.), which may cause serious privacy concerns (e.g., selling users' personal information, targeted advertising, etc.) [2].

■ Second, the notorious prerequisite for using OSN applications is continual Internet connectivity, even though mobile users are within proximity (e.g., campus, event spot, and community) and can directly exchange data through various wireless technologies (e.g., Bluetooth, Wi-Fi, etc.). This not only makes a great deal of data traffic unnecessarily go through Internet (even worse, currently, the demand–supply gap in mobile data traffic is expected to be getting only larger) but also burdens the user's data cost.

Briefly, traditional MSNs (and OSNs) lead to a shift from physical communities to virtual communities, but they fail to promote direct face-to-face interactions in the physical world. Currently, people who live or work in the same place miss opportunities to directly leverage interpersonal affinities (e.g., shared interests and backgrounds) for social interactions due to the lack of effective services that are capable of grouping like-minded people in proximity together for collaboration and socialization.

Today, modern mobile phones have the capability of detecting the proximity of other users and offer means to communicate and share data ad hoc with the people in proximity. We especially focus on mobile social networking in proximity (MSNP). In this chapter, MSNP is explicitly defined as a wireless peer-to-peer (P2P) network of spontaneously and opportunistically connected nodes that uses geoproximity as the primary filter in determining who is discoverable on the social network. This differs from location-based social networks such as Foursquare [3], which ask a user to register and actively provide his or her location and then simply broadcasts the user's location to existing friends. By enabling users to meet new people and interact with them and their locally relevant content, proximity-based social networking applications are far more engaging. Facebook has won the first stage of the social networking market evolution but may miss the rise of the proximity networks due to their focus on controlling identity. It is estimated that the emerging market of MSNP will grow to USD $1.9 billion in revenues by 2016 [4].

The trend toward social interaction though proximity social networks will drive content to the edge of networks that will be consumed by users through ad hoc

P2P local area wireless networks. This shift from wide area wireless networking to local area networking will mitigate the data burden on carrier networks while reducing their influence, creating opportunities for new players and innovative initiatives by mobile operators. Not only does this trend provide new opportunities for application vendors but it has the potential to disrupt the current social networking market and mobile landscape.

As illustrated in Figure 12.1 [5], it is should be explicitly noted that we have experienced the evolution from physical space to virtual community, and the further paradigm shift we are experiencing is from totally virtualized social communities to MSN sociality (aware of positioning and colocation). And specifically, in proximity, MSNP could be complementary to traditional MSNs. Especially, an MNSP can provide effective ways to alleviate the aforementioned two problems in MSNs: privacy concerns and continual Internet connectivity. In contrast to a centralized MSN architecture, MSNP can give the control over data back to the individuals, and no one entity in social network will have access to all personal data of the participant in social network. And moreover, local connectivity facilitates social networking without Internet access. We argue that this infrastructure-free, "crowdsourced" communication channel could be useful for real-world, proximity-based social networking, advertisement, gaming, and so on.

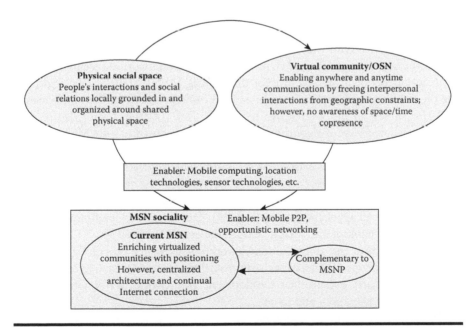

Figure 12.1 From physical space to virtual community to MSN sociality to MSNP. (With kind permission from Springer Science+Business Media: *Wireless Networks*, Survey on mobile social networking in proximity (MSNP): Approaches, challenges and architecture, 2014.)

Actually, academia [6] and mobile industry [7] have been studying the potential of mobile phones to detect social proximity and the effortless ways of communicating and sharing data with people nearby.

A new concept called proximity mobile social networks (PMSNs) was proposed in Ref. [8], which attempts to merge the growing trend of P2P services with MSNs. The vision of PMSNs is to use Bluetooth, Wi-Fi Direct, and/or P2P Wi-Fi-enabled mobile devices for people's social interactions with other users who come into their close proximity.

Similarly, through combination of the OSNs with opportunistic networks, a concept arises: local social networks (LSNs) [9]. The target of LSNs is to promote social networking benefits in physical environments in order to leverage personal affinities in the users' surroundings.

The same as the meaning of PMSN or LSN, the term "mobile social networking in proximity" first appears in Refs. [10,11], which aims to provide people with opportunities to meet new friends, to share device content, and to perform various social activities in a mobile P2P environment. Specifically, the authors proposed a mobile device–hosted, service-oriented, workflow-based mediation framework for MSNP.

However, MSNP is still in its early stage: all the aforementioned work either only provides the basic concept of MSNP or introduces preliminary basic components in MSNP application development.

In this chapter, we thoroughly summarize the primary support solutions related to MSNPs. And then, we discuss the special characteristics of MSNPs, address the underlying technical challenges, propose a networking technology and platform-independent architecture for developing MSNP applications, and provide proof-of-concept implementation of Wi-Fi Direct–based MSNP application.

A first survey on MSNs appeared recently in Ref. [12], which provided readers with a general overview of algorithms and protocols proposed to solve MSN-specific issues (community detection, content distribution, mobility modeling, security management, etc.). Various existing MSN middleware approaches were overviewed in Ref. [13]. Our previous work thoroughly characterized the basic design principles, research architecture, typical techniques, and fundamental issues in MSNs from cross-discipline and application viewpoints [14]. Our latest work preliminarily overviewed the basic concept and special characteristics of MSNPs [15].

Differently from the aforementioned work, this chapter provides a thorough survey on MSNP-specific issues: existing solutions, technical characteristics and challenges, platform-independent development framework, Wi-Fi Direct–based prototype implementation, and so on.

This chapter is organized as follows: In Section 12.2, we summarize and compare the state-of-the-art efforts of existing approaches related to MSNPs, including MSNs, mobile P2P, opportunistic networks, and several early MSNP initiatives. Section 12.3 identifies the special characteristics and open challenges in MSNPs and discusses potential solutions. Section 12.4 proposes the networking and platform independent development architecture of MSNPs, which is explicitly classified into four layers: hardware layer (networking layer), core layer, service layer, and application layer. Finally, we briefly conclude this chapter.

12.2 MSNP-Related Applications

MSNP is a wireless P2P network of spontaneously and opportunistically connected nodes that uses geoproximity as the primary filter in determining who is discoverable on the social network. Thus we argue that MSNP is mainly relevant to three fields: MSNs, mobile P2Ps, and online networks (ONs). In this section, we highlight the convergence of those technologies and the innovative use of established ones that are enabling proximity-based social networking. MSNP-related approaches are thoroughly summarized and compared through the following three dimensions: network join criteria, realization infrastructure, and supported direct wireless technologies.

12.2.1 Mobile Social Network

The nature of mobile phones makes them ideal tools to be used for social networking because people routinely carry a mobile phone with them and use it for communication. Mobile social networking is a form of social networking where individuals of similar interests or commonalities converse and connect with one another using the mobile devices. Similar to Web-based social networking, mobile social networking mainly occurs in virtual communities.

MIT's Serendipity [16] project provided a mobile profile matching and introduction service among physically proximate users. The application continuously scans for Bluetooth devices, transmits the discovered Bluetooth Indentification (BTIDs) to a central server, and alerts the matching users of the possible mutual interest.

An MSN middleware, MobiSoC [17], enabled mobile social computing application development and provided a common platform for capturing, managing, and sharing the social state of physical communities. MobiSoC is a centralized approach using simple object access protocol (SOAP) Internet for communication. All location and social information are stored on the server.

The aforementioned approaches are centralized; that is, they are primarily Internet-based. They store user data on a central server, allowing users to find friends and share data via clients running on their mobile phone. The limitations of having centralized applications are that the server may not always be reachable, breaking the communication between the server and the device, and a centralized server that may access all users' data may cause serious privacy concerns.

In literature, there exist several attempts to design decentralized MSN applications through opportunistic P2P ad hoc contacts among mobile users.

Spider [18] offers a set of services inspired by the traditional social networking sites. And moreover other services are also provided utilizing advantages of integrated technology of the mobile phones (GPS, camera, etc.). Especially, Bluetooth technology, especially, was used to discover other users within a short range and to directly exchange contents such as profiles, messages, and so on.

SCOPE is a prototype for spontaneous P2P social networking [19], which provides customized social networking applications for local users. Below the network level, SCOPE relies on 802.11 ad hoc modes and needs no infrastructure. SCOPE follows the hierarchical P2P model, in which some nodes with higher computing capability become supernodes. Supernodes form an overlay and provide the distributed data management system for the P2P social network; client nodes (CNs) connect to supernodes and rely on them for sharing their contents or for accessing shared information.

12.2.2 Mobile P2P

Recently, a number of advances in P2P systems, such as unmanaged Internet architecture (UIA) [20] and Juxtapose (JXTA) [21], have enabled seamless connectivity among users' mobile devices. These technologies create network overlays to address ubiquitous connectivity and management of device groups. Specifically, some elements could provide the basis of a platform for decentralized P2P networking that shares users' resources directly where they reside.

MyNet [22] offers a more immediate and responsive alternative to the current Web-based paradigm of personal and social networking, which allows users' distributed services and content to be accessed and shared in real time as they are produced, directly from their personal devices. MyNet built on top of UIA as its base communication platform. UIA provides strong permanent location-independent device identifiers, allows users to securely bind personal names to devices, constructs an overlay network, and offers a traditional socket API to establish connections.

Peer2Me [23] presents a pure P2P networking framework, which has been implemented in Java 2 Micro Edition (J2ME) and runs on standard mobile phones. Specifically, the framework supports management and communication of mobile ad hoc networks (MANETs) such as Bluetooth.

Photo/video sharing during an event may be one of the most popular applications so far. Some events such as sports or other occasions motivate people that share a common interest to gather together. Their proximity allows a local mobile P2P network to be used, and thus peers can share photos or videos directly (or any other type of information relevant to that event). MIT Media Lab, Information Ecology Group [24], has developed a series of opportunistic P2P proximal networking applications. For example, in CoCam [25] application, users located in the same physical space can automatically share the photos they create as well as receive photos from other users around them. CoCam does not completely eliminate the reliance on the cellular infrastructure. Although the actual content sharing is kept within the P2P links over proximal Wi-Fi, the interaction with the centralized server (through cellular interface) enables peers to join the most relevant network group and receive periodic updates on the latest status of other nodes. As for the network interface, CoCam uses the standard Wi-Fi channels. In order for CoCam to be used on off-the-shelf smartphones, it exploits the standard mobile access point feature for establishing local P2P networks.

12.2.3 Opportunistic Network

People opportunistically encounter and interact with each other during their daily lives. As long as every user carries a Wi-Fi-or Bluetooth-enabled mobile phone, each encounter would form an opportunistic network (ON). By leveraging the sensing capability of mobile phones and the communication capability of opportunistic networks, individual and group behaviors (e.g., mobility patterns, preferences, and interaction histories) can be obtained and shared during the opportunistic encounters, which might be exploited to foster future social interactions. With each individual moving from one opportunistic network to another, information can be disseminated and discussed across a sequence of opportunistic networks, which leads to the formation of so-called opportunistic mobile social networks [26].

In ONs, nodes are assumed to be mobile, and forwarding of messages occurs based on the store-carry-and-forward concept. Each node carrying a message for an intended destination evaluates the suitability of any other node it makes contact with as the next hop. Messages are thus opportunistically forwarded by exploiting nodes encounters until they reach the intended destination [27].

Routing is the most compelling challenge in ONs. A lot of algorithms for routing without infrastructure have been developed for this, which can be divided into dissemination-based and context-based algorithms. Dissemination-based algorithms are essentially forms of controlled flooding and differentiate themselves by the policy used to limit flooding. Context-based approaches usually do not adopt flooding schemes but use knowledge of the context that nodes are operating in to identify the best next hop at each forwarding step.

On one hand, context-based routing techniques [28] are generally able to significantly reduce messages duplication with respect to dissemination-based techniques. On the other hand, context-based techniques tend to increase the delay that each message experiences during delivery. This is due to possible errors and inaccuracies in selecting the best relays. Moreover, context-based techniques have higher computational costs than dissemination-based techniques.

Panisson et al. [29] improved multihop routing using social-based, opportunistic delay-tolerant strategies. The Geokad [30] uses the technique of a geographic-based table to update the peer information in such a dynamic environment. The problems they faced were peer misses and inefficient message passing due to the poor peer updating (message passing time) latency.

The Haggle project [31] is the first comprehensive architecture of autonomic and opportunistic networking, designed to enable communication in the presence of intermittent network connectivity, which allows mobile devices to exchange content with one another when they happen to come in close range contact. Haggle supports both Bluetooth and Wi-Fi connectivity. Haggle has been as a chosen platform for a variety of proximity-based applications (e.g., communications in times of disasters [32], seeking friends and localization [33], Electronic Triage Tag [34], picture sharing with mobile phones [35], and so forth).

MobiClique [36], a middleware for mobile social networking, is built on the code and functionality of the reference implementation of the Haggle network architecture, for content and profile sharing. The user's profile includes the friend list of existing social networks (Facebook is used in MobiClique). Therefore, at times Internet connection is required to synchronize the user's profile. If MobiClique users are in physical proximity and share a relationship based on the profile and the friend list, the users will be alerted and can introduce each other, exchange content, or create a friendship.

ShAir is a platform for mobile P2P resource sharing (such as data, storage, connections, and computation) without the need for Internet connections or user interventions [37], which also used Haggle as a source of inspiration. Specifically, based on ShAir platform, an Android application (MobileP2P Photo Sharing App) was developed, in which photos can be shared with those who also have this app when they happen to be physically around near a user. Users do not need to register; photos directly hop through people's devices without Internet connections.

Trifunovic et al. [38] proposed an approach named Wi-Fi-Opp, a viable and energy-efficient alternative to Wi-Fi ad hoc, to support opportunistic communications by exploiting the mobile AP mode of today's smartphones. A proof-of-concept of Wi-Fi-Opp has been implemented on stock smartphones, and it does not require root privileges.

12.2.4 Several Initiatives of MSNP

MSNP refers to the social interaction among physically proximate mobile users directly through various wireless interfaces on their smartphones or other mobile devices. MSNP has been under active research in the past few years, and several MSNP initiatives exist.

12.2.4.1 PeerDeviceNet

PeerDeviceNet [39] is both app and middleware to connect Android phones and tablets securely through Bluetooth and Wi-Fi networks. As an app, it allows a group of devices to share Web pages, contact information, pictures, videos, and other documentations. As middleware, it helps developing connected mobile apps. Developers can reuse the PeerDeviceNet connection manager to connect mobile devices for connected mobile apps or multiplayer games.

12.2.4.2 AllJoyn

Recently, Qualcomm introduced AllJoyn [6], an open-source, general networking framework that aims to simplify the development of proximity-based, device-to-device distributed applications. AllJoyn supports multiple platforms (Android, Linux, and Windows), languages (Java, C++), and networking technologies (Wi-Fi, Bluetooth). AllJoyn especially complements Wi-Fi Direct and enables ad hoc, secure, proximity-based, device-to-device communication in a simple, user-friendly manner without having to access the cloud or use an intermediary server. It enables developers to easily integrate P2P communications into applications for new types of multiscreen user experiences that span across smartphones, tablets, PCs, and TVs without having to deal with the complexities of networking in a multipeer, multitransport environment [40]. These applications provide a simplified user experience because the consumer does not have to manage network connections.

12.2.5 Summary of MSN(P) Approaches

In summary, there exist various approaches for realizing mobile social networking applications in proximity areas, which have more or less the same idea, but with different aspects for different purposes, under different conditions and circumstances, and with different assumptions. Basically, those systems can be compared and summed up with the help of the following three criteria, as shown in Table 12.1

- Network joining criteria: The presented systems can be divided into two categories: proximity-centric (i.e., only people that are in physical proximity to one another are considered to form a social network) and global-wide (i.e., in principle, all people of the world are considered to form a social network).
- Realization infrastructure: Decentralized, hybrid, and centralized. Decentralized infrastructure means the pure wireless P2P mode, in which each node quantitatively has the same right and responsibility. Centralized usually means the client/server mode, where centralized servers will be adopted for the purposes of storage, processing, and management. Hybrid roughly has two meanings: first, to utilize direct wireless P2P data exchange mode in proximity and to use centralized architecture to support global communications (hybrid-centralized server plus direct wireless P2P); second, for direct wireless P2P communications in proximity, some super peers are used to facilitate the social networking functions (hybrid-supernode plus direct wireless P2P).
- Underlying direct wireless technology: A number of technologies are emerging that facilitate this capability to pass data between two devices ranging over different distances without accessing a wide area network. Near-field communication (NFC) range is limited to a few centimeters and can quickly read small amounts of data from devices and passive tags within several centimeters (alone, it does not ensure secure communications). Infrared Data Association (IrDA)

Table 12.1 Taxonomy of MSNP-Related Solutions

Approaches		Category Criteria		
		Network Joining Criteria	Realization Infrastructure	Underlying Direct Wireless Technology
MSN	MIT Serendipity	Proximity centric	Hybrid-centralized server with direct wireless	Bluetooth
	MobiSoC	Global	Centralized	Nonsupport
	SpiderWeb	Global	Hybrid-centralized server with direct wireless	Bluetooth
	SCOPE	Proximity centric	Hybrid-supernode with direct wireless	Wi-Fi ad hoc mode
MP2P	MyNet	Global	Decentralized	Nonsupport
	Peer2Me	Proximity centric	Decentralized	Bluetooth
	CoCam	Proximity centric	Hybrid-centralized server plus direct wireless	Mobile AP mode
ON	Haggle	Proximity centric	Decentralized	Bluetooth, Wi-Fi ad hoc mode
	MobiClique	Proximity centric	Hybrid-centralized server (for profile synchronization) plus direct wireless	Bluetooth, Wi-Fi ad hoc mode
	ShAir	Proximity centric	Decentralized	Bluetooth, Wi-Fi ad hoc mode
	Wi-Fi-Opp	Proximity centric	Decentralized	Mobile AP mode

Table 12.1 (*Continued*)   Taxonomy of MSNP-Related Solutions

Approaches		Category Criteria		
		Network Joining Criteria	*Realization Infrastructure*	*Underlying Direct Wireless Technology*
MSNP initiatives	PeerDeviceNet	Proximity centric	Decentralized	Bluetooth, Wi-Fi ad hoc mode, Wi-Fi AP mode, Wi-Fi Direct
	AllJoyn	Proximity centric	Decentralized	Bluetooth, Wi-Fi ad hoc mode, Wi-Fi AP mode, Wi-Fi Direct

Source: With kind permission from Springer Science+Business Media: *Wireless Networks*, Survey on mobile social networking in proximity (MSNP): Approaches, challenges and architecture, 2014.

standards can provide data transmission speed from 2.4 Kbit/s up to 16 Mbit/s and the range can be up to 5 m. IrDA transmitters and receivers are inexpensive to implement and do not consume much battery. The main disadvantage is that IrDA transmissions require line of sight to work. Bluetooth was design for smaller devices and requires less battery, but suffers from lower transmission speeds (max 3 Mbit/s for version 2.0) and a maximum range of up to 100 m. The IEEE 802.11 standards (a, b, g, n, and y) for Wi-Fi come in various configurations with different speeds and ranges, from 2 Mbit/s and a range of 100 m indoors up to 248 Mbit/s and a range of 250 m outdoors. Specifically, the original 802.11 standard defined two modes of operation—the ad hoc mode, where stations communicate directly with each other, and the infrastructure mode, where all stations communicate through an access point that is attached to a distribution network such as the Internet. As a special application of infrastructure mode, mobile devices can be put into tethering or mobile AP mode. Although this mode is initially meant for sharing one's 3G Internet access with clients, it can always be (mis)used for client-to-client communications and even mobile AP-to-client communications, hence enabling opportunistic communications. And recently, Wi-Fi Direct technology is emerging (will be briefly introduced in Section 12.4 and Chapter 13).

Of course, this categorization has a huge impact on understanding, analyzing, and designing MSNP systems. Specifically, we can infer the following implications.

- Obviously, all MSNP systems should deal with the location of the users. Location is usually a property for being assigned to a specific social network. Some systems also introduce matching algorithms based on properties such as interests, activity, and time for assigning people to social networks.
- Proximity-centric social networks operate on local area networks and usually do not use any servers. Messages can be exchanged by routing or broadcasting the messages among the users. However, a global server (or super peer) can be useful to synchronize users' profiles and to facilitate the stability of MSNP systems.
- Proximity-centric social networking applications should support various direct wireless technologies, including Bluetooth, Wi-Fi ad hoc mode, Wi-Fi AP mode, Wi-Fi Direct, and so forth. Many problems might occur during the establishment and the maintenance of these networks. Among others, scalability and high energy consumption are still problematic because each node has to forward (route) messages to other users and share its storage and computation capabilities with other users. The message routing approach also raises important security and privacy problems. For example, the storage and management of data and lists such as friend lists are difficult.

In summary, MSN has been popular in recent years, but its communication is still limited in the virtual world without a tight integration with physical communities that are featured as direct face-to-face social interactions. ONs using direct wireless P2P technologies are good candidates to address the limitations of content sharing among uncoordinated users who may (not) know each other in a proximity area. MSNP takes advantage of real social networks and geographic proximity and, meanwhile, bypasses the growing mobile data traffic. This infrastructure-free, "crowd-sourced" communication channel could be useful for real-world proximity-based social networking, advertisement, gaming, and so forth.

12.3 Special Characteristics and Challenges of MSNPs

As described earlier, MSNP has been introduced by combining the concepts from three disciplines—mobile social networks, mobile P2P, and opportunistic networks. In MSNPs, users can fully take advantage of human interaction and physical mobility to achieve efficient and effective data delivery services. Figure 12.2 shows the special characteristics and key challenges that should be dealt with, especially in MSNPs.

12.3.1 Special Characteristics of MSNPs

Proximity-based: Collaboration is made possible by physical proximity of two or more individuals. Mobile devices can exchange content directly among one another when they happen to come in close range contact.

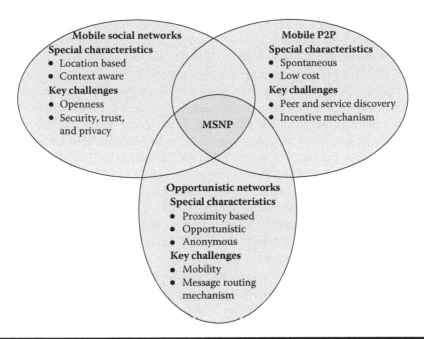

Figure 12.2 Relationship of MSNP with MSNs, mobile P2P, and ONs—its special characteristics and key challenges. (With kind permission from Springer Science+Business Media: *Wireless Networks*, Survey on mobile social networking in proximity (MSNP): Approaches, challenges and architecture, 2014.)

Opportunistic: Allows people to take advantage and make use of an opportunity that presents itself. That is, human-carried mobile devices would opportunistically communicate with each other in a "store-carry-forward" fashion without any infrastructure; data are stored, carried, and delivered by mobile devices using wireless ad hoc contact opportunities with other devices to pass the data onward, potentially over multiple hops, until it reaches its destination(s).

Anonymous: In MSNPs, initially, users are mostly anonymous and can only be discovered by people around them. Indeed, after proximity users sufficiently interact, they may want to expose more personal information and form relatively intimate social relationships. However, one of the design principles in autonomous networks, "design for user's choice" (respecting user's preference for anonymous interaction in proximity), is one of the basic characteristics of MSNPs.

Spontaneous: MSNPs should require no prior planning or preparation on behalf of the human. Usually, MSNPs are small, short-lived, and implicitly defined social networks. Relationships are created in real time and with minimal manual intervention. Briefly, spontaneous MSNP enables users to communicate, share experiences, and comment on them in situations and locations such as at events, in bars, and at schools.

Low cost: MSNPs can be built on off-the-shelf mobile devices with various wireless technologies—Bluetooth, WLAN (IEEE 802.11) ad hoc mode, and/ or Wi-Fi Direct. More important, MSNP does not rely on centralized servers and Internet connection for continuous operation. Cellular data today is often slow, expensive (especially when roaming), and not even always available (rural areas, underground transportation, popular mass events, disaster situations are just a few examples). Bluetooth or Wi-Fi can offer always available, essentially free, local connectivity. In addition, Wi-Fi offers higher bandwidths compared to the available cellular networks.

Location based: Location information is central to MSNP. Not only is content tagged with locations but MSNP uses geoproximity as the primary filter in determining who is discoverable. Besides GPS, various Wi-Fi-based or Global System for Mobile Communications (GSM) cell tower–based localization schemes have been proposed. Moreover, other components coupled with sophisticated algorithms are enabling even more precise location and proximity results. For example, application developers could analyze data from microphones on two separate user devices to matching ambient noise and determine if those devices are in the same room.

Context-aware: In MSNPs, contexts could include information about people, objects, and the surroundings, and collecting such data can be done from a number of sources. Specifically, the following contexts should be taken into account.

Social context: In MSNPs, exchanges are only possible over short distances, that is, when people come face-to-face or at least are within close physical proximity. Consequently, trading partners may be aware of whom they are trading with and be able to observe important social cues including sex, clothing, and gestures.

Usage context: The context in which mobile devices are used is different in two important aspects from using a stationary computer at home or in the office. When mobile devices are involved, user attention is a scare resource. Instead of sitting in front of a computer where we can pay full attention to the computer and its operation, handheld and mobile computers are often used in situations where our attention is occupied by demanding real-world tasks such as driving, operating a machine, or simply conversing with other people. Furthermore, with mobile terminals, time becomes a critical resource as well. If a transfer takes too long, we are less likely to complete it because we might run out of patience or we must hurry to our next appointment.

Technical context: Even if smartphones are becoming increasingly powerful, the small form factor of these devices will always introduce resource limitations compared to static computing systems. These limitations include lower processing power, smaller memory and persistent storage, slower I/O and, most critically, finite battery life. These constraints must be addressed when designing MSNP applications for mobile devices.

12.3.2 Open Issues

12.3.2.1 Mobility

Mobility means that participants of the MSNP applications may change their location during the usage of the system. It might happen that a node moves out of the scope of the network and loses the connection completely while other nodes come into reach and want to participate. Ad hoc systems should be able to work under these circumstances and even can handle the additional constraints implied by mobile nodes, such as limited bandwidth and power.

On the application level, there is a need to support changing addressing and network technologies (Bluetooth, Wi-Fi ad hoc mode, Wi-Fi Direct, etc.) in mobile environments. An intelligent support for wireless technologies is also required such as choosing an optimal multihop routing or using multilinks to transfer data.

On the contrary, the opportunistic MSNPs rely on human mobility for forwarding, which is directly related to social behaviors of people; we are more likely to meet regularly with our friends, family, or coworkers than with some strangers. Consequently, many schemes exploit social mobility patterns to make efficient forwarding decisions. For example, based on the observations that node mobility behavior is semideterministic and could be predicted once there is sufficient mobility history information, a routing approach called predict and forward was proposed and compared with a number of existing encounter-based schemes [41].

12.3.2.2 Naming and Messaging Mechanism

Different connectivity situations and applications might regard different types of names as "addressable" because of their different capabilities. For instance, to a node with geographic information and forwarding capabilities, a GPS coordinate is an addressable name; services that exploit the local connectivity to share data with a neighbor or an 802.11 MAC address as an MSNP address. Although for other nodes, a name needs to be explicitly mapped onto another address type before forwarding can occur.

One of the key problems is that how these high-level names get translated into lower-level addresses that the physical network interfaces can use for transmission. Mechanisms for mapping between name and address are therefore very important in MSNPs.

The underlying topology of an MSNP is highly dynamic. Therefore, establishing and maintaining end-to-end paths between nodes is generally infeasible. The basic principle behind most of the proposed solutions is to rely on completely decentralized hop-by-hop forwarding decisions. Specifically, it has been shown that, even in a mobile ad hoc networking environment (in multidomain network scenarios), the use of virtual coordinates or identifiers for routing and data management has several advantages compared to classical topology control techniques based on predefined addresses or geographical coordinates [42].

12.3.2.3 Peer and Service Discovery

In MSNPs, mobile peers take advantage of resources provided by mobile peers that are physically close. Because of the unpredictable movement of mobile devices, discovering resources becomes a challenge. In detail, MSNPs need algorithms through which a peer can detect the presence of neighboring peers, share configuration and service information with those peers, and be notified when corresponding peers become unavailable.

Specifically, through sending beacon and receiving probe response frames, an MSNP application can easily discover neighboring peers. However, it has no good way to further figure out which peer to establish a connection with. For example, if a game application is interested in finding all the neighboring peers that are also running the same game, it has no way to find out until after the connection is set up. Naturally, pre-association service discovery should be provided, which is meant to address this issue of filtering the peers based on the running services. With pre-association service discovery, an application can advertise a service on a peer device prior to a connection setup between the devices.

Some zero configuration (Zeroconf) networking techniques should be properly adopted in MSNP applications, which basically solve three main tasks: managing numeric network addresses for networked mobile devices (address auto configuration process); handling host names for devices (name-to-address resolution process); and offering and searching services for devices (service discovery process) [43]. Zeroconf techniques can be offered using several implementations, such as Domain Name Server (DNS)–based service discovery (Apple's Bonjour) and Universal Plug and Play (Upnp).

Finally, resource discovery must be timely (in order to detect moving peers) and efficient (so as not to overload the network). Moreover, mobile devices are resource constrained. In particular, heavyweight processing can consume too many resources of a mobile device and can also cause high latency. Such an issue can cause users to dislike participating in MSNP applications.

12.3.2.4 Incentive Mechanism

Most traditional MSNs (OSNs) do not explicitly provide specifications to ensure fairness or to provide incentives. We argue that the locus of control in creation and configuration of content in MSNPs has been shifting to the grassroots. The existence and prosperity of MSNPs will depend on achieving a critical mass of users who share their profiles, places, and real-time location information.

Generally, incentive mechanisms may include direct reciprocation methods (e.g., micropayment-based schemes that directly pay virtual currency to get others' service), indirect reciprocation methods (e.g., reputation-based incentive schemes that count on the belief that tomorrow's opponent may condition their choice on today's play), and social-based methods (e.g., users tend to respond in the most positive way when they felt that they uniquely contributed to the space).

Furthermore, physical MSNP communities can act as an incentive for sharing the limited resources of a mobile device: the incentive comes from belonging to and maintaining a community (users are usually willing to share or carry data for a friend or for someone from the same community).

12.3.2.5 Openness

A paradox with today's social networks is that although they encourage sharing of users' most intimate secrets—often with people they have never actually met—they themselves are closed, tightly controlled environments. Facebook, Foursquare, Google+, and LinkedIn require people to set up separate accounts on a centrally managed service in which every shred of information passes through the operator's virtual hands, a so-called monopoly.

The model of closed systems does make financial and business sense. It lets companies such as Facebook monetize people's collaborative endeavors. However, such centralization conflicts with MSNP's ethos of open sharing and decentralized control.

Facebook might have reached such critical mass that resistance is futile. However, the most effective way to counteract such natural monopolies is not through regulation but openness. Specifically, MSNPs should be built on a new generation of social software with open APIs, protocols, and architecture. For users, the value of an open MSNP is that it could lead to an open social platform where users in proximity are in complete control of their information and social graphs. For developers, it promises to eliminate the monopoly (e.g., Facebook) tax and to provide an easier and service-independent way of creating and distributing mobile, social applications without the need for server-side components.

Standing in the way, however, are some structural barriers to openness: (1) loss of user data privacy and control, (2) increasing difficulty for new Web services to enter the market, (3) inaccessible wireless capacity due to closed networks, and (4) network infrastructure not open to continued innovations. Stanford MobiSocial Computing Laboratory (http://mobisocial.stanford.edu/) is especially aiming to tearing down proprietary walls via the open social software that serves as a key piece of Stanford's ambitious Programmable Open Mobile Internet 2020 project [44].

12.3.2.6 Security, Trust, and Privacy

The security implications of MSNP systems, in which one can potentially track every movement of an individual as well as examine what they are doing, must be taken seriously. In an MSNP system, users may not even be aware to what devices they are connected. Someone in the next room or on the floor above may connect to someone else's mobile device and gain access to private data such as stored e-mails and meeting schedules. Thus, not only must encryption be employed to avoid eavesdropping but also robust authentication procedures need to be established for connecting trusted and nontrusted devices with each other. This, however, is made difficult by the fact that this must occur in a completely decentralized environment with little or no connection to a trusted authority. Possible solutions might use a reputation-based scheme.

Most of the MSNPs are capable of accessing the users' physical location, preferences, and social relations. Therefore, the privacy-related issues for mobile users in the MSN become crucial. Due to its profound scope and unique degree of sensitivity, privacy implementation is not an easy task. Moreover, because human interfaces are involved in MSNP, the granularity of privacy concern may differ from user to user. Specifically, profile matching (two users comparing their personal profiles) is often the first step toward effective MSNP. However, it conflicts with users' growing privacy concerns about disclosing their personal profiles to complete strangers before deciding to interact with them. Zhang et al. [45] tackle this open challenge by designing a suite of novel, fine-grained private-matching protocols without disclosing any information about their profiles beyond the comparison result.

On the contrary, some distinguished characteristics of MSNPs (e.g., giving the control over data back to the users, having no one entity access to all personal data of the participant in MSNPs, and granting users the freedom to customize its specific privacy rules, immediately verifying the validity of the source in face-to-face proximity) could significantly facilitate to solve the privacy issue. For example, a mobile social application platform, Musubi, was provided in Ref. [46], to enable users to share any data type in real-time feeds created by any application on the phone, in which all data reside on the phone (P2P approach) and public key encryption are adopted (thus leaking no user information to a third party).

We argue that the incremental three-stage privacy protection mechanism that is voluntarily chosen by individual users could alleviate the problem of privacy. First, enable users to discover one another in the proximity network. Then, build trust in an elastic network. Finally, take their friendship further by including them in their social network, revealing their true identity. In detail, the proximity network is the stage at which users are most anonymous and can only be discovered by people around them; once users begin interacting with one another in the proximity network, they enter the elastic network. The elastic network is slightly more permanent than the proximity network because it persists beyond the current time and space. In the elastic network, users begin building a relationship through a number of mechanisms, including SMS messaging, voice or video calling, and interacting with one another's content on the network. Users can also be added to buddy lists within the proximity-based social network without revealing too many personal details. Because users have to opt in to continue to interact, the elastic network is disposable and only lasts for a finite period of time if interaction diminishes. It is also possible to deploy algorithms that introduce two users who should connect based on interests and characteristics, expediting the process of moving from the proximity network to the social network. Once two users have sufficiently interacted and have built trust, they move to the more permanent social network.

In the social network, a user's full identity and contact info are revealed. At this point, the relationship is more permanent and can grow on a number of different levels, in cyberspace and in the real world.

It is should be noted that Väänänen-Vainio-Mattila et al. [47] presented the usage results of a large-scale user study of TWIN, an ad hoc social networking system that offers applications for social presence, mobile multimedia sharing, and community-based communication. Interestingly, the field study showed that privacy concerns did not arise as a significant issue in user experience.

12.4 Design Architecture of MSNP Platform

MSNP system architecture is illustrated in Figure 12.3. Generally, MSNP applications operate in ad hoc mode as the devices come within the direct communications range of each other, and through a hop-by-hop forwarding decision, the application range of MSNP can be extended to a typical proximity area (e.g., campus, event spot, or community). Moreover, through the mechanism of store-carry-forward, human mobility can help the propagation of a message. It should be explicitly pointed out that MSNP is complementary to (does not substitute for) OSNs through providing effective tools to support the direct interaction among friends and/or strangers in a proximity area. And furthermore, through sharing the wide area interface (e.g., cellular, 3G, or Wi-Fi), MSNPs can be seamlessly combined with OSNs and use various functions provided by OSNs. For example, it is possible to adopt users' OSN IDs to establish a trust relationship when they are opportunistically located in proximity.

Basically, each node executes a periodic loop that consists of three steps: (1) neighborhood discovery, (2) user identification (and authentication), and (3) data exchange [48]. The neighborhood discovery method depends on the radio technology being used. Commonly available options with today's mobile device hardware include Bluetooth device discovery or broadcast beacons on a well-known Wi-Fi Service Set IDentifier (SSID). The neighborhood can, and usually will, contain more than two devices; the system must therefore manage multiple simultaneous connections.

Upon discovery of a new device in the neighborhood, MSNP enters the identification phase where the devices open a communication link among one another to exchange user identity and profile information to some extent (according to each user's special privacy rule). Once the identification is successfully completed, the last step of the interaction is the data exchange phase in which devices exchange application level messages within each other. These messages are stored persistently on the devices within the limits of storage space and are forwarded to interest groups the user has joined. The forwarding is performed using two simple rules: (1) unicast messages (i.e., messages for a specific user) are sent either upon meeting the destination directly or are forwarded through friends of the destination, and

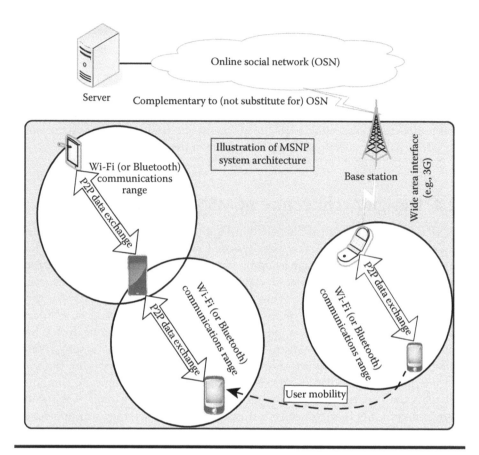

Figure 12.3   Illustration of MSNP system architecture. (With kind permission from Springer Science+Business Media: *Wireless Networks*, Survey on mobile social networking in proximity (MSNP): Approaches, challenges and architecture, 2014.)

(2) group messages are flooded within the corresponding interest group so that each member of the group will participate in the forwarding until everybody has received the data. These rules ensure that no node receives content unless it belongs to the target group of interest or is a friend of the destination. This already provides a minimum guarantee on privacy, and it helps also as an incentive mechanism. Basically, messages are removed from the system after an application-defined time-to-live (TTL) has expired. MSNP TTL could be a combination of an absolute time stamp and a maximum number of hops. Messages are deleted either when the TTL has passed or the message has taken a maximum number of hops in the network.

Because MSNPs are still in their early stages, existing approaches are tightly coupled with individual platforms. From our perspective, an ideal MSNP should be cross-platform, cross-brand, and cross-physical layer, in which participants can share their content freely without considering which platform they are using on their devices. Moreover, an ideal MSNP is highly dynamic in which mobile devices are capable of discovering the interested content for their users on-the-fly without prior knowledge about the content providers.

Figure 12.4 illustrates a schematic architecture of development platform for MSNP applications. The architecture contains four layers: hardware layer, core layer, service layer, and application layer.

The hardware layer is composed of various sensors in a mobile terminal (including camera, microphone, accelerometer, etc.) and wireless networking technologies that are used for building physical communities. As described in section 12.2, there exist a number of technologies that provide the ability for devices to directly interact in physical proximity, over different communications distances and speed. Specifically, IEEE 802.11 standard defines two modes of operation. Ad hoc mode refers to a P2P association established directly between two wireless stations. Infrastructure mode refers to associations between a group of wireless stations and an access point (AP).

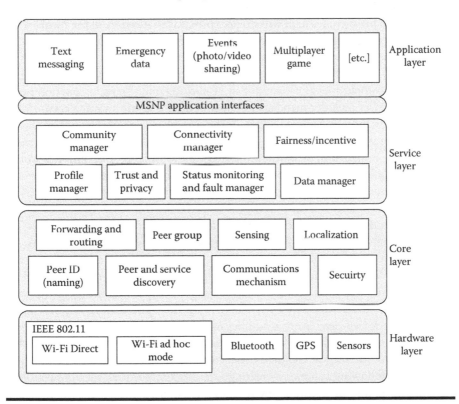

Figure 12.4 Framework of MSNP development platform; core functional components. (With kind permission from Springer Science+Business Media: *Wireless Networks*, Survey on mobile social networking in proximity (MSNP): Approaches, challenges and architecture, 2014.)

Direct device-to-device connectivity can be supported by means of the ad hoc mode of operation. However, this never became widely deployed in the market and hence presents several drawbacks when facing today's requirements (e.g. lack of efficient power saving support or extended Quality of Service (QoS) capabilities). Wi-Fi Direct is a new technology defined by the Wi-Fi Alliance that is aimed at enhancing direct device-to-device communications in Wi-Fi [49]. Unlike the previous technologies, the Wi-Fi Direct technology takes a different approach to enhance device-to-device connectivity. Instead of leveraging the ad hoc mode of operation, Wi-Fi Direct builds upon the successful IEEE 802.11 infrastructure mode and lets devices negotiate who will take over the AP-like functionalities. Thus, legacy Wi-Fi devices may seamlessly connect to Wi-Fi Direct devices. By making this decision, Wi-Fi Direct immediately inherits all the enhanced QoS, power saving, and security mechanisms. Thus, given the wide base of devices with Wi-Fi capabilities, and the fact that it can be entirely implemented in software over traditional Wi-Fi radios, Wi-Fi Direct technology is expected to have a significant impact. It is should be pointed out that Long Term Evolution (LTE) Direct is another technology being developed at Qualcomm, and it will support proximity networks of the future with higher speed and a larger communications range than Wi-Fi Direct.

Briefly, the goal of the hardware layer is to enable the researchers and developers to easily use various wireless technologies through a unified interface without having to deal with the complexities of networking in a multipeer, multitransport environment.

The core layer includes mandatory functionalities that are strictly necessary for the MSNP systems to function properly. At the core layer, our focus is on the abstraction of community formation, including naming, peer and service discovery, communications mechanisms, forwarding and routing mechanisms, peer group, sensing, localization and security mechanisms, and so forth. The dynamically formed group/community especially provides a minimum set of given services and/or shares specific interests. A community is an open set of entities. Membership is not controlled by the owner of the community (which might be one particular member or the collective of all members) but can be passed on by any member to any other entity. That is, no single authority controls membership and members can join/leave at any time.

The service level includes fundamental higher-layer services designed for flexibility and extensibility. In detail, the data manager entity is responsible for managing the data that are stored in a searchable repository. The connectivity manager is responsible for encapsulating a number of so-called connectivities representing the available network interfaces. It provides support for neighbor discovery and for managing communication channels between two peers. The community manager provides the following functions: creating interest communities, viewing community members, finding old and new friends, exchanging user profiles, and so on. Note that, in our architecture, incentive/fairness is explicitly designed as an independent component. We argue that, due to the variety of services that MSNPs should support, and given the dynamicity of MSNP interactions, in the proposed MSNP architecture, each peer should be rewarded for any service provided to the MSNP community in general and not to specific peers only. Note that our previous work shows that, based on two simple system rules, the voluntary principle (users voluntarily participate after paying an entry fee, voluntarily contribute resources, and voluntarily punish other defectors—incurring extra cost for those so-called punishers) and the entry fee (i.e., an arbitrarily small entry fee is set for each user who wants to participate in the community), the community with the feature of contributing and consuming can evolve into the state of almost full punishers who not only provide resources but who also punish the defects [50].

The application layer includes a wide variety of applications that can be developed on top of the other two layers, including (but not limited to) text messaging, dissemination of emergency data, photo/video sharing during an event, multiplayer games, and so forth. In MSNPs, text messaging can easily be implemented in direct data exchange, and if necessary, multiple hops can be used to extend the range.

Dissemination of emergency data. During emergencies, many nearby people typically try to send or retrieve information through their phones, and it may happen that the number of attempted connections exceeds the number of connections that a specific cellular network sector can accept. In MNSP-based data dissemination, the information could be shared in real time and received by anyone in proximity. The devices do not need to be connected to an AP and do not need access to the cellular network. We argue that MSNP applications should work well in those scenarios because many emergencies are local, and the data can be spread in a precise area.

Photo/video sharing during an event. Events such as sports or meetings of any kind encourage people that share a common interest to gather together. Their proximity allows a spontaneous MSNP network to be formed, and thus peers can share photos or videos (or any other type of information relevant to the event).

Multiplayer gaming. Even without an Internet connection, mobile users still can find players for a multiplayer game in proximity, such as in a train or room, and so forth.

12.5 Conclusion

We argue that the rise of proximity-based social networking lays the groundwork for long-term disruption in the mobile landscape. The trend toward social interaction though proximity social networks will drive content to the edge of networks that will be consumed by users through ad hoc P2P local area wireless networks. This shift from wide area wireless networking to local area networking will mitigate the data burden on carrier networks while reducing their influence, creating opportunities for new players and innovative initiatives by mobile operators. However, MSNP is still in its early stages; most work in literature either only provides the basic concept of MSNP or introduces preliminary basic components in MSNP application development.

In this chapter, we thoroughly overview the new shift of social networking in proximity (MSNP), which enables more tangible face-to-face social interactions in public places such as bars, airports, trains, and stadiums (in a certain time and place). Specifically, we first thoroughly introduce and compare various existing solutions related to MSNP in the current literature. And then, we discuss the special characteristics of MSNP, summarize open issues, and give potential solutions. Then, we propose a cross-platform, cross-brand, cross-networking technology architecture for developing MSNP applications.

References

1. Facebook statistics. Available online: http://www.facebook.com/press/info.php?statistics
2. He, J., D. J. Miller, and G. Kesidis. Interest-group discovery and management by peer-to-peer online social networks. Technical Report. Available online: http://www.cse.psu.edu/research/publications/tech-reports/2012/CSE%2013-001.pdf
3. Foursquare Website. Available online: https://foursquare.com/
4. Crocker, P. B. Proximity based mobile social networking: Applications and technology, making new connections in the physical world. Smith's Point Analytics, Ipswich, MA, 2011.
5. Wang, Y., A. V. Vasilakos, Q. Jin, and J. Ma. Survey on mobile social networking in proximity (MSNP): Approaches, challenges and architecture. *Wireless Networks* 2014. DOI 10.1007/s11276-013-0677-7.
6. Eagle, N. and A. Pentland. Reality mining: Sensing complex social systems. *Personal and Ubiquitous Computing* 2006; 10(4): 255–268.
7. Alljoyn: A common language for Internet of everything, Available online: https://www.alljoyn.org/
8. He, W., Y. Huang, K. Nahrstedt, and B. Wu. Message propagation in ad-hoc-based proximity mobile social network. In: Proceedings of Pervasive Computing and Communications Workshops, Mannheim, Germany, March 29–April 2, 2010.
9. Sapuppo, A. and L. S. Local social networks. *Computer Communication and Management*, Zhang Ting (ed.), Vol. 5 IACSIT Press 2011; pp. 15–22.
10. Chang, C., S. N. Srirama, and S. Ling. An adaptive mediation framework for mobile P2P social content sharing. *Lecture Notes in Computer Science* 2012; 7636: 374–388.
11. Chang, C., S. N. Srirama, and S. Krishnaswamy. Sea ling, proactive web service discovery for mobile social network in proximity. *Journal of Next Generation Information Technology* 2013; 4(2): 110–112.
12. Kayastha, N., D. Niyato, P. Wang, and E. Hossain. Applications, architectures, and protocol design issues for mobile social networks: A survey. *Proceedings of the IEEE* 2011; 99(12): 2130–2158.
13. Bellavista, P., R. Montanari, and S. K. Das. Mobile social networking middleware: A survey. *Pervasive and Mobile Computing* 2013; 9(4): 437–453.
14. Zhang, B., Y. Wang, A. V. Vasilakos, and J. Ma. Mobile social networking: Reconnect virtual community with physical space. *Telecommunication Systems Journal* 2013; 52(3): 91–110.

15. Wang, Y., J. Tang, Q. Jin, and J. Ma. Overview mobile social networking in proximity (MSNP): Applications, characteristics and challenges. In: Proceedings of the 11th IEEE/IFIP International Conference on Embedded and Ubiquitous Computing (EUC), Zhangjiajie, China, November 13–15, 2013.
16. Eagle, N. and A. Pentland. Social serendipity: Mobilizing social software. *IEEE Pervasive Computing* 2005; 4(2): 28–34.
17. Gupta, A., A. Kalra, D. Boston, and C. Borcea. MobiSoC: A middleware for mobile social computing applications. *Mobile Networks and Application* 2008; 14(1): 35–52.
18. Sapuppo, A. SpiderWeb: A social mobile network. In: Proceedings of the European Wireless Conference (EW), Lucca, Italy, April 12–15, 2010.
19. Mani, M., N. Anh-Minh, and N. Crespi. SCOPE: A prototype for spontaneous P2P social networking. In: Proceedings of the 8th IEEE International Conference on Pervasive Computing and Communications Workshops (PERCOM Workshops), Mannheim, Germany, March 29–April 2, 2010.
20. Ford, B., J. Strauss, C. Lesniewski-Laas, S. Rhea, F. Kaashoek, and R. Morris. Persistent personal names for globally connected mobile devices. In: Proceedings of Operating Systems Design and Implementation, Seattle, WA, November 6–8, 2006; pp. 233–248.
21. JXTA Community Projects. Available online: https://jxta.dev.java.net/
22. Kalofonos, D. N. et al. MyNet: A platform for secure P2P personal and social networking services. In: Proceedings of the 6th IEEE International Conference on Pervasive Computing and Communications (PERCOM), Hong Kong, China, March 17–21, 2008.
23. Wang, A. I., T. Bjørnsgard, and K. Saxlund. Peer2Me—Rapid application framework for mobile peer-to-peer applications. In: Proceedings of the International Symposium on Collaborative Technologies and Systems (CTS), Orlando, FL, May 21–25, 2007.
24. MIT Media Lab. Available online: http://www.media.mit.edu/research/groups/information-ecology
25. Toledano, E. et al. CoCam: Real-time photo sharing based on opportunistic P2P networking. In: Proceedings of the 10th IEEE Consumer Communications and Networking Conference (CCNC), Las Vegas, NV, January 11–14, 2013.
26. Guo, B., D. Zhang, Z. Yu, and X. Zhou. Enhancing spontaneous interaction in opportunistic mobile social networks. *Communications in Mobile Computing* 2012; 1(6): 1–6.
27. Boldrini, C., M. Conti, and A. Passarella. Design and performance evaluation of ContentPlace, a social-aware data dissemination system for opportunistic networks. *Computer Networks* 2010; 54(4): 589–604.
28. Pelsui, L., A. Passarella, and M. Conti. Opportunistic networking: Data forwarding in disconnected mobile ad hoc networks. *IEEE Communications Magazine* 2006; 44(11): 134–141.
29. Panisson, A., A. Barrat, C. Cattuto, W. Van den Broeck, G. Ruffo, and R. Schifanella. On the dynamics of human proximity for data diffusion in ad-hoc networks. *Ad Hoc Networks* 2012; 10: 1532–1543.
30. Marco, P., A. Michele, and Z. Francesco. GeoKad: A P2P distributed localization protocol. In: Proceedings of IEEE PERCOM Workshop, Mannheim, Germany, March 29–April 2, 2010.
31. Su, J., J. Scott, H. Pan, J. Crowcroft, E. D. Lara, C. Diot, et al. Haggle: Seamless Networking for Mobile Applications. Haggle: Seamless networking for mobile applications. In: Proceedings of Ubiquitous Computing (Ubicomp), Innsbruck, Austria, September 16–19, 2007; pp. 391–408.

32. Mokryn, O., D. Karmi, A. Elkayam, and T. Teller. Help me: Opportunistic smart rescue application and system. In: Proceedings of the 11th Annual Mediterranean Workshop on Ad Hoc Networking, Ayia Napa, Cyprus, June 19–22, 2012; pp. 98–105.

33. Papandrea, M., S. Vanini, and S. Giordano. A lightweight localization architecture and application for opportunistic networks. In: Proceedings of the World of Wireless, Mobile and Multimedia Networks & Workshops (WoWMoM), Kos, Greece, June 15–19, 2009.

34. Martín-Campillo, A., J. Crowcroft, E. Yoneki, R. Martí, and C. Martínez-García. Using Haggle to create an electronic triage tag. In: Proceedings of the Second ACM International Workshop on Mobile Opportunistic Networking, Pisa, Italy, February 22–23, 2010; pp. 167–170.

35. Nordström, E., D. Aldman, F. Bjurefors, and C. Rohner. Search-based picture sharing with mobile phones. In: Proceedings of the 10th ACM International Symposium on Mobile Ad Hoc Networking and Computing (MobiHoc), New Orleans, LA, May 18–21, 2009; pp. 327–328.

36. Pietilainen, A.-K., E. Oliver, J. LeBrun, G. Varghese, and C. Diot. MobiClique: Middleware for mobile social networking. In: Proceedings of ACM Workshop Online Social Network (WOSN), Barcelona, Spain, August 17, 2009; pp. 49–54.

37. Dubois, D., Y. Bando, K. Watanabe, and H. Holtzman. ShAir: Extensible middleware for mobile peer-to-peer resource sharing. In: Proceedings of the 9th Joint Meeting on Foundations of Software Engineering (ESEC/FSE Industrial Track), Saint Petersburg, Russia, August 18–26, 2013.

38. Trifunovic, S., B. Distl, D. Schatzmann, and F. Legendre. WiFi-Opp: Ad-hoc-less opportunistic networking. In: Proceedings of ACM MobiCom Workshop on Challenged Networks (CHANTS), Las Vegas, NV, September 23, 2011.

39. PeerDeviceNet: Secure sharing among your phones and tablets. Available online: http://www.peerdevicenet.net/

40. Stauffer, M. Connectivity solutions for smart TVs. In: Proceedings of the IEEE International Conference on Consumer Electronics-Berlin (ICCE-Berlin), Berlin, Germany, September 3–5, 2012; pp. 245–249.

41. Elwhishi, A., P.-H. Ho, and B. Shihada. Contention aware mobility prediction routing for intermittently connected mobile network. *Wireless Networks* 2013; 19(8): 2093–2108. DOI: 10.1007/s11276-013-0588-7.

42. Dressler, F. and M. Gerla. A framework for inter-domain routing in virtual coordinate based mobile networks. *Wireless Networks* 2013; 19: 1611–1626.

43. Siddiqui, F., S. Zeadally, T. Kacem, and S. Fowler. Zero configuration networking: Implementation, performance, and security. *Computers & Electrical Engineering* 2012; 38(5): 1129–1145.

44. POMI 2020 (programmable and open mobile Internet) overview. Available online: http://pomi.stanford.edu/content.php?page=about

45. Zhang, R., Y. Zhang, J. Sun, and G. Yan. Fine-grained private matching for proximity-based mobile social networking. In: Proceedings of INFOCOM, Orlando, FL, March 25–30, 2012; pp. 1969–1977.

46. Dodson, B., I. Vo, T. J. Purtell, A. Cannon, and M. S. Lam. Musubi: Disintermediated interactive social feeds for mobile devices. In: Proceedings of the 21st International Conference on World Wide Web (WWW), Lyon, France, April 16–20, 2012; pp. 211–220.

47. Väänänen-Vainio-Mattila, K., P. Saarinen, M. Wäljas, M. Hännikäinen, H. Orsila, and N. Kiukkonen. User experience of social ad hoc networking: Findings from a large-scale field trial of TWIN. In: Proceedings of the 9th International Conference on Mobile and Ubiquitous Multimedia (MUM), Limassol, Cyprus, December 1–3, 2010.
48. Yu, Z., Y. Liang, B. Xu, Y. Yang, and B. Guo. Towards a smart campus with mobile social networking. In: Proceedings of the IEEE International Conference on Internet of Things (iThings), Beijing, China, August 20–23, 2011.
49. Camps-Mur, D., A. Garcia-Saavedra, and P. Serrano. Device to device communications with Wi-Fi direct overview and experimentation. *IEEE Wireless Communications Magazine* 2013; 20(3): 96–104.
50. Wang, Y., A. Nakao, A. V. Vasilakos, and J. Ma. On the effectiveness of service differentiation based resource-provision incentive mechanisms in dynamic and autonomous P2P networks. *Computer Network* 2011; 55(17): 3811–3831.

MSN APPLICATION DEVELOPMENT

Chapter 13

Mobile Social Networking Development Platforms and Examples

13.1 Introduction

The Internet has long been used for social interaction, with some of the more popular examples being social networking applications such as Facebook, Twitter, LinkedIn, and Instagram [1–4]. These types of applications help users share digital media and have proven to be successful tools for expanding the social network. There is also a trend toward extensive use of social networking application from mobile devices [5]. The landscape of mobile platforms has seen a major evolution in the recent past. In the era of smartphones and tablets, mobile applications are providing added value to several industries including transportation, ecommerce, net banking, travel, retail, and enterprise services. Developers are exploiting the state-of-the-art functionalities of the smart devices to offer a revolutionizing user experience. In turn, they are becoming the engine for innovation. Thus, it is of prime importance for a mobile platform provider to attract more and more developers in order to boast external investment and revenue. Not only the mobile platform owners and handset manufacturers but also network service providers and chipset makers are investing heavily to develop and release software kits to reach out to the developers [6].

Programming applications for mobile phones used to be a niche task. However, the recent extensive adoption of mobile devices and especially smartphones, together with the app store mobile application distribution model, has elevated mobile application programming to a common activity. To support mobile application development, each producer of a major mobile application platform (iOS, Android, Windows Phone, etc.) provides a convenient software development kit (SDK). For example, Google's Android SDK integrates nicely with the popular Java IDE Eclipse, which makes it easy to start programming mobile phone applications. That is, programmers install an SDK on their desktop computer and develop a mobile application there just as they would develop any other application. The SDK typically contains a powerful emulator or virtual device that allows programmers to simulate how their mobile applications will behave on an actual mobile device [7].

The development of these applications is much easier than earlier mobile application development platforms but still carries some of the same complexities and issues. Although these operating systems (OSs) are rich in libraries and built-in features, they still face the heat of the market to match customer's high expectations. The basic architecture and support of OS programming language are very different from one another. Developed applications for a certain OS are not compatible for another OS [8]. The diversity of mobile platforms and the variety of SDKs and other tools pose unique challenges. These include choice of SDK, user experience, stability of framework, ease of updating, cost of development for multiple platforms, and time to market an app. Most of the developers would like to release apps for major mobile platforms (iOS, Android) and provide a consistent user experience (UX) across the platforms. Developing an app for separate mobile platforms require in-depth knowledge of them and their SDKs. This increases the cost of development and the time to market an app and decreases the ease of updating. This is where cross-platform development tools come into the picture [9].

Cross-platform tools (e.g., PhoneGap, Titanium, and RhoMobile) allow the implementation of an app and its user interface (UI) using Web technologies such as hypertext markup language (HTML) and cascading style sheet (CSS). Then the app can be built for several mobile platforms (e.g., iOS, Android, Windows Phone, and BlackBerry). This process is helpful only when a developer is willing to compromise user experience and places more importance on launching the app in several platforms to reach the maximum number of users. This approach allows an app to be developed for multiple mobile platforms at the same time. Thus, the cost of development and time to market the app is reduced [6].

This chapter focuses on the introduction of the mobile social networking (MSN) development platform. Specifically, Section 13.2 will give an overview of mobile device (hardware) and mobile operating system (software); MSN development architecture will also be discussed. Section 13.3 will take Android and iOS as representative examples to present some features of MSN applications, and then we will summarize the concerns and process when developing MSN applications. Because developing applications for separate platform requires quite a great deal of time and knowledge, cross-platform development tools will be discussed in Section 13.4. Section 13.5 presents an educational example of MSN application development, based on the Wi-Fi Direct framework. Finally, we summarize this chapter in Section 13.6.

13.2 Overview of Mobile Platform

13.2.1 Mobile Devices

Mobility can be defined as the capability of being able to move or be moved easily. In the context of mobile computing, mobility pertains to people's use of portable and functionally powerful mobile devices that offer the ability to perform a set of application functions untethered, while also being able to connect to, obtain data from, and provide data to other users, applications, and systems [10]. A mobile device has the following features—it is portable, personal, easy and fast to use, and has a network connection [11].

Clearly, cell phones, personal digital assistants (PDAs), and other manual electronic devices are converging into a single device, a smartphone, which incorporates some functions of mobile entertainment devices and other consumer electronic wireless devices. Mobile devices are not just a communication tool; they are also are a computational tool, supported from mobile networks to global cellular networks. A mobile device can be used as a wireless control instrument that can unify monitoring and control over other kinds of electronic devices.

A mobile device consists of integrated and interconnected hardware components and software. These hardware components include a microprocessor, read-only memory (ROM), random access memory (RAM), expansion storage, network interfaces, battery, and a display. A mobile device is controlled by an embedded computer system, which performs a set of functions to control any electronic device or equipment ranging from an industrial system to home use devices.

When designing an application, it is important to consider the display attributes, such as size, resolution, color depth, backlight, and power consumption. The mobile device display represents a significant amount of power consumption; therefore, it must also be taken into account in the management of global energy. Specifically, the OSs and applications should employ techniques that allow screen use with low power. In brief, the power management is a priority in mobile devices due to the constraints of battery life; therefore, it is important that hardware components and application designs use energy efficiently [12,13]. Figure 13.1 shows the hardware architecture of a mobile device [14].

13.2.2 Mobile OSs

Modern mobile OSs combine the features of a personal computer operating system with other features, including touch screen, cellular, Bluetooth, Wi-Fi,

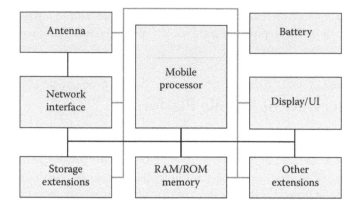

Figure 13.1 Hardware architecture of a mobile device. (Redrawn from Hernandez, I.M.T. et al. Analysis for the design of open applications on mobile devices, In: Proceedings of 2013 International Conference on Electronics, Communications and Computing (CONIELECOMP), © 2013, IEEE.)

Global Positioning System mobile navigation, camera, video camera, speech recognition, voice recorder, music player, near-field communication, and infrared blaster.

Android, iOS, Windows Phone, and BlackBerry are the most popular mobile OSs. With a collection of choices, it can be difficult for developers to decide the platform on which they will work. Table 13.1 shows the main similarities and differences of these systems [15].

Devices that operated by Android and iOS occupy most of the mobile market, although these systems have their own advantages and disadvantages. For instance, iOS provides a polished and user-friendly experience and has the best app market. However, to ensure hardware and software work well together, iOS provides a controlled smartphone experience; Android is supremely customizable with an excellent notification system and many widgets, allowing multitasking; full Google is also Android's feature. Table 13.2 shows many specific details about different systems.

13.2.3 MSN Development Platform Architecture

To support different data exchange, sharing, and delivery scenarios, different architectures are used in MSN development platforms, including centralized architecture and distributed architecture.

In centralized MSN application architecture, a centralized server is used to deliver data from server to terminal; this is a client/server (C/S) structure. Mobile users create their information and send it to the centralized server so that other users can access the online social network to receive these, as shown in Figure 13.2.

Table 13.1 Overview of Popular OSs

OS	iOS	Android	Windows Phone	BlackBerry OS
Company	Apple, Inc.	Open Handset Alliance/ Google	Microsoft	BlackBerry Ltd.
Market share [16]	13.4%	81.3%	4.1%	1.0%
Current version	7.0.4	4.4	8	10.2
OS family	Darwin	Linux	Windows CE 7/ Windows NT 8+ [17]	QNX (Unix-like)
Supported CPU architecture	ARM	ARM, MIPS, x86, I.MX	ARM	ARM
Package manager	iTunes	APK	Zune Software	BlackBerry Link
Default Web browser/ engine	WebKit	WebKit	Trident	WebKit
Official application store	App Store	Google Play	Windows Phone Store	BlackBerry World

Source: Comparison of mobile operating systems. Available online: http:// en.wikipedia.org/wiki/Comparison_of_mobile_operating_systems

A centralized architecture forms the basis for Web-based MSNs where the mobile users depend on the updates of content providers (e.g., Facebook server). The advantages of a centralized architecture include the simplicity of service implementation and the high efficiency of centralized control. However, similar to a C/S structure, a centralized MSN architecture may have a single point of failure and may experience congestion at the server when a large number of mobile users access the services at the same time [21].

In a distributed MSN architecture (shown in Figure 13.3), instead of a centralized server, mobile users communicate directly using peer-to-peer (P2P) technology such as Wi-Fi or Bluetooth on the basis of encounter/reencounter. The data flow in distributed architecture can go through other mobile users as well as access points (e.g., as the relay node).

Table 13.2   Main Differences among Mobile Operating Systems

OS	iOS	Android	Windows Phone	BlackBerry OS
Notification center	5+	Yes	No	6+
Push notifications	Yes (Apple Push Notification Service)	Yes	Yes	Yes
Multitasking	7+ [18] was limited from version 4 [19]	Yes	8+ [20]	Yes
Desktop interactive widgets	No	Yes	No; live tiles are not interactive	No
Lock screen widgets	Media player; 5+: notifications, voicemail, camera	4.2+	Media player; 8+: live apps and notifications	?
Notification view widgets	5+: Stocks and weather; 3rd party software with "jailbreak"	4.1+: Google Now and possible with 3rd party apps	No (there is no notification view)	?
Quick settings toggles	7+ or 3rd party software on jail-broken devices	2+	No	?
Screenshot	Yes	4+ also available on 3.7 or earlier with Cyanogen Mod and on certain devices (e.g., Samsung Galaxy S II)	8+	3rd party software

Source: Comparison of mobile operating systems. Available online: http://en. wikipedia.org/wiki/Comparison_of_mobile_operating_systems

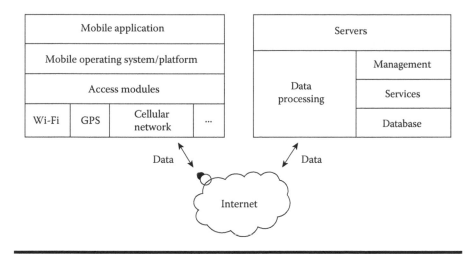

Figure 13.2 Architecture of centralized mobile social network.

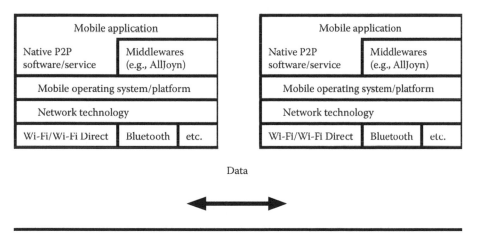

Figure 13.3 Architecture of a distributed mobile social network.

On one hand, some applications (such as Apple's AirDrop* on iOS7 or other P2P software on Android or Windows) can be built on the distributed MSN architecture. On the other hand, distributed MSN architecture can work on its own on dedicated middlewares (e.g., AllJoyn,† an open-source project led by Qualcomm). These dedicated middlewares are able to provide the necessary underlying functionalities for constructing MSN applications, including networking capability, storing the interests, identifying the other users, sharing data, and so on [21].

---

* iOS: Use AirDrop, available on http://support.apple.com/kb/ht5887
† A Common Language for the Internet of Everything-Alljoyn, Available on: https://www. alljoyn.org

13.3 Analysis of MSN Design

13.3.1 Mobile App Background

Recently, mobile apps have been rapidly evolving. Generally, they target a specific task, make heavy use of the data network, adopt C/S architecture, and have a simple delivery mechanism (i.e., Google Play Store or Apple Store). The more interesting mobile apps currently tend to use one's location and use direct P2P communications among users.

Why are mobile apps recently taking off? There are several convergent factors that make the environment attractive to mobile apps becoming pervasive. The major factors are as follows:

- Pervasive high-speed data networks: High-speed networks to support interesting, data-based applications have become relatively pervasive, particularly in metropolitan areas.
- Relatively cheap, high-performing devices: The processing and storage capacities of small devices have increased dramatically, and devices have remained affordable enough for widespread adoption.
- Introduction of easy-to-use marketplaces for apps: Apple's success with iTunes provides a platform to make apps easily available through the Apple App Store. Android's Market also supports easy availability for Android apps.
- Support for third-party mobile app development: The easier it is for others to create and publish an app (as well as to make money on them), the more likely a developer will create and publish the app.
- Underlying need for simple, targeted applications while mobile: A mobile user is often doing something else (such as trying to get to a restaurant), so many mobile apps are in support of that other activity or help one to accomplish a task while mobile. In addition, many organizations with traditional Web applications have a desire to make their applications more available and easy to use on small mobile devices where a regular Web application may be hard to use [8].

13.3.2 Android and iOS Overview

A mobile software platform is defined as the combination of an operating system for a collection of compatible mobile devices with a set of related software development libraries, application programming interfaces (APIs), and programming tools [12]. The need for mobile operating systems that enable application development has increased significantly due to the proliferation of mobile devices. The main purpose of these platforms is to create a mobile development environment where users and developers can make applications. Because Android and iOS are the most popular mobile operating systems, we will make a thorough introductory examination of them.

13.3.2.1 Android

The Android platform grew out of development with the Open Handset alliance and Google. Google purchased Android, incorporated in 2005, and made the code open source in 2008. Since then, Google has released many versions of Android, and they continue to enhance the code base to offer more features and support for a wider array of devices (such as tablets) [8].

Android is very different from the platform used in iPhone. Android is an open-source platform based on Linux with an open-source license for mobile devices. It is practically designed for a wide range of mobile systems and programming environments. Android SDK [22] is applicable in heterogeneous brands of mobile devices with well-documented APIs [23]. This platform is built on Linux kernel, native libraries, Android run time, and Android application framework. The Linux kernel core services (including hardware drivers, process and memory, security, and power management) are handled by a 2.6 kernel. Libraries running on the top of the kernel for Android include various C/C++ such as libc and SSL. Figure 13.4 shows how these are distributed in the Android platform [14,24].

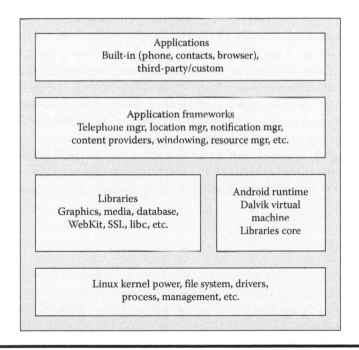

Figure 13.4 Android platform architecture. (Redrawn from Hernandez, I.M.T. et al. Analysis for the design of open applications on mobile devices, In: Proceedings of 2013 International Conference on Electronics, Communications and Computing (CONIELECOMP), © 2013, IEEE.)

Android applications are primarily written in Java and compiled into Dalvik executable (DEX) format, a custom byte code. Each application executes its own process, with its own instance of the Dalvik virtual machine. Android applications consist of a set of activities [25], which include the user interaction and may have one or more input screens. For example, selecting a contact from the internal address book is an activity. A user may flip through the contact list or may use a search box to find the specific contact number. Many actions are combined to become an activity. Activities have a well-defined life cycle and can be invoked from other activities—even activities from other applications. Besides a variety of widgets, the declarative description of UIs is also allowed in Android. XML files describe the relative layout of a UI, which simplifies internationalization and also allows rendering the UI on different screen resolutions. However, the platform's openness can bring fragmentation issues: applications developed in different versions may have problems working with different versions of the operating system [14,24].

For the Android platform, we can use Eclipse (with its plug-ins) to support Android development. In addition, at the Google I/O 2013, Google launched another Android integrated development environment (IDE) called Android Studio [26] built on IntelliJ IDEA [27] Community Edition, sharing many features with the Eclipse-based tool. Google and Android developers have provided a lot of documentation on the Android developer site. In order to actually publish the app to the Google Play Store, we will need to pay a small fee (US$25 at the time of this writing) and register so that we can publish apps on Google Play. However, to create an app, install it on actual Android-based devices, and test it, we do not need to register. In order to install our own prototype Android apps on physical devices, we only needed to adjust a few settings on each device. These are usually under the "settings" on the device and in the "applications" settings category. There you can allow "unknown sources" for apps and set "USB debugging" under the "development" item. These settings can vary a bit depending on the version of the Android operating system and the manufacturer's changes [8].

13.3.2.2 iOS

iOS is Apple's proprietary operating system that runs on the iPhone, iPod Touch, and iPad. As the parent software for iOS, Mac OS X shares the Darwin operating system foundation, making it characteristically a UNIX operating system. Apple maintains an extensive Web site to help developers. It also provides a nice simulator for iOS-based devices.

Apple provided a development kit for iOS that offered incredible tools for developers, iOS SDK. The iOS SDK contains the tools and needed interfaces to develop, install, test, and execute native applications. Native applications are built using the system frameworks and Objective-C language [28] and run directly on iOS. At the highest level, iOS acts as an intermediary between the underlying hardware and the apps that appear on the screen. The applications communicate with the hardware through a set of well-defined system interfaces that protect the application from hardware changes. This abstraction makes it easy to code applications that work consistently on devices with different hardware capabilities [14].

Figure 13.5 shows the four abstraction layers of iOS: the Core OS layer, Core Services layer, the Media layer, and the Cocoa Touch layer. The services and technologies on which all applications rely are in the lowest layers of the system; higher-level layers contain more sophisticated services and technology [14].

iOS apps were more functional, better looking, and more advanced than any other platforms. The combination gave the platform a lead on apps that other companies are still trying to close in on. The App Store, one of Apple's native applications found within iOS, is one of the largest mobile marketplaces in the world. Xcode [29] is the development environment we will be working with primarily, which is a free download from Apple's developer portal. You have to sign up for an iTunes account in order to complete the process.

In order to create apps to deploy on an actual device, you need to be able to create keys for that device and application. To do that, you need to have access to Apple to develop the keys for specific devices. (Universities can get that for free due to an agreement with Apple.) Similar to Android, to publish apps to the Apple Store, you need to pay an annual fee (US$99 at the time of this writing) and register with Apple. Apple also has a more extensive process for reviewing apps before they are released on the store [8].

Apple developed the iOS platform for the iPhone with the initial releases coming out with the iPhone in 2007. Later versions included support for the iPad and most recently (version 4.0 and greater) support multiprocessing. The strength of

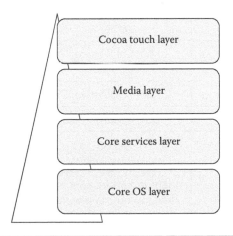

Figure 13.5 Core abstraction layers of iOS.

the iOS platform is that it has a relatively small number of versions, all supported by Apple. There are also a small number of different devices, again all supported by Apple. These factors reduce the complexity for the support a great deal [8].

13.3.2.3 Comparison of Android and iOS from Viewpoint of Developers

Smartphone platforms differ greatly in their native application development models (see Table 13.3). Android provides a mobile operating system that can run on the Linux kernel, while iOS is proprietary platform developed by Apple that is only available for Apple devices. Android employs Java, whereas iOS uses Objective-C. Java features strong typing and garbage collection, while the version of Objective-C used on iOS supports dynamic typing without garbage collection. Android and iOS differ greatly in their APIs for UI, application life cycle, and device management [24].

13.3.3 Introduction to Mobile Development: Android and iOS as Examples

Mobile application development has exploded since Apple's deployment of the iPhone and the release of Google's Android operating system. The development of these applications is much easier than earlier mobile application development platforms [8].

Building mobile applications can be as easy as opening up your IDE, throwing something together, doing a quick bit of testing, and submitting it to an App market—all done in an afternoon. Or it can be an extremely complex

Table 13.3   Android versus iOS Development

OS	iOS	Android
Official SDK platform(s)	Mac OS X using iOSSDK	Linux, Mac OS X, and Windows
Programming languages	C, C++, Objective-C	C, C++, Java
Popular IDE	Xcode	Eclipse/IntelliJ
Cost to develop on the phone	$99/year	Free
Cost to publish application on official store	Included in the cost to develop on the phone	$25 to offer it on Google Play

Source: Comparison of mobile operating systems. Available online: http://en. wikipedia.org/wiki/Comparison_of_mobile_operating_systems

process that involves rigorous upfront design, usability testing, QA testing on thousands of devices, a full beta life cycle, and then deployment in a number of different ways.

There are two elements in building mobile applications, including:

1. Considerations: There are a number of considerations when building mobile applications, especially in contrast to traditional Web or desktop applications. These considerations and how they affect mobile development will be introduced in this chapter.
2. Process: The process of software development is called the software development life cycle (SDLC). All phases of the SDLC with respect to mobile application development are also summarized.

13.3.3.1 Considerations in Mobile Development

Although developing mobile applications is not fundamentally different from traditional Web/desktop development in terms of processes or architecture, following are some common considerations.

13.3.3.1.1 Multitasking

There are two significant challenges to multitasking (having multiple applications running at once) on a mobile device. First, given the limited screen, it is difficult to display multiple applications simultaneously. Therefore, on mobile devices only one app can be in the foreground at one time. Second, having multiple applications open and performing tasks can quickly eat battery power.

13.3.3.1.2 Form Factor

Mobile devices generally fall into two categories, phones and tablets, with a few crossover devices in between. Developing for these form factors is generally very similar; however, designing applications for them can be very different. Phones have very limited screen space, and tablets, while bigger, are still mobile devices with less screen space than even most laptops. Because of this, mobile platform UI controls have been designed specifically to be effective on smaller form factors.

13.3.3.1.3 Limited Resources

Mobile devices get more and more powerful all the time, but they are still mobile devices that have limited capabilities in comparison to desktop or notebook computers. For instance, desktop developers generally do not worry about memory capacities; they are used to having physical and virtual memory in copious quantities, whereas on mobile devices you can quickly consume all available memory just by loading a handful of high-quality pictures.

In addition, processor-intensive applications such as games or text recognition can really tax the mobile central processing unit (CPU) and can adversely affect device performance. Because of considerations like these, it is important to code smartly and to deploy early and often to actual devices in order to validate responsiveness.

13.3.3.2 iOS Considerations

13.3.3.2.1 Multitasking

Multitasking is very tightly controlled in iOS, and there are a number of rules and behaviors that application must conform to when another application comes to the foreground; otherwise the application will be terminated by iOS.

13.3.3.2.2 OS-Specific Constraints

In order to make sure that applications are responsive and secure, iOS enforces a number of rules that applications must abide by. In addition to the rules regarding multitasking, there are a number of event methods, out of which app must return in a certain amount of time, otherwise it will get terminated by iOS.

Also worth noting, apps run in what is known as a sandbox [30] (see Figure 13.6), an environment that enforces security constraints restricting what the app can access. For instance, an app can read from and write to its own directory, but if it attempts to write to another app directory, it will be terminated.

13.3.3.3 Android Considerations

13.3.3.3.1 Multitasking

Multitasking in Android has two components; the first is the activity life cycle. Each screen in an Android application is represented by an activity, and there is a specific set of events that occur when an application is placed in the background or comes to the foreground. Applications must adhere to this life cycle in order to create responsive, well-behaved applications.

The second component to multitasking in Android is the use of Services. Services are long-running processes that exist independent of an application and that are used to execute processes while the application is in the background.

13.3.3.3.2 Many Devices plus Many Form Factors

Unlike iOS, which has a small set of devices, or even Windows Phone, which only runs on approved devices that meet a minimum set of platform requirements, Google does not impose any limits on which devices can run the Android OS. This open paradigm results in a product environment populated by a myriad of different devices with very different hardware, screen resolutions and ratios, device features, and capabilities.

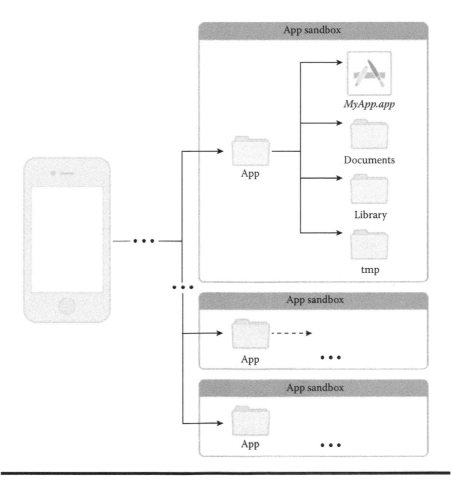

Figure 13.6   Sandbox directories in iOS.

Because of the extreme fragmentation of Android devices, most people choose the most popular five or six devices to design and test for, and then prioritize those.

13.3.3.3.3 Security Considerations

Applications in the Android OS all run under a distinct, isolated identity with limited permissions. By default, applications can do very little. For example, without special permission, an application cannot send a text message, determine the phone state, or even access the Internet! In order to access these features, applications must specify in their application manifest file which permissions they would like (see Figure 13.7).

When applications are being installed, the OS reads those permissions, notifies the user that the application is requesting those permissions, and then allows the user to continue or cancel the installation (see Figure 13.8). This is an essential step in the Android distribution model.

Figure 13.7   AndroidManifest.xml in an Android project.

Figure 13.8   Permissions at application install—Google Maps. (Available at http://source.android.com/devices/tech/security.)

13.3.3.4 Process of Mobile Development

The life cycle of mobile development is not that different from the SDLC for Web or desktop applications. Basically, there are usually five major portions of the process:

1. Inception: All apps start with an idea. That idea is usually refined into a solid basis for an application.

2. Design: The design phase consists of defining the app's user experience (UX) such as what the general layout is, how it works, and so on, as well as turning that UX into a proper UI design, usually with the help of a graphic designer.
3. Development: Usually the most resource intensive phase, this is the actual building of the application.
4. Stabilization: When development is far enough along, QA usually begins to test the application and bugs are fixed. Often times, an application will go into a limited beta phase in which a wider user audience is given a chance to use it, provide feedback, and inform changes.
5. Deployment: Often many of these pieces are overlapped; for example, it is common for development to be going on while the UI is being finalized, and it may even inform the UI design. In addition, an application may be going into a stabilization phase at the same that new features are being added to a new version. Let us discuss how each of these phases plays a part in mobile development.

13.3.3.4.1 Inception

The ubiquity and level of interaction people have with mobile devices means that nearly everyone has an idea for a mobile app. Mobile devices open up a whole new way to interact with computing, the Web, and even corporate infrastructure.

The inception stage is all about defining and refining the idea for an app. In order to create a successful app, it is important to ask some fundamental questions. For example, if you are developing an app for distribution in a public app store, some considerations are:

■ Competitive advantage: Are there similar apps out there already? If so, how does this application differ from others?

If you are intending for the app to be distributed as enterprise edition:

■ Infrastructure integration: What existing infrastructure will it integrate with or extend?

In addition, you should evaluate the usage of the app in a mobile form factor:

■ Value: What value does this app bring users? How will they use it?
■ Form/mobility: How will this app work in a mobile form factor? How can I add value using mobile technologies such as location awareness, the camera, and so on?

To help with designing the functionality of an app, it can be useful to define actors and use cases. Actors are roles within an application and are often users. Use cases are typically actions or intents.

For instance, if you are building a task-tracking application, you might have two actors: user and friend. A user might create a task, and share a task with a friend. In this case, creating a task and sharing a task are two distinct use cases that, in tandem with the actors, will inform what screens you will need to build as well as what business entities and logic will need to be developed.

If you have captured the appropriate use cases and actors, it is much easier to begin designing an application because you know exactly what you need to design, so the question becomes how to design it, rather than what to design.

13.3.3.4.2 Designing Mobile Applications

Once you have a good idea of what it is you want to design, the next step is start trying to solve the user experience (UX).

UX is usually done via wireframes or mockups using tools such as Balsamiq,* Mockingbird,† Visio,‡ or just pen and plain paper. UX mockups allow you to quickly design UX without having to worry about the actual UI design (see Figure 13.9).

Figure 13.9   User experience design using Balsamiq mockups. (Available at http://www.hksilicon.com/kb/articles/53718/Balsamiq-Mockups.)

---

* Balsamiq, http://balsamiq.com
† MockFlow, http://app.mockflow.com
‡ Visio, http://www.visio.com

Figure 13.10 Back button in user interface navigation bar. (Available at http://www.pixate.com/blog/2013-10-23-styling-navbar-pt2/index.html.)

When creating UX mockups, it is important to consider the interface guidelines for the various platforms on which you are designing. By adhering to platform-specific guidelines, you can ensure that your apps feel at home on each platform.

For example, each app has a metaphor for switching between sections in an application. iOS uses a tab bar at the bottom of the screen, Android uses a tab bar at the top of the screen, and Windows Phone uses the panorama view.

In addition, the hardware itself also dictates UX decisions. For example, iOS devices have no physical back button, so we have to design a back button in the UINavigationBar (see Figure 13.10). Furthermore, form factor also influences UX decisions. A tablet has a little larger screen, so you can fit more information; often what needs multiple screens on a phone is compressed into one for a tablet.

13.3.3.4.3 User Interface Design

Once you've nailed down the UX in your application, the next step is to create the UI design. While UX is typically just black and white mockups, the UI design phase is where colors, graphics, and so forth are introduced and finalized. Spending time on good UI design is important, and generally, the most popular apps have a professional design. As with UX, it is important to understand that each platform has its own design language, so a well-designed application may still look different on each platform (see Figure 13.11).

13.3.3.4.4 Development

The development phase usually starts very early. In fact, once an idea has some maturation in the conceptual/inspiration phase, often a working prototype is developed that validates functionality and assumptions, and helps give an understanding of the scope of the work.

Figure 13.11   User interface of iOS, Android, and Windows Phone. User interface of Windows Phone (right). (Available at http://www.cnet.com.au/windows-phone-7-339306753.htm.)

13.3.3.4.5 Stabilization

Stabilization is the process of working out the bugs in your app—not just from a functional standpoint (e.g., "It crashes when I click this button") but also with regard to usability and performance. It is best to start stabilization very early within the development process so that course corrections can occur before they become costly. Typically, applications go into prototype, alpha, beta, and release candidate stages. Different people define these differently, but they generally follow this pattern:

1. Prototype: The app is still in the proof-of-concept phase and has only core functionality or specific parts of the application are working. Major bugs are present.
2. Alpha: Core functionality is generally code-complete (built but not fully tested). Major bugs are still present; outlying functionality may still not be present.
3. Beta: Most functionality is now complete and has had at least light testing and bug fixing. Major known issues may still be present.
4. Release candidate: All functionality is complete and tested. Barring new bugs, the app is a candidate for release in the wild.

It is never too early to begin testing an application. For example, if a major issue is found in the prototype stage, the UX of the app can still be modified to accommodate it. If a performance issue is found in the alpha stage, it is early enough to modify the architecture before a lot of code has been built on top of false assumptions.

Typically, as an application moves further along in the life cycle, it is opened to more people to try it out, test it, provide feedback, and so forth. For instance, prototype applications may only be shown or made available to key stakeholders, whereas release candidate applications may be distributed to customers that sign up for early access.

13.3.3.4.6 Distribution

Once you have stabilized your application, it is time to publicly distribute it into the world. There are a number of different distribution options, depending on the platform; iOS's apps are distributed in exactly the same way.

1. Apple App Store: It allows developers to market and distribute their apps online with very little effort.
2. Enterprise Deployment: Enterprise deployment is meant for internal distribution of corporate applications that are not available publicly via the App Store.
3. Ad Hoc Deployment: Ad hoc deployment is intended primarily for development and testing and allows you to deploy to a limited number of properly provisioned devices. When you deploy to a device via Xcode, it is known as ad hoc deployment.

As for Android, all applications must be signed before being distributed. Developers sign their applications by using their own certificate protected by a private key. This certificate can provide a chain of authenticity that ties an application developer to the applications that the developer has built and released. Unlike other popular mobile platforms, Android takes a very open approach to app distribution. Devices are not locked to a single, approved app store. Instead, anyone is free to create an app store, and most Android phones allow apps to be installed from these third-party stores [31].

13.4 Cross-Platform Mobile Applications Development

In the current mobile market, there are numbers of mobile devices that run with different operating systems and languages. Developing applications in different platforms has generated many problems; a lot of time, cost, and man power is needed if there is no significant porting work [24].

Because each OS has its own language, different APIs, different integrated development environments (IDEs), and so on, cross-platform mobile development tools have been developed to meet the needs of developers with the purpose of giving them the possibility of writing the application source code once and running it on different OSs. Benefits that these tools have brought are:

- Reduction of required skills for developing applications due to the use of common programming languages
- Reduction of coding because the source code is written once and it is compiled for each supported OS
- Reduction of development time and long-term maintenance costs
- Decrement of API knowledge because with these tools it is not necessary to know the APIs of each OS, only the APIs provided by the selected tool
- Greater ease of development compared to building native applications for each OS
- Increment of market share for the corresponding business model with the advantage of raising the return on investment (ROI).

The business model includes all activities related to commercial transactions [32]. The increased usage of cross-platform mobile tools can have some effects on the respective business model of each mobile app market, such as the App Store for Apple, Google Play for Android, and so on. One of the bigger effects of these tools is to expand the application sale on more markets as much as possible with the aim of increasing the gain by both parties—business model owners and developers. Furthermore, the ROI could play a very important role, especially for companies that are looking for these tools. In fact, with these tools, developing an application and selling it on multiple markets will decrease investment costs. The reduction in capital investment for applications could encourage developers and companies to invest more in developing new applications and to register them in the respective business model. Finally, another important variation related to the business model might come from the use of these tools to support a subset of OSs [33].

13.4.1 Overview of Cross-Platform Development

Technologically, cross-platform mobile application development refers to creating mobile applications where the application logic and UI components are all primarily designed and written in common tools accepted by all platforms (e.g., the former could be JavaScript and the latter could be defined using HTML5, CSS3, etc.). All of the current leading mobile operating system platforms allow applications to display and interact with a "Web view." This Web view is an embedded browser documented and contained within each platform's SDK.

The HTML and associated JavaScript application logic are contained within a native code "wrapper." This allows what would otherwise be a Web application to be installed as an application on a mobile device's desktop. The native wrapper exposes device-level functions to the scripted portion of the application through a set of plug-ins written in the native language of the particular mobile platform, and each plug-in has an associated JavaScript API, which can expose native functions that would otherwise be inaccessible using standard HTML5 (e.g., access to the device calendar, phonebook, or built-in hardware such as the camera).

Calls from JavaScript to native codes exploit the ability to intercept navigation events and to examine the URI scheme, which can be used to indicate that the URI is an encoded method call and should not be passed on to the embedded browser to handle. The application shell can then process the encoded URI and call native methods within the application shell. Conversely, function calls can be made to the embedded browser code by injecting JavaScript at runtime into the currently loaded Web page [34].

13.4.2 Requirements and Architecture of a Cross-Platform Framework

The desirable requirements of any cross-platform framework are as follows.

- Multiple mobile platform support: The framework must support several mobile platforms. Particularly, support for Android and iOS is very essential because they have the largest share in the application markets.
- Rich user interface: Currently the cross-platform tools cannot provide rich UI as native apps. Because the success of an application highly depends on a user's experience with the interface, rich UI development should be incorporated. Support for sophisticated graphics (2D, 3D), animation, and multimedia is necessary.
- Back-end communication: Mobile devices promote an "always connected" model in which the users are sharing material in social networking sites, watching videos, communicating via live chat, and gathering information. These smooth supports for back-end communication protocols and data formats are absolute mandatory.
- Security: Applications developed by cross-platform tools are not highly secure. Proper research needs to be carried out to secure the tools and applications.
- Support for app extensions: It is necessary to install app extensions on top of existing applications such as in-app purchase/billing capability.
- Power consumption: Power consumption is an important issue nowadays with thousands of smartphones and tablets being activated daily. The generated applications must be optimized for power.
- Accessing built-in features: The tools must be able to access the built-in features of a smart device. Use of the camera, sensors, geolocalization, and other features helps to provide a better user experience.
- Open source: This attracts more application developers, and the developer community can participate in bug-fixes and further development. It should be noted that this is not a technical requirement [6].

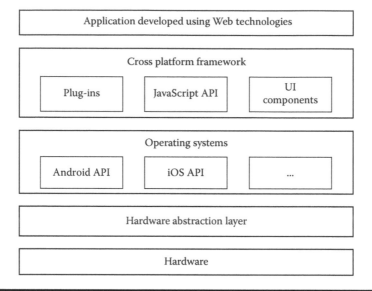

Figure 13.12 General architecture of cross-platform mobile application development. (Redrawn from Dalmasso, I. et al. Survey, comparison and evaluation of cross platform mobile application development tools, In: Proceedings of the 9th International on Wireless Communications and Mobile Computing Conference (IWCMC), © 2013, IEEE.)

To meet these requirements, a general architecture of cross-platform mobile application development has been provided, as shown in Figure 13.12. The application developer implements the business logic or the application functionalities using Web technologies. The cross-platform framework allows implementing UI and access to the storage facility and device features (sensors, camera, and contacts), which interact with a JavaScript API. The API will, in turn, interact with the native API of a mobile platform. The application is then built separately to generate the executables for different platforms. The APIs for the mobile platforms actually allow generating the respective application. Thus, the generated application can be run in a corresponding mobile device [6].

13.4.3 Comparison of Cross-Platform Mobile Development Tools

There are various types of tools used to cross compile a mobile system. *PhoneGap** is an open-source mobile development tool developed by Adobe Systems, Inc. PhoneGap allows developers and companies to build free, commercial, and open-source applications and also gives them the possibility of using any licensing combination.

---

* PhoneGap, http://www.phonegap.com

Figure 13.13 **User interface of a PhoneGap simple app. A sample app of PhoneGap. (Available at http://www.okilla.com/28/phonegap--an-open-source-mobileframework.)**

The development environment is cross-platform and permits the creation of applications for Android, Bada, BlackBerry, iOS, Symbian, webOS, and Windows Phone OSs [33]. Figure 13.13 shows the UI of a simple app running in Android.

PhoneGap is a useful solution for building mobile applications using modern Web programming languages such as HTML, HTML5, CSS, CSS3, and JavaScript, and SDKs functionality, instead of using lesser known languages such as Objective-C or others [35]. It has the benefit of bringing many advantages to skilled developers and especially of attracting Web developers [36].

Essentially, PhoneGap is a "wrapper" that allows developers to enclose applications written in known programming languages in native applications. Moreover, as a valid open-source software, it is composed of many components and extensions. PhoneGap applications are hybrid, which means that they are not purely native or Web-based. The meaning of "not purely native" comes from the layout rendering that is done via Web view instead of via the OS native language, whereas "not purely Web-based" comes from the lack of HTML support in some functions [37]. PhoneGap also offers the possibility of extending the tool by developing one's own plug-ins [33].

DragonRad * is also a cross-platform mobile application development platform; it is by Seregon Solutions, Inc., and is distributed under a commercial license. It allows developers to design, manage, and deploy mobile applications once and use them across different operating systems. The tool focuses on database-driven mobile

---

* DragonRad, http://www.seregon.com

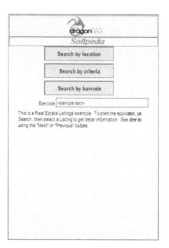

Figure 13.14   User interface of a simple DragonRad app. (Available at http://i1-win. softpedia-static.com/screenshots/DragonRAD-Designer_1.png?1349335921.)

enterprise applications with a wide range of easy database support. It provides the "drag and drop" (D&D) environment that helps developers to save programming time and to create logics. DragonRad provides its own IDE, which can be configured for different simulators such as iOS, Android, BlackBerry, and so forth. Because DragonRad has host–client architecture, it is required that the server and database are set up based on the needs of developers, but it also comes in a complete package with all server and database prerequisites such as Tomcat,* MySQL,† and so on. In brief, DragonRad is commercial tool with the support to its own language D&D, but, the possibilities of extension in terms of adding plug-ins and other support to the framework are quite limited.

DragonRad facilitates the integration and synchronization of a database system with native functions defined in mobile OSs, such as contacts, calendar, geolocation, menu, and storage. The architecture of DragonRad has three major components: DragonRad Designer, DragonRad Host, and DragonRad Client. Figure 13.14 shows a simple app of DragonRad.

RhoMobile‡ is another solution for cross-compilation; this Ruby-based mobile development manages enterprise application and data that can be used across Linux, Mac, and Windows OS. It provides a high level of productivity and portability of Web programming. RhoMobile Suite (formerly known as Rhodes Framework) uses a model view controller (MVC) pattern to do applications written with HTML, CSS,

---

* Tomcat, http://tomcat.apache.org
† MySQL, http://www.mysql.com
‡ RhoMobile, http://docs.rhomobile.com/en/4.0.0/home

**Figure 13.15 A near-field communication (NFC) app created with RhoMobile.
(Available at http://www.spritle.com/blogs/2012/04/04/create-nfc-apps-using-
rhomobile/.)**

and JavaScript using Ruby programming language. An app generator produces the
business logic of views and a controller with native applications. The applications can
be compiled to be executed natively on all mobile platforms [24]. Figure 13.15 shows
a near-field communication (NFC) app created with RhoMobile.

By comparing these frameworks, it can be observed that each has their own
strength among others. PhoneGap can support most of the different mobile platforms in the market such as iOS, BlackBerry, Android, and Windows Phone.
DragonRad is the only framework that offers the possibility to produce Web applications as well as native applications. RhoMobile is the only framework with MVC
support, which is able to write real business logic on local native applications [24].
Other available mobile frameworks include:

■ Sencha Touch*: The open-source framework built specifically to leverage
HTML5, CSS3, and JavaScript to their highest levels of power and flexibility,
such as supporting local storage, video, and audio components. The framework
also offers data integration from a variety of sources such as Asynchronous
JavaScript and XML (AJAX), JavaScript Object Notation (JSON), and YQL.

* Sencha Touch, http://www.sencha.com/products/touch/

- jQuery Mobile[*] [6]: Cross-platform and cross-device framework provides tools to build dynamic touch interfaces that will adapt to a range of device form factors. The system will include layouts (lists, detail panes, overlays) and a rich set of form controls and UI widgets (toggles, sliders, tabs).
- Titanium[†]: It has a JavaScript-based interface to native code modules included in the framework, and it can store user preferences, save data files, or implement the mobile version of a cookie using SQLite and the iPhone or Android's native file system [38].

13.5 A Simple MSN Platform Development Example

A number of technologies are emerging that facilitate the capability of passing data between two devices without accessing a wide area network over a range of different distances. Specifically, the Wi-Fi Alliance recently defined Wi-Fi Direct,[‡] which aims to effectively replace ad hoc mode. Wi-Fi Direct enables as-needed file sharing between two laptops—or between a camera and a laptop, or a laptop and a projector, and so on. For just about any situation where users might have used ad hoc mode in the past, it is now possible to use Wi-Fi Direct instead.

Wi-Fi Direct builds upon the successful IEEE 802.11 infrastructure mode and lets devices negotiate who will take over the AP-like functionalities. Wi-Fi Direct essentially embeds a software access point ("SoftAP"), into any device that must support Direct. The Wi-Fi Direct devices negotiate when they first connect to determine which device acts as an access point. The soft AP provides a version of Wi-Fi Protected Setup with its push-button or PIN-based setup. Thus, legacy Wi-Fi devices may seamlessly connect to Wi-Fi Direct devices. By making this decision, Wi-Fi Direct immediately inherits all the enhanced QoS, power saving, and security mechanisms developed for the Wi-Fi infrastructure mode in past years. Wi-Fi Direct will provide essentially the same service to end users that Bluetooth does, but it is faster and allows devices to be farther apart when communicating.

In this section, we preliminarily introduce a Wi-Fi Direct–based MSNP application development platform, shown in Figure 13.16. In terms of the functions implemented in various layers, this platform is composed of five modules: Wi-Fi Direct networking module, socket data transmission module, and several typical applications: a location application based on Google Maps, chat, and file sharing. (Due to space limitations, we have omitted introducing the application of file sharing.)

---

[*] jQuery Mobile, http://jquerymobile.com/
[†] Titanium, http://www.appcelerator.com/titanium/
[‡] http://www.wi-fi.org/discover-and-learn/wi-fi-direct

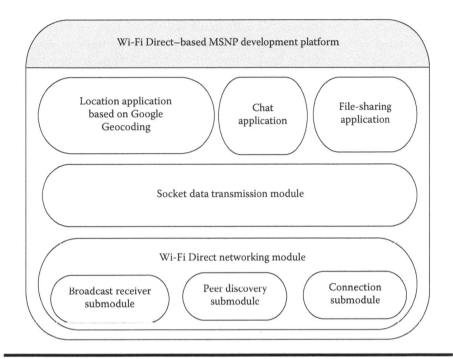

Figure 13.16 Illustration of application framework based on Wi-Fi Direct.

Briefly, Wi-Fi Direct networking module is the underlying support component; it includes three submodules and is responsible for peer discovery, connecting peer devices, and monitoring the connection status of Wi-Fi Direct P2P devices. Specifically, a Broadcast receiver sub module will monitor the results returned by Wi-Fi Direct asynchronous callback and will correspondingly react to requests from the application. The socket data transmission module encapsulates the point-to-point communication function to provide the content transmission required for the various applications superposed on it (i.e., chat and file sharing).

Location application based on Google Geocoding works as follows: First, determine the best localization method by invoking the system function provided in Android API; then, use the returned localization method to obtain the current user's accurate geographical coordinates (latitude and longitude); finally, use the reverse geocoding to get the address of the user. Moreover, we set the triggering condition of address change, and the user's address was automatically updated. Note that geocoding is in the process of converting addresses (such as "No. 66, XinMoFan Road, GuLou District, Nanjing") into geographic coordinates (such as latitude 32.06634 and longitude 118.76979). The Google Geocoding API provides a direct way to access a geocoder via an HTTP request. In addition, the service allows users to perform the converse operation (turning coordinates into addresses); this process is known as "reverse geocoding."

Technically, using Androids V4 android.net.Wi-Fi.p2p class, we built an app that can make a connection between two or more phones, and transmit data/text. app is installed on two phones in close proximity, with IDs as Android_6dob and Android_6206.

Figure 13.17 shows the screenshot of "Search Peers" in the terminal of Android_6dob. When the "Search Peers" button is clicked, terminal Android_6dob will actively find other phones within its Wi-Fi range.

As shown in Figure 13.18, the "found" other phone(s) will be listed in the fragment below the "Search Peers" button (in our scenario, the found device is the Android_6206, whose status is Available). The found peer can be selected by touching its name in the current terminal (Android_6d0b), and the button "connect" will appear. Clicking "Connect" will ask the target phone (Android_6206) to ok/refuse connection.

Once the invitee, Android_6206, accepts the Wi-Fi Direct connection request from the inviter, Android_6dob, then, on the inviter's screen, the peer's device status will changed to "Invited." Through touching and selecting the peer device, two options will occur: (1) the "Chat" button launches the chat with the selected peer device; (2) the "Disconnect" button breaks the established connection between the inviter and invitee, as shown in Figure 13.19.

As shown in Figure 13.20, once connected, both peer devices will automatically exchange their addresses (obtained through invoking Google Geocoding API) and these will correspondingly display in a proper fragment. For example, the left graph of Figure 13.20 illustrates the current peer's address (Android_6d0b) at the top

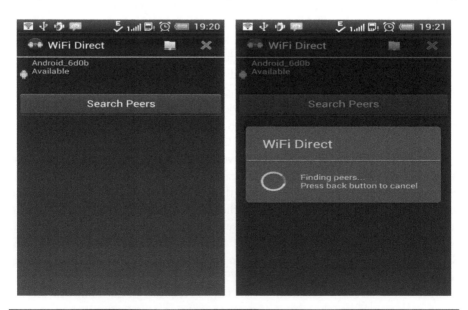

Figure 13.17   Screenshot of "Search Peers."

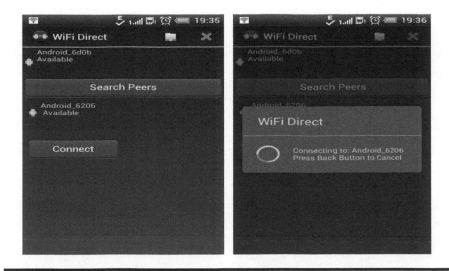

Figure 13.18 Screenshot of "Connect" to found peer device.

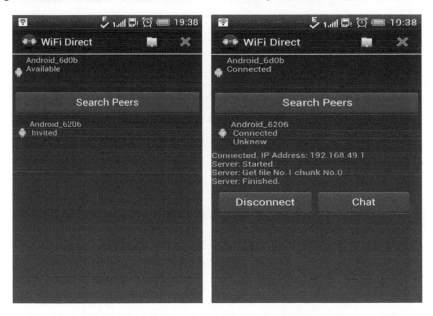

Figure 13.19 Screenshot of "Connected" status between inviter and invitee.

fragment, and its peer's (Android_6206) device address is given in the fragment below the "Search Peers" button. Similarly, the right graph of Figure 13.20 illustrates the current mobile, Android_62060b, and its peer's device addresses.

In Figure 13.19, when the button "Chat" is clicked on screen of the current peer (Android_6d0b), the bubble chat interface will appear, as shown in Figure 13.21

Figure 13.20 Illustration of exchanged address of both connected peer devices.

Figure 13.21 Bubble chat interface of current phone (left) and peer device (right).

(left graph is screen of current peer, and right graph denotes the screen of peer device). The sent messages are displayed on the right side, and the received messages are displayed on the left side. The sending (or receiving) time associated with each message is provided in the middle.

13.6 Conclusion

Mobility is now an integral part of social network. Although mobility may create data exchange difficulties, developing applications for mobile devices is no longer complicated. Popular mobile application distribution models such as the app store make it easy to get started on mobile app development and deployment. When deciding what platform to work on, developers should first consider which architecture fits their application best—centralized architecture or distributed architecture. After that, they need to decide what factors they consider important for this application, so that they can compare among numerous developing languages and tools and choose the most suitable development platform. This chapter thoroughly discusses existing main cross-platform development frameworks and considers the great benefits brought about by cross-platform development tools.

References

1. John, A., L. Adamic, M. Davis, F. Nack, D. A. Shamma, and D. D. Seligmann. The future of online social interactions: What to expect in 2020. In: Proceedings of the 17th International Conference on World Wide Web, Beijing, China, April 21–25, 2008.
2. Huberman, B., D. Romero, and F. Wu. Social networks that matter: Twitter under the microscope. *CoRR*, 2008; 14(1): 8. abs/0812.1045.
3. Papacharissi, Z. The virtual geographies of social networks: A comparative analysis of Facebook, LinkedIn and ASmallWorld. *New Media & Society* 2009; 11(1–2): 199–220.
4. Salomon, D. Moving on from Facebook using Instagram to connect with undergraduates and engage in teaching and learning. *College & Research Libraries News* 2013; 74(8): 408–412.
5. Rana, J., J. Kristiansson, J. Hallberg, and K. Synnes. An architecture for mobile social networking applications. In: Proceedings of the First International Conference on Computational Intelligence, Communication Systems and Networks (CICSYN), Indore, India, July 23–25, 2009.
6. Dalmasso, I., S. K. Datta, C. Bonnet, and N. Nikaein. Survey, comparison and evaluation of cross platform mobile application development tools. In: Proceedings of the 9th International on Wireless Communications and Mobile Computing Conference (IWCMC), Cagliari, Italy, July 1–5, 2013.
7. Nguyen, T. A., S. T. A. Rumee, C. Csallner, and N. Tillmann. An experiment in developing small mobile phone applications comparing on-phone to off-phone development. In: Proceedings of User Evaluation for Software Engineering Researchers (USER), Zurich, Switzerland, June 5, 2012.

8. Tracy, K. W. Mobile application development experiences on Apple's iOS and Android OS. *IEEE Potentials* 2012; 31(4): 30–34.

9. Ohrt, J. and V. Turau. Cross-platform development tools for smartphone applications. *IEEE Computer* 2012; 45(9): 72–79.

10. Lee, V., H. Schneider, and R. Schell. *Mobile Applications: Architecture, Design, and Development.* Prentice Hall PTR, Upper Saddle River, NJ, 2004.

11. Firtman, M. *Programming the Mobile Web.* O'Reilly Series. O'Reilly Media, Sebastopol, CA, 2010.

12. Zheng, P. and L. Ni. *Smart Phone and Next Generation Mobile Computing* (Morgan Kaufmann Series in Networking (Paperback)). Morgan Kaufmann, San Francisco, CA, 2005.

13. Ahmad, M. *Smartphone: Mobile Revolution at the Crossroads of Communications, Computing and Consumer Electronics.* CreateSpace, DBA of On-Demand Publishing LLC, 2011.

14. Hernandez, I. M. T., A. M. Viveros, and E. H. Rubio. Analysis for the design of open applications on mobile devices. In: Proceedings of 2013 International Conference on Electronics, Communications and Computing (CONIELECOMP), Cholula Puebla, México, March 11–13, 2013.

15. Comparison of mobile operating systems. Available online: http://en.wikipedia.org/wiki/Comparison_of_mobile_operating_systems

16. Android tops 81 percent of smartphone market share in Q3. Retrieved 4 November 2013. Available online: http://blogs.strategyanalytics.com/WSS/post/2013/10/31/Android-Captures-Record-81-Percent-Share-of-Global-Smartphone-Shipments-in-Q3-2013.aspx. Source: Strategy Analytics.

17. Victor, H. (2012-06-20). Windows Phone 8: The new features [Image 12: Shared Windows NT kernel, code with Windows 8]. PhoneArena.com. Retrieved 27 August 2012. Available online: http://www.phonearena.com/news/Windows-Phone-8-the-new-features_id31466#12-Shared-Windows-NT-kernel,-code-with-Windows-8

18. iOS 7: Multitasking for all apps arrives. Available online: http://www.phonearena.com/news/iOS-7-multitasking-for-all-apps-arrives_id43906

19. Head, C. (2010-06-25). Geek 101: Android and iOS multitasking compared. *PCWorld.com.* Retrieved 7 September 2011. Available online: http://www.techhive.com/article/199891/Android_multitasking.html

20. Microsoft brings true, background multitasking to Windows Phone 8—Engadget. Available online: http://www.engadget.com/2012/06/20/microsoft-brings-true-background-multitasking-to-windows-phone/

21. Kayastha, N., D. Niyato, P. Wang, and E. Hossain. Applications, architectures, and protocol design issues for mobile social networks: A survey. *Proceedings of the IEEE* 2011; 99(12): 2130–2158.

22. Puder, A. and I. Yoon. Smartphone cross-compilation framework for multiplayer online games. In: Proceedings of the 2nd IEEE International Conference on Mobile, Hybrid, and On-Line Learning, St. Maarten, Netherlands Antilles, February 10–16, 2010.

23. Puder, A. and P. Antebi. Cross-compiling Android applications to iOS and Windows Phone 7. *Mobile Networks and Applications* 2012; 18(1): 3–21.

24. Chieng, L. B., W. Y. Ting, H. H. Mohamed, M. Rafie Hj Mohd Arshad. Cross-platform mobile applications for Android and iOS. In: Proceedings of the 6th Joint IFIP Wireless and Mobile Networking Conference (WMNC), Dubai, UAE, April 23–25, 2013.

25. Gavalas, D. and D. Economou. Development platforms for mobile applications: Status and trends. *IEEE Software* 2011; 28(1): 77–86.
26. Mike, W. and D. Felker. *Android Developer Tools Essentials: Android Studio to Zipalign.* O'Reilly Media, Sebastopol, CA, 2013.
27. Assumpcao, H. O. *Getting Started with IntelliJ IDEA.* Packt Publishing Ltd, Birmingham, UK, 2013.
28. Matt, N. *IOS 7 Programming Fundamentals: Objective-C, Xcode, and Cocoa Basics.* O'Reilly Media, Sebastopol, CA, 2013.
29. Pavithra, R. and J. B. Liu. iPhone application development using Xcode. In: Proceedings of the 7th International Conference on Ubiquitous Information Management and Communication, 2013.
30. Blazakis, D. The Apple sandbox. Available online: http://www.semantiscope.com/research/BHDC2011/BHDC2011-Paper.pdf
31. Xamarin Company. Introduction to mobile development. Available online: http://docs.xamarin.com/guides/cross-platform/getting_started/introduction_to_mobile_development/
32. Hammershøj, A., A. Sapuppo, and R. Tadayoni. Challenges for mobile application development. In: Proceedings of the 14th International Conference on Intelligence in Next Generation Networks (ICIN), Berlin, Germany, October 8–11, 2012.
33. Palmieri, M., I. Singh, and A. Cicchetti. Comparison of cross-platform mobile development tools. In: Proceedings of the 16th International Conference on Intelligence in Next Generation Networks (ICIN), Berlin, Germany, October 8–11, 2012.
34. Jaramillo, D., R. Smart, B. Furht, and A. Agarwal. A secure extensible container for hybrid mobile applications. In: Proceedings of IEEE Southeastcon, Hyatt Regency Jacksonville, FL, April 4–7, 2013.
35. Fermoso, J. Seeks to bridge the gap between mobile app platforms. Available online: http://gigaom.com/2009/04/05/phonegap-seeks-to-bridge-the-gap-between-mobile-app-platforms [April 2009].
36. Myer, T. *Beginning PhoneGap.* Wrox, Wrox Press (John Wiley & Sons, Inc.), New Jersey, 2011.
37. Allen, S., V. Graupera, and L. Lundrigan. *Pro Smartphone Cross- Platform Development.* Apress, New York, 2010.
38. Smutny, P. Mobile development tools and cross-platform solutions. In: Proceedings of the 13th International on Carpathian Control Conference (ICCC), High Tatras, Slovakia, May 28–31, 2012.

Index